Lecture Notes in Computer Science 2310

Edited by G. Goos, J. Hartmanis, and J. van Leeuwen

3
Berlin
Heidelberg
New York
Barcelona
Hong Kong
London
Milan
Paris
Tokyo

Pierre Collet   Cyril Fonlupt
Jin-Kao Hao   Evelyne Lutton
Marc Schoenauer (Eds.)

Artificial Evolution

5th International Conference, Evolution Artificielle, EA 2001
Le Creusot, France, October 29-31, 2001
Selected Papers

1 3

Volume Editors

Pierre Collet
Ecole Polytechnique, Centre de Mathématiques Appliquées
91128 Palaiseau Cedex, France
E-mail: Pierre.Collet@polytechnique.fr

Cyril Fonlupt
LIL – Université du Littoral – Côte d'Opale
BP 719, 62228 Calais Cedex, France
E-mail: fonlupt@lil.univ-littoral.fr

Jin-Kao Hao
LERIA – Université d'Angers
2 Boulevard Lavoisier, 49045 Angers Cedex 01, France
E-mail: Jin-Kao.Hao@univ-angers.fr

Evelyne Lutton
Marc Schoenauer
INRIA Rocquencourt, Projet FRACTALES
Domaine de Voluceau, BP 105, 78153 Le Chesnay Cedex, France
E-mail: evelyne.lutton,marc.schoenauer @inria.fr

Cataloging-in-Publication Data applied for

Die Deutsche Bibliothek - CIP-Einheitsaufnahme

Artificial evolution : 5th international conference , evolution artificielle ;
selected papers / EA 2001, Le Creusot, France, October 2001.
Pierre Collet ... (ed.). - Berlin ; Heidelberg ; New York ; Barcelona ;
Hong Kong ; London ; Milan ; Paris ; Tokyo : Springer, 2002
     (Lecture notes in computer science ; Vol. 2310)
     ISBN 3-540-43544-1

CR Subject Classification (1998): F.1, F.2.2, I.2.6, I.5.1, G.1.6, J.3

ISSN 0302-9743
ISBN 3-540-43544-1 Springer-Verlag Berlin Heidelberg New York

Springer-Verlag Berlin Heidelberg New York
a member of BertelsmannSpringer Science+Business Media GmbH

http://www.springer.de

© Springer-Verlag Berlin Heidelberg 2002
Printed in Germany

Typesetting: Camera-ready by author, data conversion by Stefan Sossna e.K.
Printed on acid-free paper   SPIN: 10846563   06/3142   5 4 3 2 1 0

Foreword

The Evolution Artificielle cycle of conferences was originally initiated as a forum for the French-speaking evolutionary computation community. Previous EA meetings were held in Toulouse (EA'94), Brest (EA'95, LNCS 1063), N mes (EA'97, LNCS 1363), Dunkerque (EA'99, LNCS 1829), and finally, EA 2001 was hosted by the Universit´e de Bourgogne in the small town of Le Creusot, in an area of France renowned for its excellent wines.

However, the EA conferences have been receiving more and more papers from the international community: this conference can be considered fully international, with 39 submissions from non-francophonic countries on all five continents, out of a total of 68.

Out of these 68 papers, only 28 were presented orally (41) due to the formula of the conference (single session with presentations of 30 minutes) that all participants seem to appreciate a lot.

The Organizing Committee wishes to thank the members of the International Program Committee for their hard work (mainly due to the large number of submissions) and for the service they rendered to the community by ensuring the high scientific content of the papers presented.

Actually, the overall quality of the papers presented was very high and all 28 presentations are included in this volume, grouped in 8 sections which more or less reect the organization of the oral session:

1. Invited Paper: P. Bentley   gave a great talk on his classification of interdisciplinary collaborations, and showed us some of his work with musicians and biologists.
2. Theoretical Issues: Current theoretical issues concern measurement, adaptation, and control of diversity, even though connections with other disciplines are still very fruitful.   Morrison   and De Jong   introduce a unified measurement of population diversity with some interesting issues on the computation complexity of diversity measures.   Sidaner   et al. also propose a diversity measurement, which they use to analyse the way Walksat explores its search space.   Bienvenue   et al. investigate the adaptation of EA niching strategies to Monte Carlo Filtering Algorithms.   Cerruti   et al. show how an EA can be usefully exploited to tackle a hard mathematical problem related to the measure of randomness of a binary measure.   Berny   investigates the extension of a PBIL-like algorithm (more exactly a selection learning algorithm) for d-ary strings.   Brown   et al. present a very original Markov Random Field modeling of GAs, where they build an explicit probabilistic model of any fitness function. This work also seems to have some interesting connections with epistasis analysis approaches.
3. Algorithmic Issues: Devising new algorithmic issues and understanding the behavior of genetic operators and mechanisms is an important research topic in evolutionary computation.   Johnson   and Shapiro   explain the importance of selection mechanism in the case of distribution estimation

algorithms. In order to accelerate the convergence of EAs,   Abboud   and
Schoenauer   propose building and evaluating a surrogate model and introduce a surrogate mutation. To avoid stagnation in evolutionary search,
La Tendresse   et al. propose re-initializing parts of the population at given
time intervals. Dealing with noisy functions is an important topic in evolutionary computation,   Leblanc   et al. propose exploiting historical information
to devise new search strategies.

4. Applications: This section demonstrates the successful applicability of EAs
   to a broad range of problems.   Oudeyer   presents an evolutionary model of
   the origins of syllable systems. Optimizing portfolio is a challenging task.
   Korczak   et al. use artificial trading experts discovered by GA to optimize
   portfolio. Hamiez and Hao propose a scatter search approach to solve the
   graph coloring problem. By introducing an appropriate indirect representation, Bousonville   allows the application of evolutionary methods for
   solving the two stage continuous parallel ow shop problem.   Bélaidouni
   and Hao present an analysis of the search space of the famous SAT problem
   based on a measure called "density of states", and   Roudenko   et al. use
   a multi-objective evolutionary algorithm to find optimal structures for car
   front end design.

5. Implementation Issues: Until very recently, researchers in evolutionary
   computing used to design their own programs. This section concerns the use
   of tools to alleviate researchers of the task of programming.   Lutton   et al.
   present the EASEA (EAsy Specification of Evolutionary Algorithms) language and extensive tests on some famous functions.   Keijzer   et al. present
   the EO (evolving objects) library, an object-oriented framework aimed at
   building evolutionary applications.

6. Genetic Programming: Genetic Programming emerged in the 1990s as a
   very promising paradigm for automatic generation of programs.   Robilliard
   and Fonlupt   propose a way to overcome overfitting in a remote sensing
   application. Ratle   and Sebag   introduce a grammar-based GP approach,
   which uses an approach   a la PBIL during evolution, and a technique called
   boosting is presented by   Paris   et al. to improve genetic programming.

7. Constraints Handling: This section collects studies reecting ways to
   handle constraints in evolutionary computation.   Le Riche   and Guyon   provide a new insight on function penalization for constraints handling, and
   Smith   proposes to deal with constraints using the augmented Lagrangian
   penalty functions.

8. Coevolution and Agent Systems: Alternative evolutionary paradigms
   are introduced in this section.   Casillas   et al. use the coevolutionary paradigm for the learning of fuzzy-rule based systems.   Srivastava   and Kaldate   present a multi-agent simulation modeling two competing groups in
   the sphere of social and ecological resources while   Edmonds   simulates a
   foraging agent in environments with varying ecological structures.   Delepoulle   et al. give some insights on the ability of learning.   Seredy ński and
   Zomaya   report results on developing parallel algorithms for multiprocessor
   scheduling with use of cellular automata.

At this point, we would like to thank all sponsoring institutions who generously helped the Evolution Artificielle conference: the Conseil R´ egional de Bourgogne, the Universit´ e de Bourgogne, the Centre Universitaire Condorcet, the Communaut´ e Urbaine Le Creusot – Montceau, the DGA (D´ elégation G´ené-rale pour l'Armement), the INRIA (Institut National de Recherche en Informatique et Automatique), the AFIA (Association Fran   caise pour l'Intelligence Artificielle, and the CMAPX (Centre de Math´ ematiques Appliqu´ees de l'Ecole Polytechnique).

We would also like to mention all the people who donated time and energy and who therefore contributed to the success of EA 2001, namely (in alphabetical order) Val´erie Collet   (to whom we owe much of the local and financial organization as well as many of the photos), Chantal   Labeille   (secretary of the Centre Condorcet), Jean-Philippe   Rennard   (for the great web site), Nathalie Gaudechoux   (secretary of the Fractales research group at INRIA), as well as Amine Boumaza , Benoît Leblanc , Hélene Synowiecki , and Josy Liardet (for their kind help during the conference), and last but not least Alain   Blair , who generously double-registered to the conference.

January 2002   Pierre   Collet
                                               Evelyne Lutton
                                               Marc Schoenauer
                                               Cyril Fonlupt
                                               and Jin-Kao Hao

Evolution Artificielle 2001 – EA 2001

October 29-31, 2001

Universit´e de Bourgogne, Le Creusot, France

5th International Conference on Artificial Evolution

Organizing Committee

Pierre Collet (Ecole Polytechnique Paris)
Evelyne Lutton (INRIA Rocquencourt)
Marc Schoenauer (INRIA Rocquencourt) - Cyril Fonlupt (LIL Calais)
Jin-Kao Hao (LERIA Universit´ e d'Angers)

Program Committee

J.M. Alliot (ENAC Toulouse) – J.-P. Allouche (LRI Orsay)
T. B ack (NuTech Solutions GmbH) – O. Bailleux (Univ. Bourgogne)
P. Bessi ere (LIFIA Grenoble) – A. Berny (IRIN Univ. Nantes)
P. Bourgine (CREA Palaiseau) – B. Brauns chweig (IFP Rueil Malmaison)
J.-J. Chabrier (LIRSIA Univ. Dijon) – P. Collard (I3S Nice)
M. Cosnard (LORIA Nancy) – K. Deb (IIT Kanpur)
D. Delahaye (CENA Toulous e) – A. Dipanda (LE2I Univ. Dijon)
M. Dorigo (ULB Bruxelles) – R. Dorne (British Telecom London)
N. Durand (ENAC Toulouse) – M. Ebner (Univ. W urzburg)
A. Eiben (Vrije Univ. Amsterdam) – D. Fogel (Nat. Selection Inc. La Jolla)
P. Galinier (Ecole Polytechnique Montr´ eal) – C. Lattaud (Univ. Paris 5)
R. Leriche (INSA Rouen) – P. Liardet (CMI Marseille)
J. Louchet (ENSTA Paris) – J.J. Merelo (Univ. Granada)
O. Michel (Cyberbotics Ltd. Lausanne) – Z. Michalewicz (NuTech Solutions Inc.)
F. Mondada (EPFL Lausanne) – P. Preux (LIL Univ. Calais)
N. Radclie (uadstone Ltd Edinburgh) –C. Reeves (Coventry University)
D. Robilliard (LIL Univ. Calais) – E. Ronald (Ecole Polytechnique Paris)
G. Rudolph (Dortmund Univ.) – M. Sebag (LMS Paris)
M. Sipper (EPFL Lausanne) – E.-G. Talbi (LIFL Univ. Lille)
G. Venturini (E3I Univ. Tours)

Invited Talk

Why Biologists and Computer Scientists Should Work Together
P. Bentley (University College London)

Sponsoring Institutions

Conseil R´egional de Bourgogne
Universit´e de Bourgogne – Centre Universitaire Condorcet
Communaut´ e Urbaine Le Creusot – Montceau
DGA (D´el´egation G´en´erale pour l'Armement)
INRIA (Institut National de Recherche en Informatique et Automatique)
CMAPX (Centre de Math´ ematiques Appliqu´ ees de l'Ecole Polytechnique)
AFIA (Association Fran caise pour l'Intelligence Artificielle)

Table of Contents

Applications

Implementation Issues

Genetic Programming

Constraints Handling

Coevolution and Agents Systems

Why Biologists and Computer Scientists Should Work Together

Peter J. Bentley

Department of Computer Science, University College London,
Gower Street, London WC1E 6BT, UK.
P.Bentley@cs.ucl.ac.uk
http://www.peterjbentley.com/
http://www.cs.ucl.ac.uk/staff/P.Bentley/

Abstract. This is a time of increasing interdisciplinary research. Computer science is learning more from biology every day, enabling a plethora of new software techniques to flourish. And biology is now beginning to see the returns, with new models, analyses and explanations being provided by computers. The merging of computer science and biology is a hard thing to achieve. It takes a lot of effort. You have to overcome much resistance on both sides. But it's worth it.

In this paper, which accompanies the keynote presentation for Evolution Artificielle 2001, Peter J. Bentley discusses a new breed of scientist called the Digital Biologist, and why they are so important. Examples of research that benefit both fields will be provided, including swarming systems, computational development, artificial immune systems and models of ecologies. Only by working together will biology learn how nature works, and computer science develop techniques that have some of the awesome power of nature.

1 Introduction

"What do you get when you cross a computer scientist with a biologist?"

No, it's not the first line of a joke, although many computer scientists and biologists might laugh at the idea of working together. The biologists might find the idea that computers could have any relevance to biology very amusing. The computer scientists might find the idea that the natural world was related to their work quite funny too. But this is not a joke. It's a way of performing research.

So what do you get when they cross? Or to be more precise, what do you get when they collaborate? In truth, you get misunderstandings: headaches of new terminology or different meanings for existing terms, and sometimes even a complete inability to understand the words of your collaborator. You also get confusing ideas, strange motivations, different ways of performing experiments, alternative ways of interpreting the results and unlikely-sounding theories. Should you pluck up the courage to attend (or even present a paper) at the conference in your collaborator's field, you get overwhelmed with all of the above multiplied by several hundred.

P. Collet et al. (Eds.): EA 2001, LNCS 2310, pp. 3-15, 2002.

As difficult as all this sounds – and it is difficult – it's worth it. After a few weeks of learning each other's vocabulary you are able to communicate. The chances are you'll also find some fascinating new concepts along the way. The new ideas you hear will spark off exciting ideas of your own, the different motivations might suggest new applications to you. The alternative ways of performing experiments and analysing the results could suggest new ways for you to test your own work. The unlikely-sounding theories might explain something in your own field. And although you may feel a little lost in the alien territory of your collaborator's conference, you can guarantee there'll be at least one or two papers that will have your heart beating faster with excitement at the possibilities for your own work.

Many of the problems will never go away: you will probably always have different ways of thinking, different vocabularies and different motivations. But these are good things. Once you understand how your collaborator works, the differences produce far more significant and original research than you could have produced alone.

And sometimes, after computer scientists and biologists have worked together long enough, they change a little. They realise the value of using computers to model biological processes. They see the new understandings of nature and the new computational techniques that such interdisciplinary research can bring. They become digital biologists.

In this paper I argue that collaborations between biologists and computer scientists are providing the next crucial steps on the road to understanding biology and exploiting biological processes in computation. I discuss the problems of beginning collaborations and how to make them succeed. Examples of such collaborations at University College London (UCL) are provided.

2   Starting Collaborations

Scientists can be very territorial creatures who loathe venturing far from their familiar surroundings. Computer scientists are perhaps more adventurous than biologists in this respect: because computers are a means to an end, these scientists have to find something for the computers to do. This normally means finding applications or problems to solve. While computer scientists can be talented at making up theoretical problems, these are often unsatisfying and even insufficient to test their ideas. Instead they need a real application, and this is provided by industry or academics in different fields. So, many computer scientists are quite used to working with people from outside of their field. Biologists, on the other hand, tend to be more insular. They train, research and present their results only within their communities (and sometimes to the outside world via press releases). Now and again, some may get together from different fields and grudgingly compare notes, but this is less common. Perhaps more than any other field of science, biology is subdivided and segregated into a huge number of separate disciplines.

The nature of the fields means that should a computer scientist wish to learn about techniques inspired from biology or even about modelling biology, most will still only look within their own field for work performed by other computer scientists. And should a theoretical biologist decide that some computer modelling, visualisation or analysis is necessary, he is more likely to try and learn how to do the programming himself or use another biologist's software, than to talk to a computer scientist. These are fundamental barriers that are very effective in preventing collaboration. They are caused by lack of knowledge, misunderstandings and prejudices.

So how do we make the two sciences communicate? The answer is plain: educate the scientists. Spread the word about the research going on in the different fields. Let computer scientists know the value and relevance of biologists' research and let the biologists see the value of computer science.

Interestingly, one of the most successful ways of achieving such education is through popular science books. Although not necessarily written with this aim in mind, the genre of "pop science" allows a curious scientist to learn important achievements and discover current ideas in fields far from their own. Because such books are written for the general public, the terminology is drastically reduced (or at least explained a little more thoroughly than usual), overcoming the normal language barriers between fields. A number of collaborations at UCL between biologists and computer scientists (and mathematicians) were begun primarily because the biologist happened to read a pop science book (such as Kauffman's Origins of Order [1] or Bentley's Digital Biology [2]), or because the computer scientist read such a book (e.g., Dawkins's The Blind Watchmaker [3]). Other types of books also aid collaboration, for example edited collections of chapters that bring together specialists from different fields (such as Bentley's Evolutionary Design by Computers [4] or Creative Evolutionary Systems [5]).

Books are not the only trigger for collaboration. Another successful route at UCL has been the formation of special interest groups (for example, nUCLEAR: the nexus for UCL Evolutionary Algorithm Research). These meet regularly and discuss current publications or invite speakers on interesting topics. Most importantly, they focus on interdisciplinary subjects and bring together researchers from different universities as well as different fields in an informal atmosphere. A number of new collaborations and opportunities for funding at UCL have been created by these groups.

If these approaches don't appeal, there is of course the simplest of all. If you'd like to collaborate with someone in a different field, look through their web pages and publications. If they show an interest in something related to your research, just go and see them. You may get blank faces and no interest, but sometimes you may find extreme excitement and the source of an exciting new interdisciplinary research project. Again, some successful research has begun at UCL using this approach.

3 Making Collaborations Work

Once you've found a collaborator or two, you need to work out how to perform research together. As with any project with a number of researchers, the objectives need to be clearly understood and the work subdivided appropriately. With collaborations between biologists and computer scientists, this is more interesting.

From experience, these interdisciplinary projects tend to fall into four categories: biology-driven research, computing-driven research, parallel biology and computing-driven research and, rarest of all, combined biology and computing-driven research. Usually the initiator of the collaboration will determine the type of research. In more detail:

TYPE 1:   Biology-driven research. Initiated by a biologist, this form of research will focus on modelling or processing the data of real biological systems. The skills of a computer scientist or mathematician are normally crucial to ensure accurate results, but the findings will be mostly of importance to biology and not computer science.

TYPE 2:   Computing-driven research. Initiated by a computer scientist, the aims are to use the expertise and knowledge of biologists to improve existing algorithms or create new ones. The knowledge of biological processes provided by the biologist will be invaluable for the development of new computational techniques, but the results will be more significant to computer science than biology.

TYPE 3:   Parallel biology and computing-driven research . Initiated by either specialist, this type of project is two in one. It may have began as a "Type 1" project, with the computer scientist suddenly becoming inspired into developing a new algorithm. Alternatively it may have begun as a "Type 2" project, with the biologist realising that the computer could also be used to help understand some aspect of biology. Either way, two separate strands of research form, related but distinct. The results of the research will benefit both biology and computer science equally.

TYPE 4:   Combined biology and computer-driven research. Still the rarest form of research, these projects are the sole domain of the digital biologist. This type of research is usually initiated by biologists with some expertise in computing, or by computer scientists with knowledge of biology, and is a single project designed to benefit biology and computer science equally. Merging biology and computer science to this extent is difficult, but this type of research project can produce some very interesting results that would not be possible without such close collaboration.

At University College London we have much experience in all of these types of research. Indeed, whole research centres have been set up to tackle research in the ways described above. For example, the Centre for Mathematics and Physics in the Life Sciences and Experimental Biology (CoMPLEX) is a virtual group bringing together mathematicians with biologists to perform "Type 1" research. The Gatsby Computational Neuroscience Unit focuses on "Type 3" research in computational models of neurons and neural networks (but has projects of all types). The Computer Science Department, UCL, also performs all types of research mentioned above.

Choosing which type of research to perform is purely subjective. If you're a biologist uninterested in anything except using the best techniques available to solve your problems, you'll favour "Type 1" projects. If you're a computer scientist only interested in developing the best techniques using some inspiration from nature, you'll prefer "Type 2" projects. But if you're a computer scientist or biologist willing to invest a little more effort in an interdisciplinary project in the hope of far greater rewards, you may consider "Type 3" or even the challenging "Type 4" projects.

4 Getting Results

To illustrate the kinds of collaborations possible between biologists and computer scientists and show their benefits, the next sections briefly review a selection of projects begun by Peter Bentley at UCL's Department of Computer Science. These are, in order: artificial immune systems, swarms for learning, musical swarms, computational development, computational ecology, and evolving vision systems.

4.1 Artificial Immune Systems

Over the last four or five years, research performed by computer scientists Jungwon Kim and Peter Bentley has focussed on the combination of a set of biologically-inspired algorithms for the application of intrusion detection [6] (e.g., network intrusion detection, or the detection of hackers or unauthorised users in a system). These algorithms are all based on processes from the human immune system. They are known as negative selection, clonal selection and gene library evolution.[1]

Each algorithm has one aspect of our immune systems as its inspiration, and each has specific strengths. For example, the negative selection algorithm is based on the way our immune systems remove harmful antibodies from our bodies. Antibodies are generated by a variety of white blood cell known as a B-cell and help attack unwanted viruses and bacteria within us. Each B-cell produces a single, unique antibody, and a clever randomising gene expression method is employed to ensure that a huge diversity of different antibodies can be made by all the B-cells combined. Unfortunately, some B-cells produce antibodies that mistakenly attack our own 'self cells'. Luckily our immune system has a clever process (one of many) known as negative selection, which ensures that any B-cells that produce such harmful antibodies die. All that remain are B-cells that produce antibodies that do not attack self cells. In other words, negative selection tries to ensure there are antibodies for everything other than self cells.

The negative selection algorithm uses the same trick: antibodies (or detectors for some problem) are randomly generated. If the detectors are triggered by normal behaviour of the system they are supposed to protect, they are simply deleted. This leaves only the detectors that are not triggered by normal behaviour, or to put it another way, detectors for abnormal behaviour.

---

[1] Initially developed by Stephanie Forrest in "Type 2" interdisciplinary research, which has now developed into "Type 3".

In contrast, the clonal selection algorithm is based on the way B-cells are duplicated within our bodies. As B cells produce a wide diversity of antibodies, only a very small number will be effective against a particular pathogen. But our immune system is able to increase its response until there are sufficient concentrations of the correct antibody to help destroy the pathogen. It does this by cloning B-cells that make the right kind of antibody: the more of the right kind of B cells there are, the more the corresponding kind of antibody is produced. But our immune systems also have a couple of other tricks: as well as cloning the B-cells, hypermutation is used, ensuring that many slight variations of the current B-cells are produced. Should any mutated B-cell produce an even more effective kind of antibody, then it will undergo clonal selection and its solution will soon propagate through our blood streams. This is an evolutionary process being used by our immune systems: new B-cells (and the DNA within them) are evolved within us to ensure the most effective immune response to pathogens. And it doesn't end there, for our immune system also generates memory cells that are stored within us, in case the same pathogen is encountered in the future. These cells give us immunity to the disease.

The clonal selection algorithm uses these ideas: it evolves detectors for patterns of abnormal behaviour (or antigens), but evaluates them in a special way. Random groups of individuals in the population are selected and 'shown' to a single antigen. The best at detecting the antigen in the group has its fitness increased; the fitnesses of the others remain unchanged. Then another random group is picked and compared to another antigen, and so on. Finally, the fittest detectors are cloned with some mutation. The result is an evolutionary algorithm that develops niches of detectors that work together to detect a large number of different antigens.

The final algorithm under investigation was gene library evolution. This is based on the way the DNA within B-cells is generated and used. As described above, each B-cell produces a unique type of antibody, which is used to help remove unwanted viruses and bacteria. The antibodies are unique because each B-cell uses a unique (and partially random) combination of DNA fragments to specify the antibody it produces. This is a tremendously clever and complex process in itself, but there's more: the DNA fragments used to build antibodies are not completely random. Many of the fragments have been carefully evolved over millions of years to ensure that effective antibodies are usually produced. How? Through the action of the Baldwin effect. Good DNA fragments that get used in B-cells that produce good antibodies that are effective in keeping a creature alive, are more likely to be passed onto future generations than bad ones. In effect, evolution improves the capability of our immune systems to adapt to as yet unseen assailants. And research in using a genetic algorithm to evolve effective gene libraries which are used to produce a diverse and effective range of detectors has shown good results.

Through investigation of these ideas, research at UCL has shown some of the drawbacks of using small, highly abstracted processes from the immune system in isolation. For example, the basic negative selection algorithm was shown to be unable to cope with real-world network traffic data – randomly generating detectors was too inefficient. Recent work has focussed on combining the separate algorithms, and attempting to make a computer immune system that uses more of the carefully integrated processes in our own immune systems. To this end, we have developed a system which evolves useful gene libraries, which then specify useful detectors, which are kept valid by negative selection and improved through clonal selection. This integration of immune algorithms ensures that each process functions in a manner more similar to the way it was evolved in biological immune systems. The system is now being extended by Jungwon Kim at Kings College, London, to check continuously changing data (from the UK's Post Office) for fraud.

The research at UCL has not taken place without assistance. Prof. Robin Callard (an immunobiologist at the Institute of Child Health, London) has provided some invaluable support. The processes of our immune systems are highly complex and difficult to follow – it was the help of Robin that enabled us to understand the details of the processes we were interested in. We also gained his inside knowledge of which theories are most relevant and accurate about the immune system. His complete disregard of Jerne's Network Theory (which is used as the basis for other computer immune systems) was a surprise to us. Robin is no stranger to interdisciplinary research – after being inspired by a popular science book, he has worked with CoMPLEX to model the immune system. His assistance on this work (begun after we simply went to see him for a chat) has helped make this "Type 2" research project successful.

4.2 Musical Swarms

Another source of new biology-inspired algorithms has been the findings of Entomologists such as Nigel Franks, Jean-Louis Deneubourg, and Tom Seeley [rev. 2]. By studying the movement of insects such as bees, it has been discovered that the majority of the observed swarming behaviour can be produced by applying a small number of rules to every insect, or in a computer, agent. The rules cause multiple interactions of the agents with positive and negative feedback and the amplification of small random fluctuations. Together, these cause the astonishing coordination and illusion of central control so typical of swarming or flocking systems.

Exactly which rules to use in an algorithm depends on which "discoverer" you wish to follow [rev. 2]. For example, Reynolds suggests that each agent in a flock should:

1. Try to avoid colliding with any of its companions.
2. Try to move towards the centre of the flock.
3. Try and match the speed of its companions.

Alternatively, Eberhart suggests that every 'particle' in a 'particle swarm system' should also:

4. Be attracted to a 'roost' or target.

At UCL, a recent research project by Tim Blackwell and Peter Bentley investigated these ideas [7]. This research used a combination of the above rules for each agent in a swarm:

1. Try to avoid colliding with any of its companions.
2. Try to move towards the centre of the flock.
3. Be attracted to a 'roost' or target.

The work applied the swarming agents to the problem of music improvisation. Given a real-time audio input such as a saxophone or singer, the audio waveform is analysed, individual notes are identified and these are positioned into a "music space" with axies comprising pitch, loudness and start time. Because the input is constantly changing, the target continuously moves in this space. Musical agents are then allowed to swarm in the same space, each following the three rules above. As they move, their positions are used to define musical notes (for every point in the space gives a specific pitch, volume and start time). The result is a swarming behaviour that follows the target, giving the musical sensation of listening and responding, whilst the swarm's own uncertain dynamics provides novel musical ideas [7].

Although the application may be a little unusual, it did enable some interesting findings. By analysing the ability of the swarm to respond to changing targets, it was discovered that the first rule used in this work (avoid colliding with companions) played an important role in damping the oscillations of the swarm around the target. This rule is not used as standard in particle swarm optimisation research – our work suggests that making the swarm slightly more realistic will assist the ability of the swarm to search a problem space.

Again, this "Type 2" research was not performed in isolation. Assistance on insect behaviour had been provided previously by entomologist Andrew Bourke, at the Institute of Zoology in London. And this is not the only research investigating swarms at UCL.

4.3 Swarms for Learning

Another project, this time being performed by Supiya Ujjin and Peter Bentley, investigates the use of swarms for recommender systems [8]. These are software tools designed to learn the preferences of a shopper, and recommend products and services that are specifically tailored for each person. Such systems are already in use for many on-line stores. Often the user is asked to provide some feedback on products they have bought, and this information is used to work out suggestions. For example, if your feedback on one or more products is similar to the feedback provided by someone else, then it is possible that you will like other products that person likes. Indeed, given sufficient data, it is even possible to predict what your feedback might be for that product.

Currently, most recommender systems do not pay much attention to the vagaries of human beings – they do not attempt to model customer's likes and dislikes with any great sophistication. But in reality, people pay attention to specific, but different features of products. For example, my main reason for choosing a movie might be because it is science fiction, while you might choose it because it stars your favourite actor. Neither of us would be served well by a system that only suggests movies based on a general voting system.

Following work using genetic algorithms for this task, research in the early stages at UCL is examining how a swarm could search a problem space of feature weightings. These would be specific to each customer and would enable the calculation of best recommendations based on a "swarmed" feature-weight profile. The ability of swarms to cope with sparse data and continuously changing data may make this swarm intelligence-based system more effective than existing approaches.

4.4 Computational Development

For some years, I have been advocating a greater use of the mapping stage from genotypes to phenotypes in evolutionary systems [rev 9]. These views followed the discovery that for many types of problem that require complex solutions, a simple one-to-one mapping would prevent evolution from finding a result. It seemed that as the complexity of the required solution increased (e.g. requiring features such as modularity, self-similarity, symmetry, duplication, and hierarchies) so the need for a new approach to evolutionary computation increased.

Looking to nature provides the solution to this dilemma: embryology, growth, morphogenesis, or more correctly, development. Natural systems do not have a one to one mapping from gene to phenotypic effect. A highly complex process of development uses the DNA as instructions on how to build the phenotype. There is no concept of one gene specifying one feature in nature: genes only specify proteins. The proteins from one cell trigger or suppress the activation of other genes in other cells, which trigger or suppress the activation of yet more genes in other cells, and so on. At the same time, the proteins change the cells: new cells are made, existing cells are destroyed, cells are told to reshape themselves, extrude substances or even to move. Some are told to have more specialist children, which then have more specialist children again, and so on, in a process of differentiation that enables the creation of over 200 different types of cell.

Through these clever processes the most complex entities on the planet are formed: you and me. There is modularity as genes that perform similar tasks become grouped together in our chromosomes, and also as cells that perform similar tasks become grouped together as organs. There is self-similarity and duplications as genes that perform useful tasks are repeated or used repeatedly, resulting in duplicated structures such as vertebrae, ribs, or segmentation in insects. There is symmetry as the same, or similar genetic instructions are triggered on both sides of the body.

So the logical solution was to somehow incorporate development into an evolutionary algorithm. To do this required major changes to our representations: we needed genotypes that act as instructions, the use of some kind of component-based (or cell-based) representation to develop the phenotypes with, and possibly even a final phenotype representation [10]. Work at UCL began with some initial visits to see developmental biologist Paul O'Higgens. It soon developed further as Sanjeev Kumar joined UCL to work on this full time. We now have the support of eminent embryologists Lewis Wolpert and Michel Kerszberg and the research has developed from a "Type 2" project into "Type 3" work, with Sanjeev using a genetic algorithm to test Michel's theories in parallel to his own work on computational development. The main research, however, has been an investigation of how a biologically-plausible developmental system, that incorporates realistic gene-activation and suppression, protein diffusion and cellular behaviours, can aid the evolution of complex solutions. Previous work has shown indications that scalability and possibly evolvability may be greatly improved [11]. Other investigations (with Tim Gorden at UCL) include examining the potential of development-inspired methods for evolvable hardware. Work is ongoing to explore the capabilities of such systems further.

4.5 Computational Ecology

Ecologists also benefit from collaborations. For 18 months I have been working with Jacqui Dyer, an ecologist interested in the evolution of life in disturbed environments. She believes that traditional numerical models used by ecologists do not capture the behaviour of evolution with respect to environments prone to disasters such as earthquakes or fires. Such models predict that population dynamics in disturbed environments will fluctuate more strongly than those in stable environments, resulting in higher extinction rates, lower biodiversity and more simple community structures in disturbed, compared to stable environments. But these models ignore empirical data that show that many ecosystems evolve to overcome or even make use of such disasters for their survival. Frustrated by the assumptions and inaccuracies of numerical models, she approached me with the idea of developing a more realistic computational model. Early on, we decided to simplify things: we would model the evolution and responses of plants only.

With the help of undergraduate student Panash Shah, a model was created [12] (which I later rewrote and optimised for speed). The PLANTWORLD model was initially developed in order to examine the effects that the evolution of a functional response - in this case, dormancy - might have on the population dynamics of PLANTS. Each PLANT requires a single resource, moisture, which varies in availability both spatially and temporally. In addition, this implementation allows us to study the effects of two further strategies that can influence dynamics: (i) the effects of PLANT Storage Capacity ii) the effects of an alternative source of moisture, in the form of a Water Table.

Two objectives motivated the development of this system. The far-reaching objective was to attempt to develop systems that could integrate evolutionary and ecological dynamics in spatially extensive and temporally variable environments. Such an objective is prohibited in numerical models by the sheer complexity required and is only recently becoming a realistic objective in computational models. PLANTWORLD represents only the initial stages in the development of such a system, only modelling one type of agent, PLANTS, and a single resource, moisture. However, it is capable of supporting populations of 400,000 or more at the equivalent of 24000 months every hour (on a 500 Mhz Pentium III laptop computer) and uses real rainfall data to provide realistic environmental conditions. It is envisaged that other agents (herbivores, pests, etc) and variables (nutrients, light, fire, etc) will be added at later stages. In the meantime, the development of PLANTWORLD has a more immediate objective. One of the advantages of agent-based models over numerical models of population dynamics is that our agents can exhibit behaviours. Combined with evolutionary computation, such behaviours can evolve. Thus, we can examine how the evolution of traits in different environments affects the population dynamics in these environments. The immediate objective for building PLANTWORLD is therefore to examine the evolution and effects of plant dormancy on population dynamics in different spatially and temporally variable environments. The simulation is not intended to capture realistic behaviour of any specific flora but rather to test the veracity of predictions about population dynamics that arise from numerical models.

This is an example of that rare type of project, "Type 4" research - the model is the result of a close collaboration and provides fascinating results for both computer science and ecology. For example, there are no fitness functions describing what is, and what is not, fit. A PLANT merely begins as a seed, which germinates given sufficient resources. It then grows until it reaches a mature size defined by a gene, and will be fertilised by a nearby mature PLANT, producing its own seeds (with sufficient resources). At all times it follows the strategies defined in its genes, going dormant or growing during certain months. If its genes help it to survive and propagate in the environment, then those genes will be passed onto its offspring. From an evolutionary computing perspective, the model provides fascinating evidence of the evolution of different solutions to a dynamically changing and unpredictable problem. Stable niches of different types of plants evolve and coexist, from tiny, short lived "grasses" to large, long-lived "trees" that can make use of the water table below. From an ecology perspective, the model shows realistic population dynamics: interdependent cycles of population sizes emerging, or the evolution of more dynamic strategies of survival for disturbed environments.

4.6 Evolving Visual Systems

The final collaborative project I will mention here is the most recent. A couple of months ago, two people contacted me within a few days of each other: Beau Lotto, a neurobiologist at UCL's Institute of Ophthalmology, and Marcel van Gerven, a student wanting to do a Ph.D. at UCL. By some stroke of luck, both wanted to do a similar kind of research: evolve neural networks for vision recognition. I put them in touch with each other, and now we have all begun work together on what should be a fruitful "Type 4" research project.

The aim is to a test Beau's general theory of how vision evolved [13], paying particular attention to colour vision. This theory suggests the visual system perceives colour based not on the light that actually reaches the eye, but on the reflectances and different illuminances that generated the stimulus in the past. So, for example, when we see a shiny black object, we perceive it as being shiny and black, even though our eyes might be seeing something that has greys, reflections and even patches of white on its surface. We know it is black because we know that in the past, such combinations of shades mean "shininess" with specific reflections and lighting. But because our visual systems make use of past experiences of the sources of different stimuli when they process current stimuli, they can be fooled. Optical illusions demonstrate this, particularly those demonstrating that we perceive colours differently depending on which other colours are nearby.

The intention is to evolve and train neural networks such that they are capable of recognising various coloured stimuli, even when under different lighting conditions. The resulting networks will then be analysed and tested, firstly to see if they are also fooled by the same optical illusions as us, and secondly to see what kinds of neural network perform such tasks.

We are hoping that the results of this research will both help explain the evolution and functioning of our own visual systems, as well as point to new ways of developing computational visual systems in the future. Whatever we learn, the chances are it will be interesting.[2]

5 Summary:

Why Biologists and Computer Scientists Should Work Togther

In this paper I have advocated greater collaboration between biologists and computer scientists. In a field known as "evolutionary computation", one would think such views are commonplace, but in reality there are surprisingly few researchers who attempt any form of communication, let alone collaboration with their biologist counterparts. Of course biologists do not have all the answers any more than computer scientists do. However, they do often have many years of experience, knowledge and understanding that is simply ignored by most of computer science. Likewise, computer scientists have many years of expertise that is usually ignored by biologists.

Two years before his death in 1954, Alan Turing published a paper that laid the foundations of understanding for generations to come. The paper was entitled "The Chemical Basis of Morphogenesis". This advance was not in computer science like much of his previous and very famous work, but in developmental biology.

Let's not forget our roots. There have always been links between biology and computer science. By forging new ones, we can make progress in both fields at a pace greater than ever before. Digital biologists are the future.

Acknowledgements. My thanks to the following people for their assistance and for allowing me to mention our work here: Tim Blackwell, Andrew Bourke, Robin Callard, Jacqui Dyer, Marcel van Gerven, Tim Gordon, Michel Kerszberg, Jungwon Kim, Sanjeev Kumar, Beau Lotto, Paul O'Higgens, Panash Shah, Supiya Ujjin, and Lewis Wolpert.

References

1.   Kauffman, S. A. (1993). The Origins of Order: Self-Organization and Selection in Evolution. Oxford University Press.

---

[2] For more details on these and other projects, see Digital Biology [2].

2.   Bentley, P. J. (2001). Digital Biology. Hodder Headline Press, London.
3.   Dawkins, R. (1991). The Blind Watchmaker. Penguin Books.
4.   Bentley, P. J. (1999). Evolutionary Design by Computers. Morgan Kaufmann Publishers Inc., San Francisco, CA.
5.   Bentley, P. J. and Corne, D. W. (2001). Creative Evolutionary Systems. Morgan Kaufmann Publishers Inc., San Francisco, CA.
6.   Kim, J. and Bentley, P. J. (2001). Investigating the Roles of Negative Selection and Clonal Selection in an Artificial Immune System for Network Intrusion Detection. To appear in the Special Issue on Artificial Immune Systems in IEEE Transactions of Evolutionary Computation.
7.   Blackwell, T. (2001). Making Music With Swarms. M.Sc. Project Report, Department of Computer Science, University College London.
8.   Ujjin, S. and Bentley, P. J. (2001). Building a LifeStyle Recommender System. In Proc.of the Tenth International World-Wide-Web Conference. RN/01/5
9.   Bentley, P. J. (2000). Representations Are More Important Than Algorithms: Why Evolution Needs Embryology. Keynote speech, ICES2000, Edinburgh, 17-19 April 2000.
10.   Bentley, P. J. (2000). Exploring Component-Based Representations - The Secret of Creativity by Evolution? In Proc. of the Fourth International Conference on Adaptive Computing in Design and Manufacture (ACDM 2000), April 26th - 28th, 2000, University of Plymouth, UK.
11.   Kumar, S. and Bentley, P. J.(2000). Implicit Evolvability: An Investigation into the Evolvability of an Embryogeny. A late-breaking paper in the second Genetic and Evolutionary Computation Conference (GECCO 2000), July 8-12, 2000, Las Vegas, Nevada, USA.
12.   Jacqueline R. Dyer, Peter J. Bentley, Panash Shah (2001) PLANTWORLD: The Evolution of Plant Dormancy in Contrasting Environments. A late-breaking paper in the third Genetic and Evolutionary Computation Conference (GECCO) 2001.
13.   Polger, T. W., Purves, D. Lotto, B. (2000). Color Vision and the Four-Color-Map Problem. In Journal of Cognitive Neuroscience, 12(2):233-237.

Niching in Monte Carlo Filtering Algorithms

Alexis Bienven ue[1], Marc Joannides [2], Jean B´erard [3], Éric Fontenas [2], and
Olivier Fran cois[1]

[1] LMC, BP 53, 38041 Grenoble Cedex 9, France
Alexis.Bienvenue, Olivier.Francois   @imag.fr
[2] LABSAD, BP 47, 38040 Grenoble Cedex 9, France
Marc.Joannides, Eric.Fontenas   @iut2.upmf-grenoble.fr
[3] LaPCS, Universit´ e Lyon 1, 50 av. Tony Garnier, 69366 Lyon Cedex 07, France
Jean.Berard@univ-lyon1.fr

Abstract. Nonlinear multimodal filtering problems are usually addressed via Monte Carlo algorithms. These algorithms involve sampling procedures that are similar to proportional selection in genetic algorithms, and that are prone to failure due to genetic drift. This work investigates the feasibility and the relevance of niching strategies in this context. Sharing methods are evaluated experimentally, and prove to be ecient in such issues.

1 Introduction

In evolutionary computation, genetic drift is often considered as a source of premature convergence. Given a problem with multiple solutions, a genetic algorithm (GA) will at best ultimately converge to a population containing only one of these solutions. This phenomenon has been observed in natural as well as in artificial evolution, and is undesirable in many applications (e.g. multi-objective optimization). To overcome the above problem, several methods have been proposed that take their inspiration from mathematical ecology [1]. GA were developed that are capable of forming and maintaining stable sub-populations, or niches. GAs which employ niching mechanisms become capable of finding multiple solutions to a problem, within a single population [2], [3], [4]. Among these methods, the most popular is fitness sharing, that works by modifying the objective function according to the presence of nearby individuals.

Beyond the field of evolutionary computation, similar phenomena have been observed in Monte Carlo strategies such as iterated bootstrap or particle filtering [5], [6]. Such algorithms are based on selection procedures as well, and are closely connected to the traditional GA framework (see Section 2). These strategies have proved their eciency in high-dimensional nonlinear problems. In terrain navigation for instance, an aircraft measures its relative elevation sequentially, and the goal of filtering is to estimate the position and velocity of the aircraft.

Current address: TIMC, Facult´ e de M´edecine, Domaine de la Merci, 38706 La Tronche Cedex.

P. Collet et al. (Eds.): EA 2001, LNCS 2310, pp. 19–30, 2002.
c Springer-Verlag Berlin Heidelberg 2002

This problem is multimodal as several positions might correspond to a single relative elevation. While Monte Carlo strategies are theoretically able to simulate the true distribution of the aircraft position, selection often concentrates the solutions on a single mode leading to wrong decisions.

While the benefit of sharing methods has been intensively studied by the EC community, few eorts have been devoted to the other contexts. This work evaluates the feasibility of sharing methods in Monte Carlo nonlinear filtering algorithms. Section 2 presents an account on the filtering problem and Monte Carlo methods. Section 3 introduces niching strategies in sampling procedures for particle filters and discusses the choice of a sharing bandwidth. In Section 4, the algorithm is evaluated on a set of one-dimensional test problems similar to those encountered in terrain navigation. For these simple models, the solution to the filtering problems can be computed exactly. On this basis, the niching method is compared to the standard algorithm, and proves to be beneficial in this context.

2   Monte Carlo Filtering

Filtering addresses the issue of predicting an unknown signal (X_t) given noisy observations of this signal. Mathematically, the signal is modeled as a Markov process taking values in some measurable space   X :

$$X_t = F (X_{t-1}) + V_t, \quad t \quad 1, \tag{1}$$

where F is a deterministic function and (V_t) is a sequence of independent identically distributed centered random variables. More generally, such dynamics can be specified according to some Markov kernel   (x, dx) that describes the transition probabilities between successive states. The distribution of X_t is often called the prior distribution .

In the filtering problem, the signal cannot be observed directly. Instead, data are (indirectly) gathered from the observation of a second signal

$$Y_t = H (X_t) + W_t, \quad t \quad 0, \tag{2}$$

where H is usually a nonlinear function and W_t a noisy variable independent from X_t.

The filtering problem consists of making predictions about the original signal X_t given the observations Y_0, Y_1, ..., Y_t. This amounts to estimating (or computing) the conditional distribution of X_t given these observations. This distribution is called the   posterior distribution .

Filtering has an old tradition that goes back to the seminal paper by Kalman [7]. The standard approach assumes that   F and H are linear and that V_t and W_t are Gaussian random variables of known covariance matrices. In contrast, solving nonlinear ftering problems turns out to be particularly dicult, and the diculty is even increased when the signal becomes unidentifible (e.g.,   H not invertible). In such a case, the posterior distribution may be multimodal. Kalman filters would therefore lead to erroneous predictions since these methods always predict a single mode.

Kunita and Stettner [8] developed general recursion schemes that compute the exact solution of the filtering problem based on Bayes' formula. Despite their closed form, these equations are hardly of practical interest since numerical computations of high-dimensional integrals are involved.

Monte Carlo filtering is an algorithmic alternative to Kunita-Stettner recursion. It consists of a computer intensive technique, and can be useful where linear filtering fails. This method is based on a particle system approach [9], [10], [11] in which the posterior distribution is computed empirically. In this approach, a population of n particles is evolved in the signal space according to a two-stages procedure. More precisely, let $x_t = (x_1, \ldots, x_n)$ be the population at time t (x_0 is randomly initialized). The two steps are iterated as follows.

1) Prediction.   Create n new particles $x_1, \ldots, x_n$ by sampling from the transition kernel (x, dx)

$$x_i \quad (x_i, dx), \quad i = 1, \ldots, n.$$

Conditional to $x_1, \ldots, x_n$, the new particles are independent.

2) Correction.   Resample the particles according to a proportional scheme taking the observation y_t (at time t) into account:

$$x_i \quad \frac{L_t(y_t \quad H(x_i))}{\sum_j L_t(y_t \quad H(x_j))},$$

where L_t is the likelihood function of the observation noise W_t. Define the population at time $t + 1$ as being $x_{t+1} = x$.

The convergence of this algorithm to the optimal solution of the filtering problem has been proved in [10], when the population size goes to infinity.

Turning to an evolutionary computation perspective, there is a close connection between Monte Carlo filters and the simple GA without recombination. This connection has been emphasized in previous works by [10], [12]. In the above algorithm, particles can be identified as individuals in a population, where the set of phenotypes corresponds to the possible states of the signal. The first step, called prediction, is similar to the mutation step in GAs. Each individual generates an ospring by mutation from the kernel (x, dx) (and the ospring replaces its parent). In the second step, called correction, a random selection of the ospring is performed. The selection strategy is similar to the proportional selection scheme used in GAs. However, the fitness function is time-dependent as it must account for the information contained in the data at each instant. Mathematically, the fitness of ospring x_i can actually be defined as

$$f_t(x_i) = L_t(y_t \quad H(x_i)). \tag{3}$$

3 Niching in a Filtering Context

3.1 Niching Algorithms

The reason why Monte Carlo filtering methods should work is that their infinite-population models correspond to Kunita's recursion scheme precisely. However, the shortcomings of infinite-population models are well-known. By their very nature, they may not truly reect the finite-population properties that are of major interest to a practitioner. For instance, the eect of stochastic uctuations during the correction step are neglected in this approach.

The same kind of remark also holds for the traditional GA. To overcome the above problem, Goldberg and Richardson proposed a method based on sharing [1]. These methods require that the objective   fitness function be shared as a single resource among similar individuals in a population. Niching is achieved by degrading the objective function (i.e., the unshared fitness) of an individual according to the presence of nearby individuals. This type of niching requires a distance metric on the phenotype of the individuals. The objective fitness $f(i)$ of an individual i is degraded by first summing all of the shared values of individuals within a fixed bandwidth $_{sh}$ of that individual and then dividing $f(i)$ by this sum, which is known as the   niche count . More specifically, if two individuals are separated by distance $d(i, j)$, then a shared value

$$
sh(d(i,j)) = \begin{cases} 1 \quad \dfrac{d(i,j)}{sh} & \text{if } d(i,j) \quad_{sh} \\ 0 & \text{otherwise} \end{cases} \tag{4}
$$

is added to the niche count:

$$
m(i) = \sum_{j=1}^{n} sh(d(i,j)) . \tag{5}
$$

The parameters $_{sh}$ and   are chosen by the user of the niched GA based on some a priori knowledge of the fitness landscape. The parameter   is often set to one, yielding the triangular sharing function. Finally, the shared fitness is defined as

$$
f(i) = \frac{f(i)}{m(i)} . \tag{6}
$$

The actual fitness of each individual is modulated according to the density of the population around it: the fitness of isolated individuals is increased, while that of individuals in well represented regions of the search space is decreased.

3.2 Sharing Methods in Monte Carlo Filters

In this paper, we investigate the maintenance of stable sub-populations in Monte Carlo filtering algorithms using the method of sharing function. Since the available computational resources do not allow the number of particles to be arbitrary large, standard Monte Carlo flters often suer from a loss of diversity due to the stochastic nature of resampling. In real-world applications (e.g. real-time target tracking algorithms), this premature loss of diversity implies loosing the signal for some time by concentrating all individuals in a possibly wrong region of the search space. Maintaining stable niches in Monte Carlo filter is therefore a crucial point, since these niches actually correspond to existing modes of the posterior distribution. A niching procedure can be included in Monte Carlo filters as follows.

1) Prediction (unchanged).   Create n new particles $x_1, \ldots, x_n$ by sampling from the transition kernel (x, dx)

$$x_i \quad (x_i, dx), \quad i = 1, \ldots, n.$$

Conditional to $x_1, \ldots, x_n$, the new particles are independent.

2) Correction.   Resample the particles according to a proportional scheme taking the observations into account:

$$x_i \quad \frac{f_t(x_i)}{\sum_j f_t(x_j)},$$

where $f_t(x_i)$ is the shared fitness of $f_t(x_i) = L_t(y_t \quad H(x_i))$.

In implementing this algorithm, choosing the bandwidth $_{sh}$ is a critical step. Deb's rule sets this parameter by taking into account distances between peaks, and relative fitnesses. Specifically,

$$_{sh} = \min_{i,j} \quad \frac{d(x_i, x_j)}{1 \quad r_{ij}}, \tag{7}$$

where

$$r_{ij} = \min \quad \frac{f_t(x_j)}{f_t(x_i)}, \frac{f_t(x_i)}{f_t(x_j)} \tag{8}$$

and the metric d corresponds to the Euclidean Distance. In one-dimensional filtering problems, a better-supported rule is given by Silverman [13] inspired from density estimation

$$_{sh} = 0.9 \min(sd(x), iqr(x)/1.34) n^{0.2}.$$

This rule is based on the minimum of the standard deviation of x and its interquartile range divided by 1.34. Using this rule is quite natural in a niching context. Sharing indeed degrades the fitness function by dividing by the density of nearby particles and actually involves an estimation of this density. Note that a proper use of this rule requires that the sharing method be build upon a Gaussian kernel instead the triangular function. Similar rules also exist in higher dimensions.

4 Experiments

4.1 Test Problems

Evaluating the impact of the sharing method in Monte Carlo fiers is dicult in general. A number of test problems have been chosen to assess the performances empirically. The selection of six test problems is inspired from target tracking issues. Three noisy dynamical systems describe the motion of a target in one dimension. The first motion is the classical   AR(1) dynamics [7]

$$X_t = 0.9X_{t\ 1} + V_t, \quad V_t \ N \ (0,1), \tag{9}$$

where the (random) initial condition X_0 is sampled according to the Gaussian distribution N $(0, 5.26315)$ (so that the signal is stationary). The second motion is called the piecewise linear dynamics , and can be described as

$$X_t = X_{t\ 1}\quad 0.1\,\mathrm{sign}(X_{t\ 1}) + V_t, \quad V_t \ N \ (0,1) \tag{10}$$

and $X_0 = 0$. The third motion is called the   double well dynamics [14]

$$X_t = X_{t\ 1}\quad 0.04 X_{t\ 1}(X_t^2{}_1\quad 1) + V_t, \quad V_t \ N \ (0, 0.01 \quad q), \tag{11}$$

where q is a parameter set to 0 .24 in [14], and $X_0 = 0$. In addition, these three motions are observed through dierent functions. The fist observation function is a symmetric one

$$H(x) = \ x , \tag{12}$$

and the second is non-symmetric

$$H(x) = \begin{array}{l} 2x \quad \text{if } x \quad 0; \\ x/\,2 \text{ if } x \quad 0. \end{array} \tag{13}$$

The observation noise is standard Gaussian $W_t = N(0,\ ^2)$ (is often set to 1). The length of simulation runs is equal to   T = 100. Regarding the symmetric observation function, the posterior distribution is bimodal while this is not always the case for the non-symmetric function. For the six problems, the posterior distribution can be computed exactly using Kunita's recursions. Knowing the exact solution will be useful in assessing the accuracy of the filtering procedures during the experiments.

Figure 1 displays a typical trajectory from the double well dynamics (a) and the exact posterior distributions computed via Kunita's recursions under a symmetric observation function (b). A population of 20 individuals is evolved using the classical Monte Carlo procedure (c) and the niching algorithm (d).

4.2 Experimental Design

Simulation runs contain simulated trajectories of the signal motion and the corresponding observations. For each of the six models, simulations are repeated 100 times so that the performances can be evaluated statistically. In the experimental design, the following parameter settings are experimented.

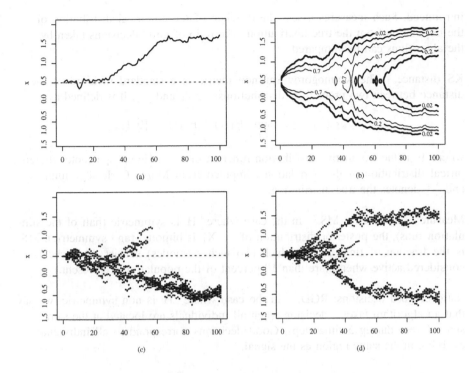

Fig. 1.   (a) A simulated trajectory from the double well dynamics. (b) Contour plot of
the posterior distribution densities. The observation function is symmetric $H(x) = x$.
(c) Classical Monte Carlo filtering plot (population size = 20). (d) Monte Carlo +
Niching simulation plot.

NS	No Sharing.
Sil	Silverman's rule for the sharing bandwidth.
Deb	Deb's rule for the sharing bandwidth.
sh	constant sharing bandwidth (values = 0.1, 1, 10).

The variable NS means that no sharing is used. The algorithm corresponds to the
classical Monte Carlo filtering method. The next levels indicate how the sharing
bandwidth has been set up: Silverman's rule, Deb's rule or constant. Except for
Silverman's rule, the triangular function is chosen (parameter   sh = 1 in Figure 2).
Other choices were tested but the results did not change significantly.

4.3   Performance Measures

As shown by Figure 1, sharing can be helpful in combating genetic drift in
Monte Carlo filtering algorithms. Indeed, the distribution of individuals seems
closer to the true distribution than the population corresponding to the classical
procedure. To quantify these observations, several performance measures can be

introduced. Such measures assess the distance of the empirical distribution of the population from the true distribution. The ratio of good decisions taken by the algorithms can be compared.

KS distance. The Kolmogorov-Smirnov metric is a standard measure of the distance between two probability distributions μ_1 and μ_2. It is defined as

$$D_{KS}(\mu_1, \mu_2) = \sup_t |F_1(t) - F_2(t)|, \quad [0, 1],$$

where F_μ is the cumulative distribution function of μ. Here, μ_1 denotes the empirical distribution of the population computed from Monte Carlo algorithms and μ_2 denotes the true distribution.

Measure of symmetry: MS. In the case where H is symmetric (half of the simulation runs), the posterior distribution of X_t is bimodal (and symmetric). MS is the fraction of time during which two niches (modes) subsist. (A niche is considered active when more than 10 percent of the population are present.)

Ratio of good decisions: RGD. In the case where H is non symmetric, we say that an algorithm takes a decision when all individuals are located at the same side of zero during 5 time steps. Good decisions correspond to all individuals evolving in the same region as the signal.

4.4   Results

This Section presents the most significant results obtained after the series of experiments. Symmetric observation functions are discussed first. Figure 2 reports the values of MS and the KS distance for the double well dynamics (DW). Similar results have been obtained for AR(1) and the piecewise linear dynamics. These results are summarized in Table 1.

Figure 2 shows that the fraction of time (MS) during which two niches subsist averages to 0.346 in the Monte Carlo filtering algorithm. This fraction increases to 0.922 when sharing is used together with Deb's rule (the best bandwidth is however $\sigma_{sh} = 1$). Simulation runs have also been performed with a population size of 100 individuals. The improvement due to sharing is more significant when the population size is small. Similar remarks can be made regarding the KS distance. For large population sizes, the goodness-of-fit seems better when Silverman's rule is used.

Table 1 reports the relative gain in using Deb's rule computed for each measure (MS and KS). This gain represents the dierence between measures with sharing / without sharing averaged over 100 runs. The results are given as percentages (MS and KS are oating point numbers between 0 and 1). Numbers in brackets represent the best ratio obtained from constant bandwidth rules (when this ratio is significantly better than Deb's rule). The star means that MS reached the maximal value (100). The following set of comments can be made about these results.

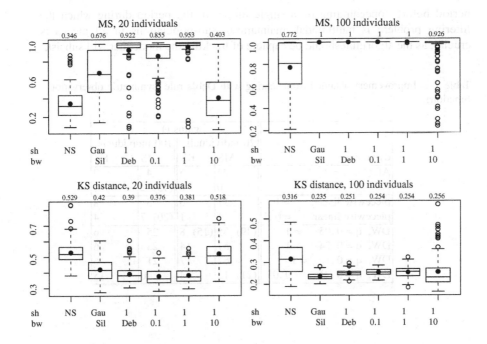

Fig. 2.   Performance measures corresponding to the double well dynamics (q = 0 .24,
= 1) and the symmetric observation function.   sh refers to the sharing function
(Gaussian or triangular) and   bw to the bandwidth parameter.

1) For this set of bimodal problems, sharing always improves Monte Carlo filters
   (the improvement may sometimes be a minor one).
2) Deb's rule is competitive (and has the advantage to be adaptive). This explains why this rule is chosen as a reference rule in Table 1.
3) Silverman's rule outperforms the other rules for large population sizes. (In
   some sense, this rule is optimal in density estimation when   n grows to infinity.)
4) Sharing is less ecient when the observation noise is small.

Turn now to the experimental results in the non-symmetric observation context. These results are summarized in Table 2 using the same notations as before.
In the non-symmetric context, the measure of symmetry has been replaced by
the ratio of good decisions (RGD). The posterior distribution is indeed multimodal only during a short interval of time, after that it concentrates on a single
mode.

In contrast to bimodal problems, high gains can hardly be expected. The aim
of sharing is maintaining individuals in niches when such niches truly exist. Note
that this method do not create artificial niches. Significant gains can nevertheless
be observed when the posterior distribution remains multimodal within a long

period before concentrating on a single mode. In the period during which the filtering problem is multimodal, maintaining sub-populations in all niches is crucial, as the algorithm should be capable of tracking the mode that will subsist.

Table 1.   Improvement obtained from sharing with Deb's rule (symmetric observation function).

dynamics	gains ()			
	20 individuals		100 individuals	
	MS	KS	MS	KS
AR (1), = 1	24	5	4	3
AR (1), = 0 .1	0	0	(15) 2	(7) 1
piecewise linear, = 1	17	6	13	8
piecewise linear, = 0 .2	1	1	(20) 7	4
DW, q = 0 .05, = 1	(49) 35	(15) 8	25	6
DW, q = 0 .24, = 1	57	14	23	6
DW, q = 0 .1, = 1	62	17	21	6
DW, q = 0 .24, = 0 .05	(21) 12	4	40	15

Table 2.   Performances of sharing for the non-symmetric observation function.

dynamics	gains ()			
	20 individuals		100 individuals	
	RGD	KS	RGD	KS
AR (1), = 1	2	-1	2	-3
AR (1), = 0 .1	(1) -5	0	-2	0
piecewise linear, = 1	(2) -4	1	0	-5
piecewise linear, = 0 .2	-2	2	4	3
DW, q = 0 .05, = 1	11	9	17	(1) -5
DW, q = 0 .24, = 1	(19) 6	6	10	(1) -12
DW, q = 0 .24, = 0 .05	-2	-1	5	-3

5   Discussion

This paper presented a new paradigm in Monte Carlo filtering algorithms: the method of likelihood sharing. While niching methods are widely accepted in evolutionary computation, the benefit of these techniques remains unexplored in several neighboring domains.

Our results give evidences that sharing methods can improve Monte Carlo filtering algorithms significantly. These methods are dedicated to problems for which posterior distributions are multimodal and standard algorithms are not ecient. Adding a sharing method allows population sizes to be reduced by a large factor, and contributes to the global eciency of the algorithm.

The method can be beneficial as well for problems that are not purely multimodal. This occurs in tracking a specific mode among several others that would be prominent after a while.

The empirical results presented in this paper have been obtained for one-dimensional problems, for which the solution can be computed by standard numerical methods. Further work is needed to extend this contribution to real-world problems (such as those arising in terrain navigation) and higher dimensional issues.

Acknowledgments.   This work is supported by the projects IMAG-SASI and AIPB.

References

1. David E. Goldberg and Jon Richardson. Genetic algorithms with sharing for multimodal function optimization. In John J. Grefenstette, editor, *Genetic algorithms and their applications : Proc. of the second Int. Conf. on Genetic Algorithms*, pages 41–49, Hillsdale, NJ, 1987. Lawrence Erlbaum Assoc.
2. Kalyanmoy Deb and David E. Goldberg. An investigation of niche and species formation in genetic function optimization. In James D. Schaer, editor, *Proc. of the Third Int. Conf. on Genetic Algorithms*, pages 42–50, San Mateo, CA, 1989. Morgan Kaufmann.
3. Carlos M. Fonseca and Peter J. Fleming. An overview of evolutionary algorithms in multiobjective optimization. *Evolutionary Computation*, 3(1):1–16, 1995.
4. Jerey Horn. *The nature of niching: Genetic Algorithms and the evolution of optimal, cooperative populations*. PhD thesis, University of Illinois at Urbana-Champaign, 1997.
5. C. Musso and N. Oudjane. Particle methods for multimodal filtering. In *Proc. of the second international conference on Information Fusion, Silicon Valley, CA, July 6-8*, pages 785–792. IEEE Press, 1999.
6. C. Musso, N. Oudjane, and F. Legland. *Improving regularized particle filters*, chapter Improving regularized particle filters. In Doucet and Gordon [11], 2001.
7. R. E. Kalman. A new approach to linear filtering and prediction problems. *Transaction of the ASME-Journal of Basic Engineering*, pages 35–45, 1960.
8. H. Kunita. Asymptotic behavior of non-linear filtering errors of markov processes. *J. Multivariate Analysis*, 1(4):365–393, 1971.
9. N. Gordon, D. Salmond, and A. Smith. Novel approach to nonlinear/non-gaussian bayesian state estimation. *IEEE proceedings, Part F*, (140):107–113, 1993.
10. P. Del Moral. Nonlinear filtering: interacting particle solution. *Markov Processes and Related Fields*, 2(4):555–579, 1996.
11. A. Doucet, J. F. G. de Freitas, and N. J. Gordon, editors. *Sequential Monte Carlo Methods in Practice*. Springer-Verlag, 2001.
12. P. Del Moral, L. Kallel, and J. Rowe. *Modeling Genetic Algorithms with Interacting Particle Systems*, chapter Modeling Genetic Algorithms with Interacting Particle Systems, pages 10–67. Springer-Verlag, Berlin, 2001.

30   A. Bienven ue et al.

13. B. W. Silverman. Density estimation for statistics and data analysis . Chapman
    Hall, London, 1986.
14. R. N. Miller, E. F. Carter, and S. T. Blue. Data assimilation into nonlinear stochas-
    tic models. Tellus , (51A):167–194, 1999.

Measurement of Population Diversity

Ronald W. Morrison [1] and Kenneth A. De Jong [2]

[1] Mitretek Systems, Inc.
7525 Colshire Drive
McLean, VA 22102-7400
ronald.morrison@mitretek.org
[2] Department of Computer Science, George Mason University
Fairfax, VA 22030
kdejong@gmu.edu

Abstract. In evolutionary algorithms (EAs), the need to eciently measure population diversity arises in a variety of contexts, including operator adaptation, algorithm stopping and re-starting criteria, and fitness sharing. In this paper we introduce a unified measure of population diversity and define its relationship to the most common phenotypic and genotypic diversity measures. We further demonstrate that this new measure provides a new and ecient method for computing population diversity, where the cost of computation increases linearly with population size.

1 Introduction

Population diversity is a key measurement in a variety of EA implementations. The question of when to stop the EA or when to re-start the EA is often based on a measure of population diversity. In fitness sharing algorithms, population diversity is used as a basis for distributing the fitness credit. The use of EAs for dynamic fitness landscapes requires measures for maintaining population diversity to ensure that the EA can detect and respond to the changes in the landscape.

Several methods for estimating population diversity have been used. They include diversity measures in both genotypic space and phenotypic space. In phenotypic space, several pair-wise and "column-based" measures (measuring the variation in values for each specific phenotypic feature) have been suggested (e.g., [1]). Genotypic measures are much more common. Principal genotypic measures are entropy (e.g., [2]), and, more commonly, pair-wise Hamming distance (e.g., [3]). Pair-wise Hamming distance H of P strings of length L is defined as:

$$H = \sum_{j=1}^{j=P-1} \sum_{j'=j+1}^{j'=P} \sum_{i=1}^{i=L} (y_{ij} \oplus y_{ij'}) \tag{1}$$

where $y_{ij}, y_{ij'} \in {0, 1}$ and the generalized notation,

$$\sum_{k=1}^{k=M-1} \sum_{k'=k+1}^{k'=M} f(x_k, x_{k'}) \tag{2}$$

P. Collet et al. (Eds.): EA 2001, LNCS 2310, pp. 31–41, 2002.

is the sum of the results of the application of $f(x_k, x_k)$ to all pair-wise combinations the members x_k and x_k of a given population of size M.

Historically, one of the major diculties in the use of pair-wise population diversity measures is that the computation of the measure is quadratic with the size of the population P for pair-wise selection:

$$\binom{P}{2} = \frac{P^2 \quad P}{2}. \tag{3}$$

In this paper we introduce a unified measure of population diversity and define its relationship between the most common phenotypic and genotypic diversity measures. We further demonstrate that this new measure provides a new and ecient method for computing population diversity, where the cost of computation increases linearly with population size. Section 2 of the paper will provide background information; Section 3 will define the diversity measure; Section 4 will relate the new diversity measure to other diversity measure in genotypic space; Section 5 will discuss the the diversity measure's relationship to other phenotypic-space measures; and Section 6 provides the conclusions and discusses future work.

2   Background

2.1   Historical Measures of Population Diversity

The most commonly used measures of population diversity include pair-wise Hamming distance in genotypic space, and column-based pair-wise distance and column variance in phenotypic space. In real-number optimization problems, phenotypic space diversity measures are often preferred over binary encoded genotypic measures. This is because, when using genotypic measures, all bit-wise diversity is treated the same, but variations at the dierent bit positions can represent signifiantly dierent levels of phenotypic diversity. Figures 1 through 9 provide illustrations of the three common diversity measures, using a simple genetic algorithm (GA), a population of 20 on a 2-dimensional, multi-modal landscape similar to that described in [4]. Gray code was used for the binary representation for the GA. Figure 1 is the initial population distribution. Figures 2 through 4 show the convergence of the population at generations 5, 16 and 20 respectively.

Figure 5 shows the pair-wise Hamming distance at each generation. Figure 6 provides the sum of the pair-wise distances of each column, and Figure 7 provides the sum of the variances of each column.

As can be seen in Figure 3, the population has lost nearly all of its diversity by generation 16. The three diversity measures provide somewhat dierent views of this loss of diversity, with the column variances (Figure 7) most clearly indicating population convergence, while the low-order bit dierences cause the genotypic space pair-wise Hamming distance measure (Figure 5) to indicate more diversity than is present in phenotypic space.

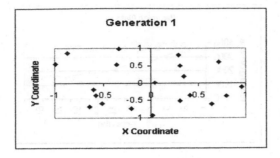

Fig. 1.   Population at Generation=1

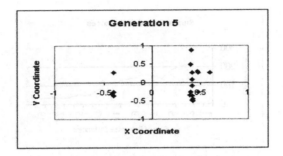

Fig. 2.   Population at Generation=5

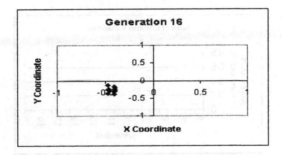

Fig. 3.   Population at Generation=16

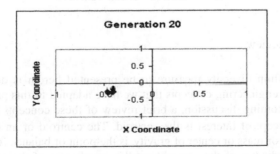

Fig. 4.   Population at Generation=20

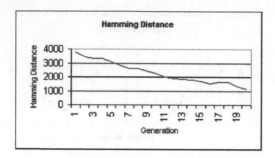

Fig. 5.   Population Pair-wise Hamming Distance

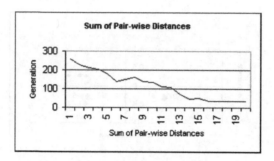

Fig. 6.   Sum of Pair-wise Distances

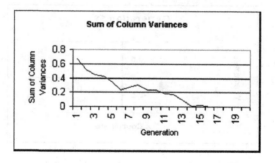

Fig. 7.   Sum of Column Variances

2.2   Concept Review

The new population diversity measure to be presented herein is derived from some traditional engineering concepts that we have adapted to this problem. To facilitate the upcoming discussion, a brief review of these concepts is provided.

The first concept of interest is the centroid. The centroid of an object, also called the center of mass or center of gravity, is the point of balance for the entire object. The coordinates of the centroid are the coordinates of the midpoints of the mass distribution along each axis.

The second concept of interest is the moment of inertia. Moment of inertia is a term used in many engineering problems and calculations. Just as mass is the relationship between force and acceleration according to Newton's second law, moment of inertia is the relationship between torque and angular acceleration. The moment of inertia indicates how easily an object rotates about a point of rotation. In any object, the parts that are farthest from the axis of rotation contribute more to the moment of inertia than the parts that are closer to the axis. Conceptually, when the point of rotation is the centroid of an object, the moment of inertia is a measure of how far the mass of the object is distributed from the center of gravity of the object. The engineering moment of inertia for a point mass is defined as:

$$I = mr^2 \qquad (4)$$

where: I is the usual symbol for moment of inertia, m is the mass, and r^2 is the square of the distance to the point of rotation.

3   A New Measure of Diversity

Our new measure of population diversity is based on extension of the concept of moment of inertia for measurement of mass distribution into arbitrarily high dimensionality spaces for the measurement of EA population diversity.

Extended into N-space, the coordinates of the centroid of P equally weighted points in N-space, $C = (c_1, c_2, c_3, \ldots c_N)$, are computed as follows:

$$c_i = \frac{\sum_{j=1}^{j=P} x_{ij}}{P} \qquad (5)$$

where x_{ij} and c_i is the ith coordinate of the centroid.

Continuing with P equally-weighted points in N-space, we define the moment-of-inertia based measure of diversity of these points about their centroid is:

$$I = \sum_{i=1}^{i=N} \sum_{j=1}^{j=P} (x_{ij} - c_i)^2. \qquad (6)$$

As will be shown in later sections, this measurement of population diversity is closely related to commonly used measures of both genotypic diversity and phenotypic diversity, providing a single diversity measurement method for use in both situations. The principal advantage of this measure of diversity is that, in comparison with traditional methods of computing pair-wise population diversity which are quadratic in population size, P, this method is linear in P. Specifically, for an N-dimensional problem with a population size of P, computation of the coordinates of the centroid requires N times P additions and N divisions. Computation of the moment of inertia around the centroid is then N times P subtractions plus N times P multiplications plus N times P additions. Total computational requirements for the centroid-based moment of inertia, therefore are $4(NP) + N$ calculations, making it a computationally ecient diversity measure.

4   Relationship to Diversity Measures in Genotypic Space

Genotypic diversity of EAs is most often measured using pair-wise Hamming distance, but the population diversity is much more eciently computed using the new moment of inertia method.

When applying the moment of inertia calculation in the context of binary strings, each bit is assumed to be an independent "spatial" dimension. Under these circumstances, the coordinates of the centroid, ($c_1, c_2, c_3, \ldots, c_L$), of P bit strings of length L are computed as:

$$c_i = \frac{\sum_{j=1}^{j=P} x_{ij}}{P} \tag{7}$$

and the moment of inertia about the centroid is:

$$I = \sum_{i=1}^{i=L} \sum_{j=1}^{j=P} (x_{ij} - c_i)^2. \tag{8}$$

It turns out that by transitioning from discrete mathematics to continuous mathematics, it can be shown that the moment of inertia as described in equation (8) is equal to the pair-wise Hamming distance divided by the population size.

Theorem 1.   For $y_{ij} \in 0, 1$:

$$\sum_{i=1}^{i=L} \sum_{j=1}^{j=P} \sum_{j'=j+1}^{1 \le j' \le P} y_{ij} \oplus y_{ij'} = P \left[\sum_{i=1}^{i=L} \sum_{j=1}^{j=P} (y_{ij} - c_i)^2 \right] \tag{9}$$

where:

$$c_i = \frac{\sum_{j=1}^{j=P} x_{ij}}{P}.$$

Verbally: the pair-wise Hamming distance for P bit strings of length L is equal to the L-space moment of inertia of the population computed around the centroid of the population times the population size. In short, the pair-wise Hamming distance is the binary case of the centroid moment of inertia.

Proof. [1]
First we will examine the right hand side of the theorem:

$$P \sum_{i=1}^{L} \sum_{j=1}^{P} (y_{ij} - c_j)^2 = P \sum_{i=1}^{L} \sum_{j=1}^{P} (y_{ij} - \frac{\sum_{j=1}^{P} y_{ij}}{P})^2$$

$$= P \sum_{i=1}^{L} \sum_{j=1}^{P} (y_{ij}^2 - 2y_{ij} \frac{\sum_{j=1}^{P} y_{ij}}{P} + \frac{1}{P^2} (\sum_{j=1}^{P} y_{ij})^2)$$

---
[1] Proof based on suggestions by Chris Reedy, Mitretek Systems.

$$= P \left[\sum_{i=1}^{L} \sum_{j=1}^{P} y_{ij}^2 - \frac{2}{P}\left(\sum_{j}^{P} y_{ij} \right)^2 + \frac{1}{P^2}\left(\sum_{j=1}^{P} \sum_{j=1}^{P} y_{ij} \right)^2 \right]$$

$$= P \left[\sum_{i=1}^{L} \sum_{j=1}^{P} y_{ij}^2 - \frac{2}{P}\left(\sum_{j=1}^{P} y_{ij} \right)^2 + \frac{1}{P}\left(\sum_{j=1}^{P} y_{ij} \right)^2 \right]$$

$$= P \sum_{i=1}^{L} \sum_{j=1}^{P} \left[y_{ij}^2 - \frac{1}{P}\left(\sum_{j=1}^{P} y_{ij} \right)^2 \right] = P \sum_{i=1}^{L} \sum_{j=1}^{P} y_{ij}^2 - \sum_{i=1}^{L} \sum_{j=1}^{P} \left(y_{ij} \right)^2. \tag{10}$$

To examine the left hand side of the theorem, let's first examine the properties of the quantity:

$$\sum_{i=1}^{L} \sum_{j=1}^{P} \sum_{j=1}^{P} \left(y_{ij} - y_{ij} \right)^2 = \sum_{i=1}^{L} \sum_{j=1}^{P} \sum_{j=1}^{P} y_{ij}^2 - 2 \sum_{i=1}^{L} \sum_{j=1}^{P} \sum_{j=1}^{P} y_{ij} y_{ij} + \sum_{i=1}^{L} \sum_{j=1}^{P} \sum_{j=1}^{P} y_{ij}^2$$

$$= 2 \sum_{i=1}^{L} \left[P \sum_{j=1}^{P} y_{ij}^2 - \left(\sum_{j=1}^{P} y_{ij} \right)^2 \right]. \tag{11}$$

Examined dierently, and changing notation for convenience, such that:

$$\sum_{i=1}^{L} \sum_{j=1}^{P} \sum_{j=j+1}^{P} . \tag{12}$$

Noticing that:

$$\sum_{i} \sum_{j} \sum_{j} \left(y_{ij} - y_{ij} \right)^2 = \sum_{i} \sum_{j} \sum_{j} \left(y_{ij} - y_{ij} \right)^2$$

$$+ \sum_{i} \sum_{j} \sum_{j=j} \left(y_{ij} - y_{ij} \right)^2 + \sum_{i} \sum_{j} \sum_{j} \left(y_{ij} - y_{ij} \right)^2 \tag{13}$$

and since:

$$\sum_{i} \sum_{j} \sum_{j=j} \left(y_{ij} - y_{ij} \right)^2 = 0 \tag{14}$$

then:

$$\sum_{i} \sum_{j} \sum_{j} \left(y_{ij} - y_{ij} \right)^2 = \sum_{i} \sum_{j} \sum_{j} \left(y_{ij} - y_{ij} \right)^2 + \sum_{i} \sum_{j} \sum_{j} \left(y_{ij} - y_{ij} \right)^2 \tag{15}$$

so, by symmetry:

$$\sum_i \sum_j \sum_j (y_{ij} - y_{ij})^2 = 2 \sum_i \sum_j \sum_j \sum_j (y_{ij} - y_{ij})^2. \qquad (16)$$

Combining (11) and (16):

$$2 \sum_i \sum_j \sum_j \sum_j (y_{ij} - y_{ij})^2 = 2 \sum_i [P \sum_j y_{ij}^2 - (\sum_j y_{ij})^2] \qquad (17)$$

so that:

$$\sum_i \sum_j \sum_j (y_{ij} - y_{ij})^2 = \sum_i [P \sum_j y_{ij}^2 - (\sum_j y_{ij})^2]. \qquad (18)$$

Since, for $y_{ij} \in 0, 1$, the left hand side of the theorem:

$$\sum_{i=1}^{i=L} \sum_{j=1}^{j=P} \sum_{j=j+1}^{j=P} y_{ij} \oplus y_{ij} = \sum_i \sum_j \sum_j (y_{ij} - y_{ij})^2, \qquad (19)$$

so that combining (10), (18) and (19):

$$\sum_i [P \sum_j y_{ij}^2 - (\sum_j y_{ij})^2] = P \sum_i \sum_j y_{ij}^2 - \sum_i (\sum_j y_{ij})^2 \qquad (20)$$

shows that the pair-wise Hamming distance is equal to the moment of inertia around the centroid times the population size.

4.1   Explanation and Example

The moment of inertia computational method for computing pair-wise Hamming distance works because all coordinates are either 0 or 1. This means that $x^2 = x$ and x times x is equal to x or x or both. As a simplified example of how this computational method is used, consider a population of six strings ($P = 6$), each three bits long and having values $y_{\text{gene,individual}}$ equal to:

$$y_{11} = 1, y_{21} = 1, y_{31} = 1$$

$$y_{12} = 0, y_{22} = 0, y_{32} = 0$$

$$y_{13} = 1, y_{23} = 1, y_{33} = 0$$

$$y_{14} = 1, y_{24} = 0, y_{34} = 0$$

$$y_{15} = 0, y_{25} = 1, y_{35} = 0$$

$$y_{16} = 1, y_{26} = 0, y_{36} = 1.$$

The coordinates of the population centroid are:
$C_1 = \frac{4}{6} = \frac{2}{3}$, $C_2 = \frac{3}{6} = \frac{1}{2}$, $C_3 = \frac{2}{6} = \frac{1}{3}$.
The population size times the moment of inertia around the centroid

$$P\left[\sum_{i=1}^{i=L}\sum_{j=1}^{j=P}(y_{ij} - c_i)^2\right] \tag{21}$$

is computed as:

$$6[(1-\tfrac{2}{3})^2 + (0-\tfrac{2}{3})^2 + (1-\tfrac{2}{3})^2 + (1-\tfrac{2}{3})^2 + (0-\tfrac{2}{3})^2 + (1-\tfrac{2}{3})^2$$

$$+(1-\tfrac{1}{2})^2 + (0-\tfrac{1}{2})^2 + (1-\tfrac{1}{2})^2 + (0-\tfrac{1}{2})^2 + (1-\tfrac{1}{2})^2 + (0-\tfrac{1}{2})^2$$

$$+(1-\tfrac{1}{3})^2 + (0-\tfrac{1}{3})^2 + (0-\tfrac{1}{3})^2 + (0-\tfrac{1}{3})^2 + (0-\tfrac{1}{3})^2 + (1-\tfrac{1}{3})^2]$$

$$= 6\left(\frac{12}{9} + \frac{6}{4} + \frac{12}{9}\right) = (8+9+8) = 25$$

which is the same value as the pair-wise Hamming distance for this population.

The computational eciency of the moment of inertia method of computing pair-wise Hamming distance makes a considerable dierence at population sizes normally encountered in evolutionary computation. For a bit string length of 50 and a population size of 1000, the number of computations necessary for calculation of the pair-wise Hamming distance by the moment of inertia method is two orders of magnitude less than that required by usual computational methods. Even adjusting for the fact that the moment of inertia method involves oating-point calculations, whereas Hamming distance calculations can be made using integer or binary data types, the moment of inertia method for computing pair-wise Hamming distance is considerably more ecient.

5 Relationship to Diversity Measures in Phenotypic Space

For an individual dimension, the moment of inertia measure is closely related to the calculation of statistical variance:

$$\sigma^2 = \frac{\sum(X - \mu)^2}{N} \tag{22}$$

diering only in the use of population size in the calculation.

The moment of inertia diversity measure, therefore, when applied in phenotypic space for real numbered parameters is related to the sum of the column variances. It should be noted, however, that when using the moment of inertia population diversity measure for real-numbered parameters, just as when combining traditional column-wise phenotypic diversity measures across columns, attention must be paid to individual parameter scaling. When searching a space, it is important to realize the impact of search-space size on the problem to be solved and understand the resolution (granularity) with which the search for a solution is to be conducted. For example, in a real-numbered convex-space optimization problem, the search space is defined by the ranges of real-numbered parameters. If the range of parameter   A is twice as large as that of parameter B, at the same granularity of search, the search space is twice as large along dimension A as along dimension B. In dierent cases the resolution of interest might be defined as a single percentage of the range, and this percentage might be equally applicable to all parameters. In this case, all parameters should be scaled equally. The moment of inertia calculations can be transformed to equally scale all parameters merely by dividing all parameter values by the parameter range. As long as the parameters are scaled so that they have an equal granularity of interest, the moment of inertia calculations provide an ecient method for measuring population diversity.

It is possible to envision circumstances where it would be desirable to compare the diversity of two dierent-sized populations on the same problem. In these cases, scaling the diversity by the population size would then be appropriate. When scaled in this manner, the moment of inertia diversity measure for the real parameter problem is equal to the sum of the column-wise variances of the individual parameters.

Figure (8) shows the moment of inertia diversity measure for the example problem used for Figures (1) through (7). Comparing Figure (8) to Figure (6) illustrates that, in addition to being more computationally ecient than the pair-wise column distance measure, the moment of inertia measure more dramatically portrays the population loss of diversity by generation 16 than does the pair-wise distance measure.

6   Conclusions and Future Work

We have introduced a unified method for computing population diversity that is equally useful for measuring diversity for both real-parameter populations and binary populations. Closely related to variance for real-parameter populations, and pair-wise Hamming distance for binary populations, moment of inertia diversity provides a single method of computing population diversity that is computationally more ecient than normal pair-wise diversity measures for medium and large sized EA problems.

The insight into the measurement of population diversity presented here leads to further questions and opens opportunities for other investigation. One area for further investigation relates to whether a suitable Levenshtein-distance [5] version of moment of inertia diversity measurement could be derived, to create more computationally ecient methods of measuring diversity in populations of

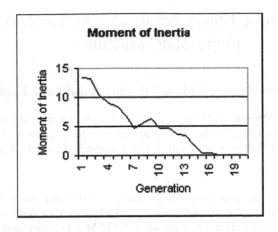

Fig. 8.   Moment of Inertia Diversity

unequal string lengths. Another area for further research relates to the use of this measure for investigating EA performance. For example, if the population points in N-space are not equally weighted, but are provided "mass" in accordance with their fitness, could the fitness-weighted moment of inertia (the "detected" fitness landscape) and the population moment of inertia (the EA's response to the landscape detection) be used as a measure of EA performance? These and other questions await further investigation.

References

1. DeJong, K.: An Analysis of the Behaviour of a Class of Genetic Adaptive Systems. Ph.D. Thesis, University of Michigan (1975)
2. Mori, N., Imanishi, S., Kita, H., Nishikawa, Y.: Adaptation to Changing Environments by Means of the Memory Based Thermodynamic Genetic Algorithm. In: Proceedings of the Seventh International Conference on Genetic Algorithms, Morgan Kaufmann, (1997) 299-306
3. Horn, J.: The Nature of Niching: Genetic Algorithms and the Evolution of Optimal, Cooperative Populations. Ph.D. Thesis, University of Illinois-Champaign (1997)
4. Morrison, R. De Jong, K.: A Test Problem Generator for Non-stationary Environments. In: Proceedings of Congress on Evolutionary Computation. IEEE (1999) 2047-2053
5. Sanko, D. Kruskal, J., Time Warps, String Edits, and Macromolecules: the Theory and Practice of Sequence Comparison. CSLI Publications, Stanford, California, (2000)

Prediction of Binary Sequences by Evolving Finite State Machines

Umberto Cerruti [1], Mario Giacobini [2], and Pierre Liardet [3]

[1] Dipartimento di Matematica, Universit   a degli Studi di Torino, 10100 Torino, Italy.
[2] Institut d'Informatique, Universit´   e de Lausanne,1015 Lausanne, Switzerland.
[3] CMI, Universit´ e de Provence, 13453 Marseille Cedex 13, France.

Abstract. This paper explores the possibility of using the evolution of a population of finite state machines (FSMs) as a measure of the 'randomness' of a given binary sequence. An FSM with binary input and output alphabet can be seen as a predictor of a binary sequence. For any finite binary sequence, there exists an FSM able to perfectly predict the string but such a predictor, in general, has a large number of states. In this paper, we address the problem of finding the best predictor for a given sequence. This is an optimization problem over the space of all possible FSMs with a fixed number of states evaluated on the sequence considered. For this optimization an evolutionary algorithm is used: the better the FSMs found are, the less 'random' the given sequence will be.

1   Introduction

The use of random binary sequences has become more and more frequent in many areas of applied mathematics, such as numerical integration, cryptography or statistics. From this viewpoint we can see growing interest in statistical algorithms able to give a measure of the   randomness of numerical sequences. In this direction, we propose an evolutionary algorithm that searches for automata able to predict a binary string better than randomly. The underlying idea is to use the evolutive ability of prediction of the algorithm to get measures of the randomness of the sequence. It is interesting to relate our method to the following one which is strongly connected to inverse problems for finite automata considered in [1]: by iterating a substitution it produces a string which is close (according to the Hamming distance for example) to a given one.

In order to detect regularities of sequences, we will use finite state machines (FSMs).The first investigation in this direction for binary sequences is due to M. O'Connor [10]. Generalizations to any symbolic sequences have been intensively studied by Broglio and Liardet [2] in the case of Mealy FSMs. In Section 2 we will outline their main results, and we will show how these results can be considered for both the Mealy and the Moore FSM models. Such a generalization will allow us to use their theorems in our case study, the Moore FSMs. Then we define the notion of intrinsic prediction skill: it represents the expected value of the prediction for a finite state machine which performs the best prediction during the initial phase.

P. Collet et al. (Eds.): EA 2001, LNCS 2310, pp. 42–53, 2002.
c Springer-Verlag Berlin Heidelberg 2002

In the third section, the problem will be presented: the search for the best predicting FSM in the space of all possible FSMs with a fixed number of states can be seen as an optimization problem. This observation justifies the use of an evolutionary algorithm to perform the search: this algorithm will be described in Section 4. Sections 5 and 6 will be devoted to the results found during the experimental phase. Two dierent cases are considered: the prediction of pseudo-random sequences, and the prediction of a special sequence obtained by the nuclear decay of krypton-85.

The conclusions that can be drawn from these experiments together with the future work we intend to do will be presented in Section 7.

2   Prediction by Automata

The concept of random sequences has not yet found a unique mathematical formulation and underlies the kind of problem we are working on. One of the main questions the researchers are trying to answer is if an eective process which recognizes randomness exists. In [2] the possibility of recognizing such random sequences with the help of deterministic finite state machines is studied. The authors introduced the prediction ratio of a   q-ary sequence. Since in our research we dealt only with binary sequences, we present in this section their results for $q = 2$ (see also [10]). Let $A = \{0, 1\}$ be the binary alphabet, let A^* be the set of finite strings of symbols from A, and let $u = u_1 u_2 \ldots u_n$ be any element of A^*. It is natural to think that if u is random, then no automaton is able, after reading $u_1, u_2, \ldots, u_{k-1}$ $(k \leq n)$ successively, to guess u_k with a probability of success exceeding $1 / 2$.

2.1   Finite State Machines

Independently defined in 1955 by G. H. Mealy and in 1956 by E. F. Moore, finite state machines are automata that, reading an input sequence, produce an output sequence. Both models can be seen as directed labelled finite graphs whose vertices are called states (one of them is called initial), and where a labelled directed edge between two dierent states indicates the passage of the automaton from the first to the second state when the labelling symbol is read. The dierence between the two models lies in the output function   : in the Mealy model   has domain S, the set of states, while in the Moore model its domain is $S \times A$, where A is the set of edges.

In this paper, a finite state machine M is defined by giving a string $M = \langle S, s_0, , \rangle$, where S is the set of states, s_0 is the initial state, $= \{ _a ; a \in A \}$ is the set of transition functions $_a : S \to S$ (one for each letter of the input alphabet, that determines the transition from one state to another one when the input bit a is read), and the output function   that, as mentioned above, has the two dierent domains in the two models. The underlying labelled graph of M is furnished by the set of labelled edges $e = (s, a, s') \in S \times A \times S$ (from s to s' with label a) verifying the relation $s' = _a(s)$, so that each edge e is completly determined by the couple (s, a). Consequently, in the Moore model, the output function will be defined directly on the set $S \times A$.

The results of [2] that we will present in the next section concern the Mealy model FSMs. It can be easily shown (see D. Cohen [5]) that these two different models are equivalent: given a finite state machine belonging to a family there exists another one belonging to the other family that produces, reading the same input strings, the same output sequences if the first output symbol is omitted in the Mealy model. This equivalence naturally extends all results in [2] to the Moore FSMs, our case study.

2.2 Predictors

We define a predictor P as a map $P : A^* \to A$. Given any word $w \in A^*$, we say that P predicts the letter $a \in A$ knowing the word w, if $P(w) = a$.

Let us consider $w = w_1 w_2 \ldots w_n$ a string of symbols of A of length n and set $w[k] := w_1 w_2 \ldots w_{k-1}$, for $k \in \{1, 2, \ldots, n+1\}$. Following this definition, $w[1]$ is the empty word that we denote by ϵ. It is easy to construct a predictor P able to perfectly predict the word w: it is sufficient to choose $P(\epsilon) = w_1$ and, for all $i \in \{1, 2, \ldots, n\}$, $P(w[i]) = w_i$.

With any finite state machine M (according to the Mealy model), we associate a predictor P_M, called automatic, defined by

$$\forall w \in A^*, \; P_M(w) = \lambda_{w_n} \circ \lambda_{w_{n-1}} \circ \cdots \circ \lambda_{w_1}(s_0)$$

and $P(\epsilon) = \lambda(s_0)$. In the case of the Moore representation, the corresponding output function μ is given by $\mu(s, a) = \lambda(\delta_a(s))$ and the automatic predictor P_M is then defined for any non-empty word w by

$$P_M(w) = \mu(\delta_{w_{n-1}} \circ \cdots \circ \delta_{w_1}(s_0), w_n)$$

with $P_M(w_1) = \mu(s_0, w_1)$. From this formula, P_M is not yet defined at the empty word but there is no loss of generality to fix arbitrarily (but definitively) the value $P_M(\epsilon)$.

Let P be a predictor; for any binary word $w = w_1 w_2 \ldots w_n$ of length n, the prediction ratio $\rho(P, w)$ of w relative to P is by definition

$$\rho(P, w) := \frac{1}{n} \sum_{j=1}^{n} \delta(P(w[j]), w_j)$$

where the function $\delta(\cdot, \cdot)$ is the Kronecker symbol, that is to say $\delta(a, b) = 1$ if $a = b$ and $\delta(a, b) = 0$ otherwise. If P derives from a finite state machine M we simply denote $\rho(M, w)$ for $\rho(P_M, w)$.

The definition of $\rho(P, \cdot)$ is extended to any infinite string w by setting

$$\rho(P, w) := \limsup_n \rho(P, w[n]).$$

A predictor P is said to be good for a certain sequence w if its prediction ratio is strictly greater than $1/2$, i.e., if P predicts the sequence better than randomly. The purpose of this paper is to find, given a binary sequence w, a good automatic predictor; the aim is to use automata in order to detect regularities in w.

Let $T = (S, s_0,)$ be a fixed automaton and, for any output function for T, let P be the automatic predictor given by the FSM $(S, s_0, ,)$. Now we define $(T, w) := \max (P, w)$ where the maximum is taken over all output functions . From a result of Broglio and Liardet [2], given any word $w_1 \ldots w_N \in A$ there exists a good automatic predictor among the P, i.e., $(T, w) \geq 1/2$. A natural question is to identify cases corresponding to $(T, w) = 1/2$. To see this we consider the Mealy model. In fact, the equality $(T, w) = 1/2$ leads to $(P, w) = 1/2$ for all output functions and this is equivalent to a mixing of properties about the walk on the labelled graph of T and derived from w. More precisely, set $s_n = w_n \ldots w_1(s_0)$. Then we have

Proposition 1. $(T, w) = 1/2$ if and only if for any $(s, a) \in S \times A$ the equality

$$ \{1 \leq n \leq N ; s_{n-1} = s, u_n = a \} = \frac{\{1 \leq n \leq N ; s_{n-1} = s \}}{2} \tag{1} $$

holds.

Proof. Assume that $(T, w) = 1/2$. Using a constant output function $_0$ we first have that both the letters of 0 and 1 occur in w the same number of times, and if we change the value of $_0$ at the state s we immediately derive that the number of integers n ($1, \ldots, N$) such that $w_n = 0$ and $s_{n-1} = s$ is the same as the number of integers n with $w_n = 1$ and $s_{n-1} = s$. This proves (1). Reciprocally, assume (1). Then, for any output function the prediction ratio can be written as follows:

$$ N (P, w) = \sum_{s \in S} \sum_{\substack{1 \leq n \leq N \\ s_{n-1} = s}} ((s), u_n). $$

Therefore,

$$ N (P, w) = \frac{1}{2} \sum_{s \in S} \{1 \leq n \leq N ; s_{n-1} = s \} = \frac{N}{2} $$

as expected.

Notice that the length of w satisfying (1) is necessarily even. Moreover, the proposition can easily be extended to the case of words over q symbols.

2.3   Intrinsic Prediction Skill

If we consider infinite binary words (i.e., elements of A) the above definitions and results maintain their validity, when the limit sup operator is used. In this case, it is proved in [10] and [2] that normal binary sequences are those whose prediction ratio of any automatic predictor is equal to $\frac{1}{2}$.

Following [2], an infinite binary string $u = u_1 u_2 u_3 \cdots$ is said to be exactly predicted by a predictor P if there exists N_P such that for all $n \geq N_P$ the equality $P(u[n]) = u_n$ holds. If, given an infinite word, there exists an automatic predictor that exactly predicts the sequence, then u is ultimately periodic and its period is at most equal to the cardinality of the definition domain of the underlying output function. Such a period will be called a period-word of P. The converse of this result is also evident: for any ultimately periodic (binary) word, there exists an automatic predictor that exactly predicts the word. Notice that for any given automatic predictor, it is easy to construct an infinite binary string u which is not ultimately periodic but $(P, u) = 1$.

The following result derives from a theorem in [2] which gives the general structure of infinite words such that $(P, u) = 1$.

Proposition 2. Let P be an automatic predictor and let u be an infinite binary word such that $\lim_n (P, u[n]) = 1$. Then, there exist two increasing sequences $(m_k)_k$ and $(n_k)_k$ of integers, a constant C and a sequence of period-words W_k of P such that for all k, $m_k \leq n_k \leq m_k + C$, $n_k \leq m_{k+1}$,

$$u_{n_k} \cdots u_{m_{k+1} - 1} = (W_k)^{s_k}$$

for a suitable integer s_k and $\lim_k \frac{k}{n_k} = 0$.

We have now presented the theoretical basis, elaborated in [2], that supports our research. Now, for a good interpretation of the prediction ratio found by the evolutionary computation techniques, we need an agreed definition of a prediction skill theoretically estimated for a random sequence. The question we posed was: "a random sequence should not have any good automatic predictor, so, in the case of a random sequence, what is the greatest prediction ratio we could find in the space of all the automatic predictors defined by FSMs with a fixed number of states?"

We know that, in the case of a Moore FSM with n states, the maximal period p_{max} of a sequence produced by this automaton is equal to the number of edges of the automaton (i.e., $p_{max} = 2n$). If we consider a random binary sequence of length $l \geq 2n$, we can suppose that there is an automatic predictor with n states which exactly predicts a subword of length at most equal to the maximal period of a sequence produced by this automaton (this fixes definitely the FSM) but the remaining part of the word must be predicted on average with a ratio of $1/2$. Consequently, we introduce the so-called intrinsic prediction skill (ips), as

$$\mathrm{ips} = \cdots_{max} = \frac{\frac{2n}{2} + 2n}{\cdots} = \frac{1}{2} + \frac{n}{\cdots}.$$

The following result explains and justifies the theoretical interest of this definition. Let M_n be the set of Moore FSM with n states and let $S : A^{2n} \to M_n$ such that for any $w \in A^{2n}$ the prediction ratio $s(w) := (S(w), w)$ is maximum. We extend S to any infinite binary word u by setting $S(u) = S(u[2n + 1])$. The map S will be called a selection rule and in general S is not uniquely determined.

Proposition 3. Let   denote the uniform probability on the set of binary strings of length   . Then

$$\int_A (S(w), w) d \quad (w) \quad ips + O\left(\frac{n^{1/3}(\quad 2n)^{2/3}}{}\right).$$

Proof. By definition of   and $S(w)$ the integral can be written as

$$I_S := \int_{A^{2n}} \left(\quad_{A^{2n}} (S(w), w)\right) d \quad (w). \tag{2}$$

Therefore

$$I_S \quad \frac{2n}{} + \frac{2n}{} \max_M \int_{A^{2n}} (\quad (M, u)) d \quad_{2n}(u), \tag{3}$$

where the maximum runs on the set of Moore FSMs M with n states. We know that the integral under the maximum converges to 1 $/2$, but from the proof of this result given in [2] we can derive the estimate $\int_{A^m} (\quad (M, u)) d \quad_m(u) = \frac{1}{2} + O\left(\frac{n}{m}^{1/3}\right)$.

The value of the integral I_S can be estimated more accurately to obtain

$$I_S = \frac{1}{2} + \frac{n(2 \quad_S \quad 1)}{} + O\left(\frac{n^{1/3}(\quad 2n)^{2/3}}{}\right)$$

where

$$\quad_S := \int_{A^{2n}} \quad_S(w) d \quad_{2n}(w)$$

is the expectation of $\quad_S$. This is due to introducing $\frac{2n}{}\quad_S$ in (3). On the other hand, the corresponding lower bound for I can be obtained in a similar way by taking the minimum over Moore FSMs in place of the maximum. In practice, the binary strings we study are finite but long enough to justify, as above, the introduction of the set A of all infinite words endowed with the uniform Bernoulli measure   . But it is of interest to consider another Borel measure   on A , also assumed to be invariant under the shift operator $: w_1 w_2 w_3 \quad w_2 w_3 w_4 \quad$.

In this more general situation, our algorithm can be adapted to build a selection rule S in order to maximize the integral $\quad_S() = \int_A (S(w), w) d \quad (w)$ which is $\quad_S$ if $=$. To be complete, we have to compute for any $M \quad M_n$ the expectation

$$J(M, \quad) := \int (M, u) d \quad (u).$$

As a consequence of [2], the Bernoulli case corresponds to $J(M, \quad) = \frac{1}{2}$. It will be of interest to study measures   such that $J(M, \quad)$ does not depend on M and n. The Markovian case seem to be a good candidate.

3   The Problem

In 1951 John von Neumann wrote: "Anyone who considers arithmetical methods of producing random digits is, of course, in a state of sin" (cited in Knuth [9]). In fact, a numerical sequence produced arithmetically to simulate randomness will always be ultimately periodic. But it is this kind of sequences which are used for most applications like calculus, computer programming, cryptography or games.

If we consider a purely periodic sequence, it is easy to construct an automatic predictor that exactly predicts it. Such a predictor can be defined by an automaton with a number of states equal to the length of the period of the sequence. Unfortunately, from a practical viewpoint, it is very hard to find, for a sequence with long period, the equivalent automaton in the space of all the possible automata. Moreover the eventual solution of this problem will hardly be implemented. But does there exist a predictor, defined by an automaton with a number of states smaller than the period of the sequence, able to predict in a 'good enough' way the sequence? This problem is obviously an optimization problem: we look for the FSM that better predicts a given sequence in the space of all the automata with a fixed number of states.

4   The Algorithm

To solve the problem explained in the previous section, we propose an evolutionary algorithm similar to the Generated Simulated Annealing proposed by Goldberg [8]. Such a choice is due to the fact that some initial experiments showed a limited inuence of the crossover operator on the evolution process in the usual genetic algorithms. Moreover the studies on the convergence of this kind of algorithm, done by Raphael Cerf ([3] and [4]), guarantee the absolute convergence to an optimal solution.

Binary sequences of dierent degrees of randomness' constitute the various environments in which the evolution processes have to take place. As we already said, the individuals subject to the evolution processes are Moore FSMs with a fixed number of states. Each FSM is used as a predictor for a given binary sequence: the output sequence, compared to the input one, gives the prediction skill (fitness) of the FSM for the considered binary sequence.

To determine a finite state machine $T = \langle S, s_0, \rangle$, we have chosen a binary representation. First we suppose that $S = \{1, \ldots, n\}$, $s_0 = 1$ and set $\ell := \inf\{i \in N; n \leq 2^i\}$. Then, T is coded by a binary string $A = A_1 \ldots A_k$ of length $2n(\ell + 1)$ with $A_k = B_k b_k C_k c_k$, where B_k, C_k are strings of length ℓ, and b_k, c_k are letters. By construction B_k (resp. C_k) corresponds to the classical binary expansion of $\delta_0(k)$ (resp. $\delta_1(k)$) completed to the left by a suitable number of 0 to obtain a string of length ℓ. Finally, $b_k = \partial(k, 0)$ and $c_k = \partial(k, 1)$. It is easy to see that such a representation furnishes a one to one correspondance between the set of all possible chromosomes and the set of all FSMs.

As in the Generated Simulated Annealing, only two genetic operators have been used: a mutation operator for the variation phase, and a selection operator for the selection phase. In any generation cycle, each bit of the chromosome of each member of the M parents population is mutated with a fixed probability. Therefore, after the reproduction phase, the algorithm has produced, besides the parent population, an ospring population of the same size.

The choice of the mutation probability determines the ratio between exploration of the search space and exploitation of the information already contained in the population. The first experimental results have shown that the algorithm possesses a good convergence character when this parameter is set between 0   .05 and 0 .2. It is clear that a probability close to 0   .2 increases the exploration character of the algorithm, and a probability close to 0   .05 increases the exploitation action. From our experiments, the best results have been obtained with a probability varying in that interval during the evolution process.

The selection phase is similar to that of an (m + m)-evolution strategy: in the population formed by the   m parents and the   m mutated osprings, the   m individuals with the best fitness pass as parent population in the next generation, and the rest are deleted.

There are two principal characteristics of this algorithm: the produced evolution process is non-regressive and hyper-elitist. In fact, the selection operator is such that in the worst case (i.e., no better individual is produced in the reproduction phase) the new population will be equal to the parent population. An evolutionary process results in which the best individuals are always kept in the next population (hyper-elitism) and the average fitness function over the populations is an increasing function (non-regressive evolution).

5   Prediction of Pseudo-Random Sequences

In the first experimental phase we chose as environments three sequences produced by binary linear feedback shift registers with 8, 11 and 12 cells. This choice was made principally for three reasons. First of all the sequences produced have, if the generating polynomial is irreducible, long periods (for a shift register with k cells, the period can be of length 2^k 1). A second reason is that these sequences furnish very good results with the usual randomness tests (see Knuth [9] for more details). For example, if we consider the sequence produced by an 8-cell shift register with irreducible polynomial $f(x) = x^8 + x^7 + x^2 + x + 1$, we have:

- entropy: 0 .999968 (optimal value 1 .0);
- compression: 0 (optimal value 0);
-    χ^2 distribution: 0 .73 (a random sequence would have a greater value 75 of the time);
- arithmetical mean: 0 .4967 (optimal value 0 .5);
- correlation coecient: 0.02 (optimal value 0 .0).

The last reason is that for these generators the automata that exactly predict the sequences have a too large number of states. Therefore, an exhaustive search of a good predictor is computationally unfeasible for large   k (40). In practice, in our cases, the equivalent automata would have 2 ^{8} 1, 2^{11} 1 and 2^{12} 1 states respectively, while the evolved automata have 8, 16 and 32 states. In this experimental phase, two dierent kinds of relations were studied:

1. Let us consider the environment sequence of length 80000 produced by a 12-cell shift register. What is the relation between the length   of the sequence presented to the automata for the evolutionary process and the prediction skills found by the algorithm?

   For a sequence of period length   p = 4095, populations of automata with 8, 16 and 32 states (of size 500, 1000 and 2000 respectively) were evolved. The mutation probability was set to 0.2 for the first 2000 cycles and to 0.05 for the remaining 400 cycles. The results are set out in Table 1.

Table 1.   Results of the evolutions of 8, 16 and 32 state FSMs in environments of lengths   = 400,   = 2000,   = 4100,   = 8200 and   = 41000. Notations: ips denotes the intrinsic prediction skill,   m is the maximal fitness found, and   r the evaluation of the best individual on the rest of the sequence.

	8 states	16 states	32 states
= 400	ips = 52 m = 65.09 r = 51.32	ips = 54 m = 68.33 r = 51.69	ips = 58 m = 72.07 r = 51.75
= 2000	ips = 50.40 m = 57.47 r = 54.16	ips = 50.80 m = 57.07 r = 54.56	ips = 51.60 m = 60.16 r = 54.99
= 4100	ips = 50.19 m = 56.47 r = 56.48	ips = 50.39 m = 57.89 r = 57.91	ips = 50.78 m = 58.42 r = 57.94
= 8200	ips = 50.09 m = 56.36 r = 56.36	ips = 50.19 m = 57.91 r = 57.93	ips = 50.39 m = 57.47 r = 57.47
= 41000	ips = 50.01 m = 56.17 r = 56.20	ips = 50.03 m = 58.44 r = 58.53	ips = 50.07 m = 58.68 r = 58.73

2. Let us consider three environment sequences with dierent linear complexities: one of period   p = 255 (8-cell shift register), one of period   p = 2047 (11-cell shift register), and one of period   p = 4095 (12-cell shift register). What is the relation between the linear complexity of the environment sequence (i.e., the number of cells of the generator) and the evolved prediction skill? The three sequences, with length equal to the period (for the results of the previous point, a length sucient for a good evolution), are presented to the same populations of FSMs considered in the previous experiments with the same parameters. The results are shown in Table 2.

Table 2.   Results of the evolutions of 8, 16 and 32 state FSMs for pseudo-random sequences of period lengths   p = 255, p = 2047 and   p = 4095. ips, m, and r have the same meaning as above.

	8 states	16 states	32 states
p = 255	ips = 51 m = 70.25 r = 70	ips = 52 m = 71.50 r = 72	ips = 54 m = 72.5 r = 72
p = 2047	ips = 50.13 m = 58.90 r = 58.23	ips = 50.26 m = 58.67 r = 58.42	ips = 50.53 m = 56.97 r = 56.88
p = 4095	ips = 50.19 m = 56.47 r = 56.48	ips = 50.39 m = 57.89 r = 57.91	ips = 50.78 m = 58.42 r = 57.94

6   Prediction of a 'Random' Sequence

After the encouraging results about prediction of pseudo-random sequences, we tested our algorithm on a sequence with a more random character. It is known that half of a mass of krypton-85 decays in rubidium-85 in 10.73 years, but it is not known which atoms are involved in the decay, or the lengths of the successive moments of the decay process. During a discussion with John Nagle in 1985, John Walker ([11]) had the idea of using the random character of this process to construct a random sequence. Considering the string T_i of the time intervals between two successive nuclear decays, one can define the binary sequence B_i by: $B_i = 1$ if T_i $T_{i 1}$ and $B_i = 0$ otherwise. This sequence possesses very good results from the most common statistical tests:

- entropy: 0 .999975 (optimal value 1 .0);
- compression:0 (optimal value 0);
- ^{2} distribution: 0 .04 (a random sequence would have a greater value 75 of the time);
- arithmetical mean: 0 .5029 (optimal value 0 .5);
- correlation coecient: 0.003941 (optimal value 0 .0).

The algorithm (with the same parameter setting as the previous experiments and a population of 200 automata) found the results shown Table 3.

7   Conclusions and Future Work

The results of this research are not considered to be definitive: many other experiments must be done, in order to explore the FSM prediction capability. In fact, from these first experiments, we can only draw partial conclusions that assure us about the validity of the investigation undertaken:

Table 3.   Results of the evolutions of populations of automata with 4, 5, 6 and 7 states
for the random sequence krypton-85 of length   = 400.

	4 states	5 states	6 states	7 states
= 400	m = 58	m = 58.25	m = 59.50	m =60
	ips = 50.5	ips = 50.62	ips = 50.75	ips = 50.87

- the evolved prediction skills are always better than the intrinsic prediction skills: we have thus always found automatic complete predictors that predict the sequences 'better than randomly';
- to evolve good prediction skills it is sucient to give half of the sequence period as the environment in the evolutions;
- even if it is evident that the space of all possible FSMs with a certain number of states contains sub-spaces isomorphic to spaces of FSMs with a smaller number of states, the bigger the FSMs are, the better the evolved prediction skills are;
- the evolved prediction skills are in inverse proportion to the period length of the considered sequence;
- the evolution of FSMs prediction skills seems to be directly linked to the linear complexity of the sequences considered.

These last two conclusions make us hope that the evolution of FSMs could be used as a measure of the randomness character of a binary string. More experiments have still to be done for a correct evaluation of parameter setting in the algorithm: the relation between the populations and the search space sizes, and the mutation probability setting require deeper investigations.

Even if these results are encouraging, further investigations on the use of FSMs in random tests for binary sequences should be made. In fact, the FSM prediction could be used in any search for regularities in numerical sequences: for example, during the COIL Summer School 2000 (Limerick, Ireland) the possibility of this approach in some bio-informatics problems, such as the modelling of gene promoter sequences in DNA, was explored.

Acknowledgements.   This research began as a university degree diploma in the Department of Mathematics of the University of Torino (Italy). Most of the experimental results were obtained during a cooperation with Telsy Elettronica e Telecomunicazioni ([6]), a firm specialized in the development of cryptographic systems: we especially want to thank Guglielmo Morgari for his help and for the final implementation of the algorithm. This research was completed as a master's thesis ([7]) in the Centre de Math´ ematiques et d'Informatique of the University of Provence (Marseille, France). We also want to thank Marco Tomassini and Leonardo Vanneschi of the University of Lausanne (Lausanne, Switzerland) for useful discussions and remarks on the final version of this paper.

References

1. Allouche, J.-P., Leblanc, B., Lutton,   É.: Inverse problems for finite automata: a solution based on genetic algorithms.   Lecture Notes in Computer Science (Artificial Evolution, 1997, Eds. Hao J.-K., Lutton E., Ronald E., Schoenauer M., Snyers D.) 1363 (1998) 157–166
2. Broglio, A., Liardet, P.: Prediction with automata. In   Symbolic Dynamics and its Applications, Contemporary Mathematics   135 (1992) 111–124
3. Cerf, R.: The dynamics of mutation-selection algorithms with large population sizes. Annales de l'Institut Henri Poincar´   e, Probabilit´ es et Statistiques 32 -4 (1996) 455–508
4. Cerf, R.: Asymptotic convergence of genetic algorithms.   Advances in Applied Probability 30 -2 (1998) 521–550
5. Cohen, D.: Introduction to Computer Theory.   John Wiley and Sons, New York (1991)
6. Giacobini, M.: A randomness test for binary sequences based on evolutionary algorithms. In Proceedings of the 1999 Genetic and Evolutionary Computation Conference Workshop Program   , Annie S. Wu Ed., Orlando (1999) 355–356
7. Giacobini, M.: Recherche de r´ egularit´es dans des suites binaires pseudo-al´ eatoires au moyen des algorithmes ´ evolutionnaires.   Master's Degree, Universit´ e de Provence, Marseille (2000)
8. Goldberg, D.: A note on Boltzmann tournament selection for genetic algorithms and population-oriented simulated annealing.   Complex Systems 4 (1990) 445–460
9. Knuth, D. E.: The Art of Computer Programming II.   Addison-Wesley Publishing Company, New York (1969)
10. O'Connor, M. G.: An unpredictability approach to finite state randomness.   J. Comp. System Sciences 37 (1988) 324–336
11. Walter, J.: Hot Bits: Guenuine Random Numbers, Generated by Radioactive Decay. http://www.fourmilab.ch/hotbits/   .

Extending Selection Learning toward Fixed-Length   d-Ary Strings

Arnaud Berny

arnaud.berny@free.fr

Abstract. The aim of this paper is to extend   selection learning , initially designed for the optimization of real functions over fixed-length binary strings, toward fixed-length strings on an arbitrary finite alphabet. We derive selection learning algorithms from clear principles. First, we are looking for product probability measures over   d-ary strings, or equivalently, random variables whose components are statistically independent. Second, these distributions are evaluated relatively to the expectation of the fitness function. More precisely, we consider the logarithm of the expectation to introduce fitness proportional and Boltzmann selections. Third, we define two kinds of gradient systems to maximize the expectation. The first one drives unbounded parameters, whereas the second one directly drives probabilities,   a la PBIL. We also introduce composite selection, that is algorithms which take into account positively as well as negatively selected strings. We propose stochastic approximations for the gradient systems, and finally, we apply three of the resulting algorithms to two test functions, OneMax and BigJump, and draw some conclusions on their relative strengths and weaknesses.

1 Introduction

Population-Based Incremental Learning (PBIL) [1] is an early example of selection learning over binary strings. It is an adaptive algorithm that periodically updates a probability vector which defines a search distribution (instead of a search population as with genetic algorithms). A PBIL iteration has three steps. First, a finite population is sampled from the search distribution. Then, the best two strings, according to the fitness function to maximize, are selected from the population and averaged. Finally, their average is used to update the probability vector with a linear equation. PBIL is a weak optimizer in the sense that it offers no guaranty of convergence to a global optimum, even if it may outperform genetic algorithms in some cases. However, it has some interesting features: its complexity is low and it is robust, by which we mean that it may be applied to time-varying or noisy fitness functions.

Extending selection learning involving strings on an arbitrary finite alphabet is useful because it will avoid coding issues which arise when one symbol of an arbitrary finite alphabet is replaced with its equivalent number of bits. Servais et al. [14] have extended PBIL by ordering the alphabet and have proposed a specific update rule. Lo and Hsu [8] have proposed a discrete time algorithm in

P. Collet et al. (Eds.): EA 2001, LNCS 2310, pp. 54–64, 2002.

which the probability of each symbol of each component depends on the best fitness encountered for strings in such a configuration. In the context of reinforcement learning, Meuleau and Dorigo [9] have established connections between Ant Colony Optimization and stochastic gradient. In the context of genetic programming, Ratle and Sebag [11] have proposed a multiplicative update rule for an extended PBIL algorithm applied to the choice of rules of context-free grammars.

We will not put any order on the alphabet and we will emphasize the derivation of update rules which essentially relies on the computation of the gradient of an average fitness. Doing this, we will extend the work initiated in [2,5,4]. We study continuous time gradient systems which we discretize in order to obtain adaptive algorithms. We also explain the link between fitness proportional or Boltzmann selection and the gradient approach. Moreover, in order to derive an extended PBIL algorithm from a gradient, it is necessary to apply what we call a stabilization technique. We have used it in the binary case [5] and we provide its counterpart in the general case.

Sec. 2 presents some preliminary notions such as statistical optimization and the unit simplex. Sec. 3 presents a first selection learning algorithm which drives unbounded parameters of the search distribution rather than probabilities. It explains how the selection operator derives from the gradient of some statistical criterion. Sec. 4 presents a second selection learning algorithm which is closer to PBIL than the previous one and directly drives probabilities. Sec. 5 gives stochastic approximations for both algorithms. Sec. 6 presents early experiments with dierent selection learning algorithms maximizing OneMax and BigJump fitness functions generalized to   d-ary strings.

2   Preliminary Notions

2.1   Statistical Optimization

Let $f : E \rightarrow R$ be a function to maximize, also called fitness function, over some finite set E. The idea behind statistical optimization is to replace the optimization problem $\max(f, E)$ with the problem $\max(J, M(E))$, where $M(E)$ is the set of probabilities over E and J is a statistical criterion. More precisely, for $\lambda \in M(E)$, we define $J(\lambda)$ as the expectation $E_\lambda(f)$ of f relatively to λ, that is $J(\lambda) = \sum_{x \in E} \lambda(x) f(x)$. It can be proved that essentially nothing is lost from the original problem when solving the new one [5]. More precisely, if λ is J-optimal, then it only charges points of the search space E which are f-optimal.

2.2   Product Probability Measures

Let $E = L^n$ be the set of strings of length n on some finite alphabet L of size d. For simplicity and without any loss of generality, we will identify L with the set $1, \ldots, d$. To keep the new problem $\max(J, M(E))$ tractable, we restrict ourselves to the set of product probability measures λ over $E = L^n$ such that $\lambda(x) = \prod_{i=1}^n \lambda_i(x_i)$, where x is a string of E with components x_i, $1 \leq i \leq n$, and each π_i is a multinomial probability over the alphabet Λ. Put another way, the components of random variables with such a distribution are statistically independent . Most of what is new in this paper relies on the higher cardinality of the alphabet, and can be studied in the one dimensional case $n = 1$, by which we mean strings reduced to one component.

2.3   Unit Simplex

A non degenerate multinomial probability over Λ is completely defined by d positive real numbers summing to 1, or equivalently a point in the unit simplex . Let v denote a vector of $\mathbb{R}^d$ with components v_j, $1 \le j \le d$. The unit simplex S^{d-1} is the set $\{ v \in (\mathbb{R}_+)^d : \sum_{j=1}^d v_j = 1 \}$ which is a $d-1$ dimensional differentiable manifold. The tangent space of S^{d-1} at any of its points is the set $\{ v \in \mathbb{R}^d : \sum_{j=1}^d v_j = 0 \}$. We will call the points of the unit simplex probability vectors.

3   Gradient Systems over Unbounded Parameters

3.1   Computing the Gradient in One Dimension

Let $f : \Lambda \to \mathbb{R}$ be some fitness function. We define the statistical criterion $J : S^{d-1} \to \mathbb{R}$ such that, for all probability vector v, $J(v) = \sum_{j=1}^d v_j f(j)$ or $E(f)$, where the expectation is relative to the probability defined by v.

We introduce a dierentiable mapping between S^{d-1} and $\mathbb{R}^{d-1}$ that has an inverse which is also dierentiable. Let $\phi_d : S^{d-1} \to \mathbb{R}^{d-1}$ be such that if $u = \phi_d(v)$, then $u_j = \log(v_j/v_d)$, for all $1 \le j \le d-1$. Observe that u is unbounded, on the contrary to v. The mapping ϕ_d can be easily inverted with the formulas $v_j = e^{u_j} / (1 + \sum_{k=1}^{d-1} e^{u_k})$ and $v_d = 1 / (1 + \sum_{k=1}^{d-1} e^{u_k})$.

With the criterion J and the mapping ϕ_d, we define the function $J_d : \mathbb{R}^{d-1} \to \mathbb{R}$ by the composition $J_d = J \circ \phi_d^{-1}$. This is close to the approach of [9], except that we maintain only $d-1$ parameters instead of d. Since the function J_d is dierentiable, we can introduce the gradient system in $\mathbb{R}^{d-1}$ $\dot{u} = \nabla J_d$, where $\nabla J_d = (\partial J_d/\partial u_1, \ldots, \partial J_d/\partial u_{d-1})$. The motivation behind such an approach is that the singular points of the dynamical system are those at which the gradient vanishes, which is a necessary condition for local optimality. However, we will not study the connection between stability and optimality; we consider the gradient system as a heuristic procedure.

Using the chain rule, the partial derivative of J_d w.r.t. u_j can be written $\sum_{k=1}^d \partial J/\partial v_k \cdot \partial v_k/\partial u_j$. From the expression of the simplex mapping, we have $\partial v_k/\partial u_j = -v_j v_k$ if $k \ne j$, and $v_j(1 - v_j)$ otherwise. After substitution, we express the partial derivative as $v_j(f(j) - E(f))$ which reminds us of models in population genetics [10] and reinforcement learning [5]. However, we prefer another expression which will allow us to give a stochastic approximation of the dynamical system. We can write the partial derivative as the expectation $E(e_j f)$, where e_j is the function defined by $e_j(k) = 1_{k=j} - v_j$, for all $1 \le k \le d$, and is called the eligibility.

3.2 Computing the Gradient in Arbitrary Dimension

In order to completely define a product probability measure over $E = L^n$, we need a collection of n probability vectors v_i, $1 \le i \le n$, with components v_{ij}, $1 \le j \le d$. With this collection, we define the probability $\pi(x) = \prod_{i=1}^n \pi_i(x_i)$ where $\pi_i(x_i) = v_{ix_i}$. For all probability vector v_i, let $u_i = \sigma_d(v_i) \in R^{d-1}$ be its image under the simplex mapping.

In order to compute the gradient of J_d, we first express it as an expectation, $\nabla J_d = \sum_{x \in E} f(x) \nabla \log \pi(x) \pi(x)$ or simply $E(f \nabla \log \pi)$. From the factorization of π, we have $\partial \log \pi(x)/\partial u_{ij} = \partial \log v_{ix_i}/\partial u_{ij}$. With the expression of the simplex mapping, we arrive at $\partial \log v_{ix_i}/\partial u_{ij} = 1 - v_{ij}$ if $x_i = j$, and $-v_{ij}$ otherwise. Just as in Sec. 3.1, we combine the previous results into an eligibility function $e_{ij}(x_i) = 1_{x_i = j} - v_{ij}$. The partial derivative of J_d can then be written as the expectation $E(e_{ij} f)$.

3.3 Selection Learning

We have considered the maximization of the expectation of some fitness function relatively to a probability over the search space. Since we are interested in population-based incremental learning or selection learning, we follow the guidelines introduced in [5,3] and define another statistical criterion which is the logarithm of the expectation of the fitness function. We will show that its maximization, which is equivalent to the maximization of the expectation itself, leads to the application of fitness proportional selection to the probability π seen as an infinite and implicit search population.

Let $J = \log E(f)$ be the logarithmic statistical criterion. By the chain rule, we find that $\nabla J_d = E(f \nabla \log \pi)/E(f)$ or simply $\tilde{E}(\nabla \log \pi)$, where the last expectation is taken relatively to $\tilde{\pi}$, the probability which results from the fitness proportional selection of the probability π. For all point x of the search space, we have

$$\pi(x) \xrightarrow{\text{selection}} \tilde{\pi}(x) = \frac{f(x)\pi(x)}{E(f)} .$$

The probability of sampling x from $\tilde{\pi}$ is proportional to its probability of being sampled from π times its fitness, which models the effect of proportional selection on π. Observe that we have to restrict to positive functions.

Finally, using the eligibility function, the partial derivative of the criterion can be written $\tilde{E}(e_{ij}) = \tilde{E}(1_{x_i = j}) - v_{ij}$, where the random variable is x_i. The effect of the corresponding dynamical system is to increase the probability of selected symbols by moving each probability toward the frequency of its symbol in the selected distribution.

We can derive Boltzmann selection in the same manner as proportional selection: it suffices to replace f by $e^{\beta f}$, where $\beta \ge 0$ controls the selective pressure. The statistical criterion is then $\log E(e^{\beta f})$ and the expression of the selected probability is $\tilde{\pi}(x) = e^{\beta f(x)}\pi(x)/E(e^{\beta f})$. Boltzmann selection is interesting for two reasons. First, we can relax the positivity constraint on f since $e^{\beta f}$ is positive. Second, the limit case $\beta \to \infty$ leads to PBIL-like algorithms.

3.4   Composite Selection

We show how both positive and negative selection operators can be combined at the same time in a gradient system. Let f be a positive fitness function. Observe that maximizing $E(f)$ is equivalent to minimizing $E(1/f)$. This comes from the fact that the points where f is maximal are those where its inverse $1/f$ is minimal. Thus, maximizing $\log E(f)$ is equivalent to maximizing $\log E(1/f)$, and also equivalent to maximizing their sum. From this observation, we propose the statistical criterion $J(\ ) = \log E(f) \quad \log E(1/f)$. When f can take values of both signs, which we assume from now on, we replace f with e^f. From the search distribution , define two selected distributions, $_+$ which is equal to , and $(x) = e^{f(x)} (x) \quad E(e^{f})$. The partial derivative of the composite criterion J_d can then be written $E_+(1_{x_i = j}) \quad E (1_{x_i = j})$, where the first expectation is relative to $_+$ and the second one is relative to . The negative part of the vector field, that is the expectation relatively to , plays the role of a repoussoir which is an example from which the algorithm tries to move away. The term "repoussoir" has been introduced in [13]. [12] also describes a population algorithm which maintains both a repoussoir and an attractor, switching from one behavior to the other whenever the fitness does not increase for a given number of iterations.

4   Stabilized Gradient Systems over Bounded Parameters

In this section, we present another way to derive selection learning for d-ary strings which will lead to algorithms close to PBIL, that is algorithms which directly update probability vectors (bounded parameters). The main dierence from Sec. 3 lies in the mapping which does not bound probability vectors. Thus, we will have to bound them a posteriori by stabilization of a gradient system, which means to introduce artificial zeros in the vector field. We only consider the one dimensional case since its generalization presents no diculty.

Let $H^{d\ 1}$ be the hyperplane $v \quad R^d : \sum_{j=1}^d v_j = 1$ which is a $d\ 1$ dimensional dierentiable manifold with the same tangent space as the unit simplex. Let $_d : H^{d\ 1} \quad R^{d\ 1}$ be the mapping such that if $u = _d(v)$, then $u_j = v_j$, for all $1 \quad j \quad d\ 1$. $_d$ can be easily inverted with the formula $v_d = 1 \quad \sum_{k=1}^{d\ 1} u_k$. Let $J_d = J \quad {}_d^{1}$, where J is the logarithmic criterion, and consider the gradient system over J_d. We only constrain the probability vector v to stay in $H^{d\ 1}$ rather than in $S^{d\ 1}$. Just as in Sec. 3.2, we focus our attention on the eligibility. Using the mapping $_d$, we find $\log (k)/u_j = 1/v_d$ if $k = d$, $1/v_j$ if $k = j$, and 0 otherwise. The corresponding (unconstrained) dynamical system is then $u_j = E\ 1_{k=j} /v_j \quad E\ 1_{k=d} /v_d$, where k is the random variable.

In order to constrain v inside the unit simplex, we multiply each component j of the vector field by the quantity $v_j v_d$ which vanishes when $v_j = 0$ or $v_d = 0$. In the binary case $d = 2$, this transformation reduces to multiplying the only component of the vector field by $v_1(1 \quad v_1)$, since $v_2 = 1 \quad v_1$ [5]. We obtain

$u_j =$ $_{jd}$, where $_{jd} = v_d E (1_{k=j})$ $v_j E (1_{k=d})$. Observe that for all $j = d$, $_{dj} =$ $_{jd}$. Using the mapping $_d$, the dynamics of v is given by the system $v =$ $_d$, where the last component of the vector field is $_{dd} =$ $_{j=d}$ $_{jd}$.

At this point, we have derived a vector field $_d$ which depends on the choice of the reference. Thus, we propose to   symmetrize the vector field by averaging it over all the possible references and computing the new vector field $= \sum_{k=1}^{d}$ $_k$. If we examine   componentwise, we find $_j = \sum_{k=1}^{d}$ $_{jk} =$ 2 $_{jj}$. With this new vector field, the resulting dierential equation is then $v_j = E (1_{k=j})$ v_j for all 1 j d. The last equation may be modified to take into account composite selection. However, its stochastic approximation would not bound probabilities. In consequence, we will study stochastic approximations for composite selection only in the case of unbounded parameters.

5   Stochastic Approximation

We have to design algorithms which correspond to the dynamical systems we have studied in Secs. 3 and 4. Two kinds of approximation are required. First, we have to discretize the time parameter. Second, we have to compute expectations of random variables which are sums over the entire search space, and thus cannot be exactly computed in a reasonable amount of time. In this paper, we will focus on selection learning, although Monte Carlo integration and reinforcement learning can also be applied.

Selection learning means that we use the metaphor of natural selection, like genetic algorithms do, to update the probability vectors, directly or indirectly if we use unbounded parameters. All selection learning algorithms share the same template. First, a finite population is sampled from the distribution defined by the probability vectors. Then, a fraction of this population is selected and used to update the probability vectors.

5.1   Unbounded Parameters

Let us describe the algorithm:

1. Sample a population of N independently and identically distributed (i.i.d.) strings, (x^k), 1 k N, whose common law   is defined by n probability vectors v_i, 1 i n;
2. Apply Boltzmann selection to (x^k) and let (y^k), 1 k P, be the selected population;
3. For each component i and each symbol j, let p_{ij} be the number of selected strings whose ith component is j;
4. For each component i and each symbol $j = d$, update u_{ij} with the equation $u_{ij}(t+1) = u_{ij}(t) +$ $(p_{ij}/P$ $v_{ij})$, where 0 is the learning gain;
5. Recompute the probabilities with the exponential mapping.

The stochastic approximation itself consists in replacing the expectation $E (1_{x_i = j})$ by the ratio p_{ij}/P. This can be decomposed into two approximations. First, we approximate the theoretical selected population   by the empirical measure μ = $\sum_{k=1}^{N} w_k \, \delta_{x^k}$, where the weight w_k of the string x_k is $e^{f(x^k)} / \sum_{l=1}^{N} e^{f(x^l)}$, and δ_a is the Dirac measure at $a \in E$. Second, the expectation of $1_{x_i = j}$ relatively to μ is replaced with its average over the selected population (y^k) whose strings are i.i.d. with common law μ. Hence the counting ratio p_{ij} / P.

In the case of composite selection, we introduce the population (z^k), $1 \leq k \leq$, of negatively selected strings using Boltzmann selection. Let q_{ij} be the number of negatively selected strings whose ith component is j. The new update rule is then $u_{ij}(t+1) = u_{ij}(t) + \lambda (p_{ij}/P - q_{ij}/)$.

5.2 Bounded Parameters

The algorithm for bounded parameters is similar to that for unbounded parameters. Step 4 needs to be modified. For each component i and each symbol j, directly update v_{ij} with the equation $v_{ij}(t+1) = v_{ij}(t) + \lambda (p_{ij}/P - v_{ij})$, where $0 \leq \lambda \leq 1$ is the learning gain. Observe that the sum over j of all increments vanishes and that no probability vector can escape the unit simplex, provided that it lies in it before it is updated. This results from the convexity of the unit simplex. It is also clear that such a stability property does not hold when composite selection is applied. There is no Step 5 since the probability vectors are directly updated.

In the limit case , and when $P = 1$, the fittest string in the population (x^k) is selected with probability 1. Therefore we get an extended PBIL algorithm. When $\lambda = 1$, we get a generalized Univariate Marginal Distribution Algorithm (UMDA) [10] which does not take into account $v_{ij}(t)$ to compute $v_{ij}(t+1)$.

6 Experiments

We have adapted OneMax and BigJump [10] functions to d-ary strings. Both functions are maximal at some arbitrary string s. For all string x, let m be the number of components where the target string s and x match. OneMax(x) is simply m, whereas BigJump(x) = n if $m = n$, 0 if $n - t \leq m < n$, and m otherwise, where the threshold t controls the hardness of the function. In all experiments, the size of strings is $n = 40$ and the size of the alphabet is $d = 4$, which is similar to DNA.

We have studied three algorithms which follow the guidelines of Sec. 5. They dier in the choice of parameters (bounded or unbounded), and in the selection operator (Boltzmann selection, composite or not):

1. Bounded parameters and Boltzmann selection
2. Unbounded parameters and composite Boltzmann selection
3. Unbounded parameters and Boltzmann selection

For all three algorithms, the number of iterations is set to 15000, the learning gain $\lambda = 0.001$, the population size $N = 50$, the selective pressure $= 0.5$, and the size of the selected population $P = 5$. For the second algorithm (composite selection), the size of the negatively selected population is   = 5. We are interested in the best fitness encountered at each iteration which we average over 30 runs for all algorithms. We have applied all three algorithms to OneMax (or BigJump with   t = 0), and BigJump with   t = 5 and t = 10.

The results are shown in Fig. 1. The algorithms are ordered according to their speed of convergence (iteration): Alg. 1 is the fastest, then Alg. 2, and finally Alg. 3. In the case t = 0, all three algorithms found the optimum. In the case t = 5, Alg. 3 could not find the optimum at all. In the case   t = 10, both Alg. 2 and 3 only discovered local optima. Alg. 1 found the optimum more than 40 of the time. In the case of unbounded parameters, composite selection improves on positive selection alone.

Alg. 2 and 3 are slower (real time) than Alg. 1 since they require a call to the simplex mapping, which involves the exponential function, each time the unbounded parameters are updated.

Fig. 1(d) shows the inuence of the learning gain on the dynamics of a selection algorithm with bounded parameters maximizing BigJump with   t = 5. The optimal value is   = 0 .05, higher values leading to local optima, and lower ones leading to slower dynamics. In particular, the   d-ary equivalent of UMDA, which corresponds to   = 1, is clearly suboptimal.

7 Conclusion

We have extended selection learning toward fixed-length strings of   d-ary symbols. The basic tool for this work is the gradient of a statistical criterion defined as the logarithm of the expectation of the fitness function. The definition of a gradient relies on the dierentiable structure of the unit simplex. We did not assume any order between symbols for each string component. We have proposed two dynamical systems for maximizing the statistical criterion. The first one is a pure gradient system over unbounded parameters. The second one is a stabilized gradient system over bounded parameters (the probabilities themselves). The eect of both systems can be interpreted from an evolutionary point of view, since they involve fitness proportional or Boltzmann selection. Composite selection, or put another way, positive and negative selection at the same time, follows from the combination of two symmetrical statistical criteria. Composite selection requires unbounded parameters. We have proposed stochastic approximations for selection learning algorithms. In particular, a generalized PBIL algorithm can be obtained from stabilized gradient systems over bounded parameters.

We have compared three selection learning algorithms with test functions of increasing hardness (OneMax and BigJump functions). The PBIL-like algorithm (with bounded parameters) is the fastest of them. Algorithms with unbounded parameters are improved if composite selection is used. Also, incremental update rules seem crucial to avoid being trapped in local optima too quickly, which has happened with an UMDA-like algorithm.

Fitness proportional selection derives from the definition of the logarithmic criterion. Other selection schemes are also of interest (deterministic, tournament)

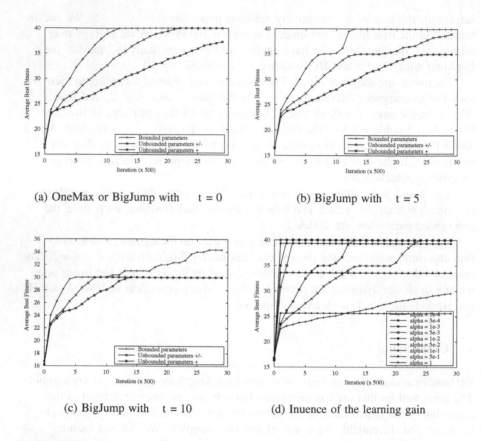

(a) OneMax or BigJump with   t = 0   (b) BigJump with   t = 5

(c) BigJump with   t = 10   (d) Inuence of the learning gain

Fig. 1.   Comparison of three selection learning algorithms with fitness functions of increasing diculty.

and may be combined with update rules to define a whole class of selection learning algorithms. However such algorithms cannot be interpreted as gradient systems maximizing the expectation of the fitness function.

As in the binary case, much work remains to be done. We have initiated a stability analysis of strings in the binary case, linking the Hamming graph to the properties of the dynamical system. Such an analysis seems possible in the general case, using generalized Hamming graphs.

The main theoretical problem, in our opinion, is the definition of a precise criterion to decide whether or not a given fitness function is globally optimizable with the continuous time systems we have presented or their stochastic approximations. Observe that a needle in a haystack can be maximized with a continuous time system, but not by its stochastic approximation (at least in an ecient way). Isolation of optimum is one cause of failure of PBIL-like algorithms, but it is probably not the only one.

It is also desirable to know to what extent a stochastic approximation tracks its continuous time deterministic model, in particular with respect to the population size. Such a connection has been initiated in [16] and further developed in [15] for Evolution Strategy algorithms. The main dierence from our model and other PBIL-like algorithms is that an ES-algorithm updates points in the search space instead of probability parameters, in which case the analysis may be more dicult.

From a practical point of view, as done in the binary case in [6], empirical studies should determine the importance of the selection operator and its parameters for a wide class of fitness functions. Graph coloring seems to provide natural test functions [8]. It is not clear however whether selection learning can be eciently applied to strings on alphabets with high cardinal.

References

1. S. Baluja and R. Caruana. Removing the genetics from the standard genetic algorithm. In A. Prieditis and S. Russel, editors,   Procee dings of the 12th International Conference on Machine Learning   , pages 38–46. Morgan Kaufmann, 1995.
2. A. Berny. Statistical machine learning and combinatorial optimization. in [7].
3. A. Berny. An adaptive scheme for real function optimization acting as a selection operator. In X. Yao and D.B. Fogel, editors,   First IEEE Symposium on Combinations of Evolutionary Computation and Neural Networks   , pages 140–149, San Antonio, May 2000.
4. A. Berny.   Apprentissage et optimisation statistiques, application   a la radiot´elé- phonie mobile . PhD thesis, Universit´ e de Nantes, 2000. in french.
5. A. Berny. Selection and reinforcement learning for combinatorial optimization. In M. Schoenauer et al., editors,   Parallel Problem Solving from Nature VI   , Lecture Notes in Computer Science, pages 601–610, Paris, September 2000. Springer-Verlag.
6. A. Johnson and J. Shapiro. The importance of selection mechanisms in distribution estimation algorithms. In   Artificial Evolution , Le Creusot, France, October 2001.
7. L. Kallel, B. Naudts, and A. Rogers, editors.   Theoretical Aspects of Evolutionary Computing . Natural Computing Series. Springer-Verlag, 2001.
8. C.-C. Lo and C.-C. Hsu. An annealing framework with learning memory.   IEEE Trans. on Systems, Man, and Cybernetics part A   , 28(5):648–661, September 1998.
9. N. Meuleau and M. Dorigo. Ant colony optimization and stochastic gradient descent. Technical report, IRIDIA, December 2000.
10. H. M uhlenbein. Evolutionary algorithms: from recombination to search distributions. in [7].
11. A. Ratle and M. Sebag. Avoiding the bloat with stochastic grammar-based genetic programming In   Artificial Evolution , Le Creusot, France, October 2001.
12. D. Robilliard and C. Fonlupt. A shepherd and a sheepdog to guide evolutionary computation? In C. Fonlupt, J.-K. Hao, E. Lutton, E. Ronald, and M. Schoenauer, editors, Artificial Evolution , Lecture Notes in Computer Science, pages 277–291. Springer-Verlag, 1999.
13. M. Sebag and M. Schoenauer. A society of hill-climbers. In   Proc. IEEE Int. Conf. on Evolutionary Computation   , pages 319–324, Indianapolis, April 1997.

14. M. P. Servais, G. de Jaer, and J. R. Geene. Function optimization using multiple-base population based incremental learning. In   Proc. Eight South African Workshop on Pattern R ecognition , 1997.
15. G. Yin, G. Rudolph, and H.-P. S   chwefel. Analyzing (1 ,) Evolution Strategy via stochastic approximation methods.   Informatica , 3(4):473–489, 1995.
16. G. Yin, G. Rudolph, and H.-P. S   chwefel. Establishing connections between evolutionary algorithms and stochastic approximation.   Informatica , 6(1):93–116, 1995.

Markov Random Field Modelling of Royal Road Genetic Algorithms

D.F. Brown [1], A.B. Garmendia-Doval [2], and J.A.W. McCall [1]

[1] School of Computer and Mathematical Sciences, The Robert Gordon University,
St Andrew Street, Aberdeen AB25 1HG, Scotland.
db,jm @scms.rgu.ac.uk
[2] RiboTargets, Granta Park, Abington CB1 6GB, Cambridgeshire, England.
beatriz@ribotargets.com

Abstract. Markov Random Fields (MRFs) [5] are a class of probabalistic models that have been applied for many years to the analysis of
visual patterns or textures. In this paper, our objective is to establish
MRFs as an interesting approach to modelling genetic algorithms. Our
approach bears strong similarities to recent work on the Bayesian Optimisation Algorithm [9], but there are also some signifiant dierences.
We establish a theoretical result that every genetic algorithm problem
can be characterised in terms of a MRF model. This allows us to construct an explicit probabilistic model of the GA fitness function. The
model can be used to generate chromosomes, and derive a MRF fitness
measure for the population. We then use a specific MRF model to analyse two Royal Road problems, relating our analysis to that of Mitchell
et al. [7].

1 Introduction

Markov Random Fields (MRFs) [5] are a class of probabalistic models. They
have been applied for many years to the analysis of images, particularly in the
detection of visual patterns or textures. They use probabalistic information to
characterise particular pixel values in terms of their neighbours. More generally,
MRF theory may be used for analysing spatial or contextual dependencies.

In this paper we investigate a MRF approach to modelling genetic algorithms
(GAs). Our motivation is that the bit patterns in chromosomes represent textures
generated more or less imperfectly by the GA fitness function.

The paper is structured as follows. In Sect. 2 we provide a description of
a GA problem as an instance of the Labelling Problem, well known in image
analysis. One approach to solving this problem is Markov Random Field modelling, and we provide a description of this in Sect. 3. We also prove a theorem
showing that every GA encoding can be modelled as a Markov Random Field
with respect to some neighbourhood relation on the chromosome alleles. The
theorem also establishes an explicit relationship between the MRF parameters
and the GA fitness function. In Sect. 4, we focus on a particular MRF model,
the Ising model. We define the model with its associated MRF parameters, and

show how these can be used to generate an optimal chromosome with respect to the model, using a zero-temperature Metropolis method. In Sect. 5 we describe experiments applying the Ising model to the Royal Road genetic algorithm. We consider two encodings with preferred schemata of dierent lengths. MRF parameters are used to define a MRF fitness measure that is distinct from standard measures of population fitness used in GAs. In Sect. 6 we analyse the results and compare MRF fitness with standard measures of population fitness. We compare our analysis with that of Mitchell et al. [7], and find that the operation of the Royal Road GA is explicable in terms of the MRF theory. In particular we can characterise hitchhiking in terms of the MRF parameters. Finally, in Sect. 7 we conclude by relating our findings to recent work on Probabilistic Model Building GAs and setting out a future direction for the MRF approach.

2   The Labelling Problem for Genetic Algorithms

An important goal of image analysis is the detection of particular features in images, e.g., points, lines, edges and, more generally, visual patterns or textures. Typically this is accomplished by looking at localised collections of pixels and their grey-scale values. The key features of the detection process can be encapsulated in the Labelling Problem, which is stated as follows.

Let L be a set of labels, and let A be a set of locations to which labels can be assigned. A map $c : A \to L$, which assigns a particular label to each location is called a labelling. The general Labelling Problem is to determine an optimal labelling with respect to a particular set of criteria.

The purpose of this section is to formulate the Labelling Problem for Genetic Algorithms. Assume we have a problem, G, to which genetic algorithms are to be applied. Assume that G has an encoding consisting of chromosomes of length n with an associated fitness function f. The objective is to search for a chromosome that maximises f.

Let A denote the set of chromosome alleles, and let L denote the set of possible allele values. A particular chromosome c represents the assignment of an element of L to each element of A. In other words, each chromosome is a labelling $c : A \to L$. Each labelling c has a fitness value $f(c)$, and we wish to find a chromosome that maximises fitness. Thus we have the Labelling Problem for G:

$$\text{Find a labelling } c : A \to L \text{ which maximises } f(c) . \qquad (2.1)$$

Note that the notion of encoding here is quite general, encompassing bit-string, other finite alphabet and oating-point encodings. The theorem presented in the following section is therefore applicable to a wide range of genetic algorithms.

3   Markov Random Field Approach

In this section, we describe a MRF approach to the labelling problem. We begin with some notation and definitions.

Given a set A of locations, we can define a neighbourhood system N on A by specifying for each location, k a set of neighbouring locations, N_k. We use the notation $a \sim b$ to indicate that location b is a neighbour of location a. Neighbourhood is a symmetric relation, that is $a \sim b \Leftrightarrow b \sim a$.

Given a neighbourhood relation $\sim$ on A, we define a clique to be a subset of A satisfying:

1. is a singleton k for some $k \in A$, or,
2. $a \sim b$ for all locations a, b .

Informally, a clique is either a single location or a set of mutual neighbours. We denote the set of all cliques by K.

We write c_k to represent $c(k)$ the assignment of a label to location k. More generally, if $B \subseteq A$ we will write c_B to denote the restriction of c to B.

Markov Random Field modelling [5] regards locations as random variables taking values in the set of labels, and assigns a non-zero probability $P(c)$ to each labelling $c \in C$. Thus we have:

1. $P : C \to [0, 1]$, a probability for each labelling,
2. $P(c) > 0$, the Positivity Condition and,
3. $\sum_{c \in C} P(c) = 1$, the sum of the probabilities is 1.

Note that, in the case of oating point encodings, P will be a probability density and the summation will become an integral.

P must also be consistent with the neighbourhood system in that the value of the label in a particular location is conditional only upon the values of the labels in the neighbouring locations. This is expressed mathematically as:

4. $P(c_k \mid c_{A \setminus k}) = P(c_k \mid c_{N_k})$ for all labellings c (the Markovianity Condition).

This last condition relates directly to the interaction between allele values in chromosome fitness evaluation, as we shall see below.

To each clique , we may associate a function $V : \to (\ ,\ )$. We call these clique potential functions , and they encapsulate information about related locations.

The Hammersley-Cliord Theorem (HCT) [1] states that, for any Markov Random Field with probability function P, there is a (non-unique) formulation:

$$P(c) = \frac{e^{U(c)/T}}{Z}, \text{ for all labellings } c \in C . \tag{3.1}$$

Here Z is the normalising constant,

$$Z = \sum_{s \in C} e^{U(s)/T} .$$

U is called the energy function and is defined by:

$$U(c) = \quad V \ (c) \ . \tag{3.2}$$

T is a temperature coecient which will remain constant for the purposes of this paper. Without loss of generality, we will set T equal to 1.

Equation (3.1) shows that the probability of a particular labelling is completely determined by the values of the potential functions. Conversely, if a set of labellings satisfies a probability distribution of this form, then the HCT states that there is a MRF with probability function P. The important consequence of this is that P must then satisfy the Markovianity Condition.

This preamble now leads us to a useful result for genetic algorithms.

Theorem 1.   Let G be a GA encoding with fitness function f. If f (c)   0 for all chromosomes c, then G defines a Markov Random Field with respect to some neighbourhood system N.

Proof. We define the joint probability function to be:

$$P(c) \quad \frac{f(c)}{Z} \ . \tag{3.3}$$

Here, $Z = \ _s \ _c f(s)$ is the sum of all chromosome fitnesses. Clearly:

1. P : C   [0, 1]
2. $_c \ _C P(c) = 1$

We define a neighbourhood system on A by setting $N_k = A$ for each k A. (This means that any collection of alleles will form a clique.) We assign clique potential functions as follows:

$$V \ = \quad \begin{matrix} 0, & \text{if} & = A \\ \ln \frac{1}{f(c)}, & \text{if} & = A \end{matrix}$$

Putting U(c) =   V (c), we obtain:

$$P(c) = \frac{e^{\ U(c)}}{Z}, \text{for all chromosomes } c \quad C.$$

We can now invoke the converse to the HCT to deduce that   G defines a Markov Random Field with respect to our chosen neighbourhood system.

In general, the MRF defined by   G will not be unique, and there may be other neighbourhood systems and associated potential functions that can be similarly defined. For any such system, we obtain the following expression for chromosome fitness in terms of a sum of potential functions:

$$\ln f(c) = \quad V(c). \tag{3.4}$$

The neighbourhood system used in the proof is not very interesting, as it contains no more information than is already contained in f. What will be of more interest is finding MRFs based on neighbourhood systems where the cliques consist of only a few interacting alleles. The potential functions for such MRFs will then encapsulate localised information about the fitness function. In particular, they can be used to characterise highly-fit schemata.

In the next section, we present a simple model using cliques containing only one or two chromosome alleles.

4   The Ising GA Model

In a simple GA and in many practical applications, chromosomes are encoded as a bit string. In other words, the set of labels $L = 0, 1$. Where the bit-string is of length n, it is well-known that there are 2^n labellings. In what follows, we let $A = 1, \dots, n$ and define a neighbourhood system N to be the set of neighbourhoods N_k, one for each allele k, where $N_k = k \quad 1, k, k+1$.

For notational, computational and theoretical convenience, we interpret these numbers modulo n. Thus the alleles are numbered from 1 to n, running left to right, and each allele has as its neighbours the two alleles immediately adjacent to it. This includes the first and last alleles, which are neighbours of each other.

There are two types of clique for this neighbourhood system. First, there are n singleton cliques (1-cliques) of type k. Second, there are n 2-cliques containing two neighbouring locations. These have the form $k, k+1$ for each $k \quad A$. In the case $n = 3$ only, there is a 3-clique $1, 2, 3$ because here all alleles are mutual neighbours. However, since GAs with chromosomes of length 3 are intrinsically uninteresting, we may safely assume $n \quad 3$, and that there are precisely $2n$ cliques.

Using the notation of Sect. 3 and identifying the assignment with the value, we may represent a chromosome literally as:

$$c = c_1 c_2 \dots c_n.$$

The Ising model [4] is defined by assigning clique potential functions as follows:

$$U(c) = \quad {}_k c_k + \quad {}_{k,k+1} c_k c_{k+1}. \tag{4.1}$$

The   and   coecients are real numbers. Each set of coecients defies a MRF (though a particular MRF can be determined by more than one set of coecients). We will refer to the   and   coecients as MRF parameters. Substituting from (3.4) we obtain, for each chromosome   c:

$$\ln f(c) = \qquad \sum_k {}_k c_k + {}_{k,k+1} c_k c_{k+1} \qquad . \tag{4.2}$$

Each chromosome c provides a unique equation of the form (4.2). These are linear equations in the MRF parameters. Applying this to a suciently large set of chromosomes (2n), we obtain an over-specified system of linear equations. For example, taking all possible chromosomes we would obtain 2^n equations in 2 n variables. In practice, a GA will sample only a small population of chromosomes at any one time. If the population size is 2n however, a best-fit solution of (4.2) provides an estimate of the MRF parameters. The parameters obtained from a population thus provide a probabilistic model of the fitness function, based on the population sampled.

Given a set of MRF parameters, it is then possible to generate approximations to optimal labellings using a zero-temperature Metropolis method [6], which can be described as follows:

1. Generate a chromosome c at random.
2. For N iterations, repeat:
  2.1. Flip the label on allele k chosen at random to obtain a labelling c .
  2.2. Set U = U(c) U(c).
  2.3. If U 0, set c = c .
3. Terminate with answer c.

From (4.1), U can be determined explicitly from the following formula:

$$U = (c_k \quad c_k)({}_k + {}_{k-1,k} c_{k-1} + {}_{k,k+1} c_{k+1}) . \tag{4.3}$$

This gives an algorithm for generating chromosomes that approximate the optimum. First we generate sucient chromosomes to calculate MRF parameters ${}_k$, ${}_{k,k+1}$. We then iterate the Metropolis method until it converges. The resultant chromosome will be an estimate for the optimum.

There is a close relationship between the parameter values and the generated chromosomes due to the form of the energy function (4.1). Broadly speaking, a negative value of ${}_k$ indicates that the optimal chromosome is likely to have an allele k labelled 1, in order to minimise the energy function. Conversely, a positive value of ${}_k$ indicate s a 0 on that allele in the optimal chromosome. The ${}_{k,k+1}$ parameters indicate a binding strength between neighbouring alleles. A negative value ${}_{k,k+1}$ of indicates that the same label should be attached to both neighbours, whereas a positive value indicates opposite labels.

Mitchell et al. [7] use the Royal Road functions to investigate schema processing and recombination. As part of this work, they compare the performance of a GA against a Random-Mutation Hill-Climbing algorithm (RMHC) on the Royal Road problem. We recall RMHC for comparison with the Metropolis method:

1. Generate a chromosome c at random.
2. For N iterations, repeat:
  2.1. Mutate an allelle k chosen at random to produce c .
  2.2. If f (c) f (c), set c = c .
3. Terminate with answer c.

The relationship between RHMC and the Metropolis method is given by (4.2) and the realisation that:

$$f(c) \quad f \quad (c) \quad U \quad 0 . \tag{4.4}$$

In the next section we present experiments designed to investigate schema processing using MRF modelling. We follow the approach of Mitchell et al. in examining GA performance on Royal Road problems.

5   Experiments

The chromosomes for a Royal Road function consist of bit strings of length n, and the fitness function f is defined using a list of preferred schemata, which we will denote B_i. The fitness $f(c)$ of a chromosome c is defined to be:

$$f(c) = 1 + \sum_i {}_i {}_i(c), \text{ where } {}_i(c) = \begin{cases} 1, \text{ if } c \quad B_i \\ 0, \text{ otherwise} \end{cases} \tag{5.1}$$

The purpose of G is to determine the optimal chromosome.

Note that this is a slight adaptation of the fitness function used in [7]. We have used our own notation, and added 1 to the fitness function to avoid zero fitness values (adding any small amount would do). Note also that ${}_i$ is a coecient used in the definition of the fitness function. Its meaning is distinct from the MRF parameters ${}_{k,k+1}$.

We investigated two Royal Road problems and performed the same set of experiments for each. The first experiment is to generate a population of chromosomes at random, compute MRF parameters for this population, and then apply the Metropolis method to those parameters to generate an optimal chromosome. The second experiment is to apply the same process to a population that is mostly random but has been "seeded" with chromosomes containing some of the preferred schemata. The final experiment is to apply the same process to populations of chromosomes that have been generated by successive generations of the GA. (Note that in this last experiment, MRF parameters cannot be calculated once the GA population has converged on only a few distinct chromosomes. This is because the system of equations becomes under-specified). In each run of the experiment, 5 chromosomes were generated from the MRF parameters derived from the resulting population.

We use the expression "MRF fitness" to mean the fitness of these generated chromosomes. As we are generating 5 chromosomes, we have a best MRF fitness and an average MRF fitness by taking the best, respectively the average, of the fitnesses of the chromosomes generated from the MRF parameters. Since we generate the MRF parameters from a population, MRF fitness can be regarded as a measure of population fitness. As such it is distinct from measures such as best or average fitness of the chromosomes that belong to the population. As we shall see from our experiments, MRF fness can dierentiate chromosome populations in a way that the usual measures cannot.

Our GA is implemented in Standard ML using the functional GA framework described in [3]. In particular, this framework provides explicit control over random seeds [2]. All experimental runs were generated using separate seeds.

The first Royal Road problem is that presented in [7]. The chromosome length is 64, and the preferred schemata and coecients can be seen in Fig. 1.

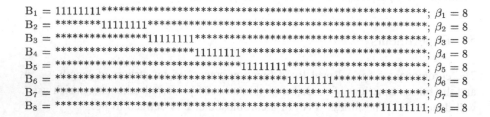

Fig. 1.   Preferred schemata and coecients for chromosome length 64.

Each chromosome has a fitness of 8 m + 1, where m   0, 1, 2, . . . , 8 . The optimal chromosome labels each allele with 1 and has a fitness of 65.

In the random experiment, we performed 10 runs with populations of 200 chromosomes generated at random. We performed two seeded experiments consisting of 10 runs in which 175 chromosomes were generated at random, and 25 were seeded. The 25 seeded chromosomes are created by generating a random chromosome and then setting one (respectively two) of the schema blocks to 1s, so that the fitness value is either 9 (respectively 17) or higher. Finally, in the GA experiment, we made 10 runs of the genetic algorithm using a population size of 200 chromosomes. We used each successive population to calculate MRF parameters, and used these in turn to generate chromosomes via the Metropolis method. After some generations, the GA converged to a point where the system of equations (4.2) was underspecified, and so MRF parameters could not be calculated. We retained the MRF fitness for the last generation of the GA for which MRF parameters could be obtained. The results for all four experiments are presented in Table 1.

The second Royal Road problem has a chromosome length of 20, and the preferred schemata and coecients can be seen in Fig. 2.

Each chromosome has a fitness of 4 m + 1, where m   0, 1, . . . , 5 . The optimal chromosome labels each allele with 1 and has a fitness of 21.

In the random experiment, we performed 10 runs with populations of 100 chromosomes generated at random. The seeded experiment consisted of 10 runs in which 75 chromosomes were generated at random, and 25 were seeded. The seeded chromosomes were created by generating a random chromosome and then setting two of the schema blocks to 1s, so that the fitness value was either 9 or higher. Finally, in the GA experiment, we made 10 runs of the genetic algorithm

Table 1.   Experimental results for the 64-bit Royal Road function.

Experiment	Population fitness			MRF fitness	
	Fitness = 1	Mean fitness	Best fitness	Mean fitness	Best fitness
Random	97	1.3	9	1.6	9
Seeded (1)	85	2.6	17	3.7	17
Seeded (2)	85	3.7	25	6.6	17
GA	58	11.5	25	1.6	9

$$B_1 = 1111; \qquad\qquad \beta_1 = 4$$
$$B_2 = 1111; \qquad\qquad \beta_2 = 4$$
$$B_3 = 1111; \qquad\qquad \beta_3 = 4$$
$$B_4 = 1111; \qquad\qquad \beta_4 = 4$$
$$B_5 = 1111; \qquad\qquad \beta_5 = 4$$

Fig. 2.   Preferred schemata and coecients for chromosome length 20.

using a population size of 100 chromosomes. We used each successive population to calculate MRF parameters, and used these in turn to generate chromosomes via the Metropolis method. After some generations, the GA converged to a point where the system of equations (4.2) was underspecified, and so MRF parameters could not be calculated. We retained the MRF fitness for the last generation of the GA for which MRF parameters could be obtained. The results for all three experiments are presented in Table 2.

Table 2.   Experimental results for the 20-bit Royal Road function.

Experiment	Population fitness			MRF fitness	
	Fitness = 1	Mean fitness	Best fitness	Mean fitness	Best fitness
Random	71	4.8	13	15.3	21
Seeded	54	5.5	13	12.8	21
GA	52	8.1	13	5.0	9

6   Results

Each experiment yields a population for which dierent characterisations of population fitness are derived. Tables 1 and 2 show, for each experiment, the percentage of the population of chromosomes that have a minimum fitness of 1.

These are each averaged over 10 runs. Also, we show the mean population fitness averaged over 10 runs, and the median over 10 runs of the best population fitnesses. Finally, we also show the corresponding mean and best MRF fitnesses.

We hypothesise that MRF parameters are able to detect the presence of preferred schemata in a population, and that the Metropolis method can be used to generate chromosomes that contain them. Thus we would, in general, expect MRF fitness to improve on population fitness.

We consider the random and seeded experiments first. In the random and seeded populations, the mean MRF fitness is better than population mean fitness in all cases. The best MRF fitness agrees closely with the best population fitness. There is a dierence however between the two sets. From the 20-bit data, the best MRF fitness is far greater than the best population fitness. In fact in each case the true optimal chromosome was generated. From the 64-bit data, however, the best MRF fitness is only equal to the best population fitness for the random and first seeded experiment. It is less than the best population fitness (by the value of one preferred schema) for the second seeded experiment, though it does correctly reect the number of preferred schemata with which those populations were seeded.

These results are consistent with the hypothesis that preferred schemata are being detected. The dierence between the two sets lies in the dierent lengths of preferred schemata. Recall from Sect. 4 that the Metropolis method builds low-energy chromosomes using allele labels and binding energies within cliques. The largest cliques in the Ising model are of length 2.

A preferred schema of length 4 is built from 3 overlapping 2-cliques. The least squares estimate must assign these the correct binding energies in order to build the schema. The precise probability of a particular preferred schema being generated at random is $\frac{1}{16}$. This is a parallel process, and so one would expect each preferred schema to be represented 6 times in a random population of 100. This explains why the results from the random and seeded experiments on the 20-bit problem are so similar. It also explains why all preferred schemata are detected in each case.

It requires 7 overlapping 2-cliques to be assigned the correct binding energies to build a preferred schema of length 8. The precise probability of a particular preferred schema being generated at random is $\frac{1}{128}$. In a random population of 200, one would expect each preferred schema to be represented 1.5 times, and would not expect chromosomes to contain 2 preferred schemata. Thus in the second set of results, the random population is clearly distinguished from the seeded populations. The Metropolis method constructs one or two schemata from the information present, but is unable to construct more schemata due to the diculty of obtaining a suitable coincidence of 2-cliques.

We now consider the experiments on evolved GA populations. The data from the GA runs are markedly dierent for both problems. These data are derived from evolved populations that have a higher mean fitness than the random or seeded populations. However, for both problems, the mean MRF fitness is comparable to that of a random population. Also, the best MRF fitness is usually worse than, and certainly no better than, those generated from the random and seeded populations. It is consistently worse than the best population fitness in the evolved populations. This suggests that the information about preferred schemata in the evolved GA population is deceptive in some way.

In [7], Mitchell et al. demonstrate that the Royal Road GA is hampered by hitchhiking, the process by which poor schemata are propagated by being present in highly-fit chromosomes that quickly come to dominate a population. One can in fact see quite explicitly from (4.2) that the MRF parameters characterise this phenomenon. A chromosome that contains one Royal Road schema will quickly dominate a population where a high percentage of the chromosomes contain no preferred schemata. The low energy of this particular labelling will bias the values of all the parameters in (4.2), not just those that correspond to the cliques responsible for the low energy value. Thus the parameter values resulting from the least squares approximation will reect this bias in the energies assigned to particular cliques. When a chromosome is generated from these parameters, labellings that are poor with respect to the Royal Road function will nevertheless give low-energy solutions with respect to the MRF parameters, and are thus likely to be generated. Therefore we have a population with high population fitnesses but low MRF fitnesses. The low MRF fitnesses in our GA experiments signal that hitchhiking is taking place.

7   Conclusion

In this paper we have demonstrated how a mathematical link can be defined between a GA fitness function and a MRF model of fitness, derived from the fitnesses of a population of chromosomes. Our Theorem shows that such a link can always be defined for a very general class of GA encodings. In practice, the strength of this link depends on the choice of neighbourhood system, and the clique potential functions. Technically, the neighbourhoods form a sub-base for a topology on the allele set. The extent to which this topology is compatible with epistatic interaction between alleles will govern the ecacy of the MRF model. For the Royal Road problems, the Ising model has performed well, detecting preferred schemata and signalling hitchhiking in a converging GA.

It should be noted that our approach is distinct from the work done in modelling the evolution of a GA using Markov chains. We do not model the evolution. There is a close link however in that the MRF model is an approximation to the fitness function. Therefore MRF models could potentially be used to estimate transition probabilities in Markov chain studies of specific encodings.

The MRF approach bears strong similarities to the Bayesian Optimisation Algorithm (BOA) developed in [8,9], and indeed to the wider class of Probabilistic Model-Building Genetic Algorithms (PMBGA) [10]. These approaches also build probabilistic models that characterise highly-fit schemata, and use these models to generate better chromosomes. However there are some significant differences. The BOA uses a directed acyclic graph structure to model interactions between dierent allele values; the concept is one of parenthood rather than neighbourhood. The MRF model does not assume that dependency is directed. Another signi6ant dierence is that the BOA searches the space of possible models using a greedy algorithm. The best model found is then used to generate the next BOA population. Contrastingly, the MRF model uses Least Squares to fit parameters to a fixed set of clique potential functions.

A major problem with simple genetic algorithms is that recombination is accomplished through operators that are broadly fixed in how they operate on chromosomes. This operation is not always compatible with the true interaction between dierent allele values, and sub-optimal evolution phenomena such as hitchhiking are the result. There is a considerable body of recent research aimed at using PMBGAs to address these problems. MRF models may possibly be used, in a similar way, to improve the eectiveness of mutation and crossover operators during GA runtime.

References

1. J. Besag. Spatial interaction and the statistical analysis of lattice systems (with discussions). Journal of the Royal Statistical Society , 36:192–236, 1974.
2. D. F. Brown, A. B. Garmendia-Doval, and J. A. W. McCall. A genetic algorithm framework using Haskell. In Procee dings of the 2nd Asia-Pacific Conference on Genetic Algorithms . Global Link Publishing, May 2000.
3. D. F. Brown, A. B. Garmendia-Doval, and J. A. W. McCall. A functional framework for the implementation of genetic algorithms: comparing Haskell and Standard ML. In S. Gilmore, editor, Trends in Functional Programming , volume 2, pages 27–37, Portland, Oregon, 2001. Intellect Books.
4. H. Derin and P. A. Kelly. Discrete-index Markov-type random fields.   Procee dings of the IEEE , 77:1485–1510, 1989.
5. S. Z. Li. Markov Random Field Modelling in Computer Vision . Springer, 1995.
6. N. Metropolis. Equations of state calculations by fast computational machine. Journal of Chemical Physics , 21:1087–1091, 1953.
7. M. Mitchell, J. H. Holland, and S. Forrest. When will a genetic algorithm outperform hillclimbing? In J. D. Cowan, G. Tesauro, and J. Alspector, editors,   Advances in Neural Information P   rocessing Systems 6 . Morgan Kaufmann, 1994.
8. M. Pelikan and D. E. Goldberg. Research on the Bayesian Optimization Algorithm. Technical Report 2000010, Illinois Genetic Algorithms Lab, UIUC, Urbana, IL, 2000.
9. M. Pelikan, D. E. Goldberg, and E. Cant'u–Paz. BOA: The Bayesian Optimization Algorithm. In W. Banzhaf et al., editor,   Procee dings of the Genetic and Evolutionary Computation Conference GECCO-99   , volume I, pages 525–532, San Fransisco, CA, 1999. Morgan Kaufmann Publishers.
10. M. Pelikan, D. E. Goldberg, and F. Lobo. A survey of optimization by building and using probabilistic models. Technical Report 99018, Illinois Genetic Algorithms Lab, UIUC, Urbana, IL, 1999.

Measuring the Spatial Dispersion of Evolutionary Search Processes: Application to Walksat

Alain Sidaner [1], Olivier Bailleux [2], and Jean-Jacques Chabrier [1]

[1] LIRSIA,
Université de Bourgogne,
9 avenue A. Savary,
B.P. 47870, 21078 Dijon Cedex, France
[2] CRIL,
Université d'Artois,
rue de l'Université,
S.P. 16, 62307 Lens Cedex, France

Abstract. In this paper, we propose a simple and ecient method for measuring the spatial dispersion of a set of points in a metric space. This method allows the quantifying of the population diversity in genetic algorithms. It can also be used to measure the spatial dispersion of any local search process during a specified time interval. We then use this method to study the way Walksat explores its search space, showing that the search for a solution often includes several stages of intensification and diversification.

1 Introduction

The notion of evolutionary algorithm covers many search heuristics derived from the natural process of species evolution. Genetic algorithms are typical examples of evolutionary algorithms [10]. Immediately inspired by the Darwinian theory of species evolution, they handle a population of   chromosomes (each of them obtained by encoding a point in the search space) by applying repetitively mutation, crossing over and selection operators.

Many studies invoke the notion of population diversity in order to analyze and/or improve the convergence of genetic algorithms [6,8,11]. Typically, the population diversity is a measure of spatial dispersion using a metric on the chromosomes space [1].

Like genetic algorithms, local search procedures are based on the concept of cumulative selection. These procedures are proved to be very ecient for solving such problems as the search for a model of a CNF formula in propositional logic (i.e. SAT certification). In particular, the Walksat procedure [14,9], which will be studied in section 3, is a major reference in this domain from several years.

This paper provides two contributions. First, we propose a simple and ecient method for measuring the spatial dispersion of a set of points in a metric space. This method allows the quantifying of the population diversity in genetic

P. Collet et al. (Eds.): EA 2001, LNCS 2310, pp. 77–87, 2002.

algorithms at a given time or during several generations. It can also be used to measure the spatial dispersion of any local search process during a specified time interval. We then use this method to study the way Walksat explores its search space, showing that the search for a solution often includes several stages of intensification and diversification. With some hard instances, the process tends to focus on regions of the search space without a solution.

In section 2, we propose a definition of a spatial dispersion measure of a multi-set of points, which can be applied both to genetic algorithms and to local search procedures. This measure is based on the notion of proximity between a point and a multi-set of points in a metric space.

In section 3, we apply our dispersion measure to the study the way Walksat explores its search space. We present some experimental results based on a sample of 100 random 3CNF formulae that are typical of the ones that are usually used for evaluation and performance comparison of the solving algorithms for SAT. Our results clearly show that during the search for a solution, Walksat alternates several stages of diversification (increasing of the spatial dispersion) and intensification (decreasing of the spatial dispersion). We also study the average distance between the points Walksat visits in the search space and the nearest solutions of these points according to the proximity measure previously mentioned. Thanks to a theoretical property of the dispersion measure introduced in section 2, we show that Walksat often focuses on an area that does not contain any solution. The more dicult the SAT instance is to solve, the farther the process tends to stay from the nearest solution, as if some deceptive regions of the search space wanted to attract it. The study of such deceptive regions is otherwise one of the perspectives we will develop in the last section, after a short synthesis of the experimental results.

2   A Measure of Spatial Dispersion

In this section, we propose a measure of the spatial dispersion of a multi-set of points in a metric space. When this space is the set of binary strings associated with the Hamming distance, this measure is very ecient in terms of time and space complexity.

2.1   Notations

Let F be a multi-set containing $f_1, \ldots, f_p$.

Let $\mathrm{nocc}(f_i)$, $1 \le i \le p$, be the number of occurrences of f_i in F.

The value $\sum_{i=1}^{p} \mathrm{nocc}(f_i)$ will be denoted $|F|$.

For any function g with domain included in F, $\sum_{x \in F} g(x)$ denotes $\sum_{i=1}^{p} \sum_{j=1}^{\mathrm{nocc}(f_i)} g(f_i)$.

For any $\alpha, \beta \in \{0,1\}$, $\mathrm{diff}(\alpha, \beta)$ denotes the value $\max(\alpha, \beta) - \min(\alpha, \beta)$.

For any $x \in \{0,1\}^n$, for any $i \in 1..n$, x_i denotes the ith component of x.

For any $(x,y) \in \{0,1\}^n \times \{0,1\}^n$, $h(x,y) = \sum_{i=1}^{n} \mathrm{diff}(x_i, y_i)$ denotes the Hamming distance between x and y.

2.2   Definitions and Properties

Definition 1.   Let (E, h) be a metric space. Let x be an element of E and Y be a multi-set of elements of E.

We call $H(x, Y) = \frac{1}{|Y|} \sum_Y h(x,)$ the proximity between x and Y.

We call average point of Y any point x in E that minimize $H(x, Y)$.

We call dispersion of Y the proximity between Y and any average point of Y, that is min $H(x, Y), x \in E$.

The dispersion of a multi-set Y is then defined as the average distance between the points of Y and any point in E that minimizes this average distance. Clearly, a null dispersion happens only when Y collapses to the same point with any number of occurrences. We can also remark that if all the points in Y are in a ball with radius d then the dispersion of Y is lower than or equal to d.

The following theorem allows the computation of a lower bound of the distance between two points, given their proximity to a same multi-set.

Theorem 1.   Let (E, h) be a metric space, Y be a multi-set of points of E and $,$ be two points in E such that $H(, Y) = H(, Y) + $, with 0. Then $h(,)$.

Proof.   By definition $H(, Y) = H(, Y) + $ (0)

Then $\frac{1}{|Y|} \sum_{x \in Y} h(x,) = \frac{1}{|Y|} \sum_{x \in Y} h(x,) + $

Hence $= \frac{1}{|Y|} \sum_{x \in Y} h(x,) \frac{1}{|Y|} \sum_{x \in Y} h(x,) = \frac{1}{|Y|} \sum_{x \in Y} (h(x,) h(x,))$

Now $x \in E, h(x,) + h(,) h(x,)$

Then, $x \in E, h(,) h(x,) h(x,)$

Hence $\frac{1}{|Y|} \sum_{x \in Y} h(,) \frac{1}{|Y|} \sum_{x \in Y} (h(x,) h(x,))$

Finally, we have $h(,)$.

Definition 2.   Let Y be a multi-set of points of $0,1 ^n$. We call dispersion vector of Y the vector $W = (w_1, \ldots, w_n)$ such that i $1..n, w_i = \frac{1}{|Y|} \sum_{x \in Y} x_i$.

Each component w_i is then the proportion of strings in Y with 1 at rank i.

Theorem 2.   Let (E, h) be the metric space such that $E = 0,1 ^n$ and h is the Hamming distance. Let Y be a multi-set of elements of E and x be an element of E. Then $H(x, Y) = \sum_{i=1}^{n} ((1 \quad x_i) w_i + x_i (1 \quad w_i))$.

Proof.   By definition, $H(x, Y) = \frac{1}{|Y|} \sum_Y h(x,) = \frac{1}{|Y|} \sum_Y \sum_{i=1}^{n} di (x_i, _i)$

Then $H(x, Y) = \frac{1}{|Y|} \sum_{i=1}^{n} \sum_Y di (x_i, _i)$

$$H(x, Y) = \frac{1}{|Y|} \sum_{i \in 1..n/x_i=0} di\,(x_i, {}_i) + \frac{1}{|Y|} \sum_{i \in 1..n/x_i=1} di\,(x_i, {}_i)$$

$$H(x, Y) = \frac{1}{|Y|} \sum_{i \in 1..n/x_i=0} {}_i + \frac{1}{|Y|} \sum_{i \in 1..n/x_i=1} (1 - {}_i)$$

$$H(x, Y) = \sum_{i \in 1..n/x_i=0} \frac{1}{|Y|} {}_i + \sum_{i \in 1..n/x_i=1} \left(1 - \frac{1}{|Y|}\right) {}_i$$

$$H(x, Y) = \sum_{i \in 1..n/x_i=0} w_i + \sum_{i \in 1..n/x_i=1} (1 - w_i)$$

The result follows.

Corollary 1. Let Y be a multi-set of binary strings with a dispersion vector W. An average point of Y can be obtained by choosing for each bit at rank i the value 1 if w_i > $1/2$, the value 0 if w_i < $1/2$, and a value 0 or 1 if $w_i = 1/2$. Moreover, the dispersion of Y can be easily computed with the formula:

$$disp(Y) = \sum_{i=1}^{n} \min(w_i, 1 - w_i).$$

Clearly enough, computing the dispersion of a multi-set of k binary strings of n bits requires time $O(kn)$. Furthermore, if the multi-set is given as a sequence where each string differs from the next one to the value of at most one bit (the rank of which is known), the time complexity can be reduced to $O(k + n)$.

3 Application to Walksat

3.1 Random Instances, Peak of Difficulty, Walksat

SAT is the decision problem related to the satisfiability of a Conjunctive Normal Form (CNF) Boolean formula. A formula (or instance) is a conjunction of clauses, a clause is a disjunction of literals, a literal is a propositional variable or its negation. An interpretation (or configuration) is a truth assignment to the variables. A model (or solution) is an interpretation that satisfies the formula in the usual sense in propositional logic. In this paper, we only consider the problem of certification (i.e. search for a solution) of satisfiable instances.

A 3SAT instance is a CNF formula where each clause contains at most three literals. The 3SAT problem is NP-Complete [3]. A random 3SAT instance is a formula each clause of which is obtained by drawing 3 distinct variables at random and then negating each of them with probability 0.5 [2]. The difficulty of the certification of the random 3SAT instances is maximum when the ratio of number of clauses to number of variables is near 4.25 [4]. This value is called difficulty peak.

The Walksat procedure tries to solve a SAT instance by repeating local search runs: starting from a random initial configuration, it chooses a variable according to some heuristic and then flips this variable. This sequence choice and modification is repeated until a solution is found or a maximum number of flips is

achieved. The user sets the maximum number of ips. We will call   trajectory any
sequence of configurations visited by Walksat during one search.

Several heuristics are available for Walksat. In our experiments, the SKC
heuristic was used: pick a clause that is falsified (i.e. unsatisfied) by the current
configuration; if there is a ip which doesn't make a currently satisfied clause
unsatisfied, pick any such variable from this clause, otherwise pick a variable
that minimizes the number of clauses that are true in the current state but that
would become false if the ip were made [9].

We define the   diculty   of a satisfiable instance as the average number of
ips Walksat/SKC needs to reach a solution. Because of the cost of the proposed
experiments, we limited ourselves to instances of 100 variables and 425 clauses.

3.2   A First Example of Trajectory

From a sample of 100 random satisfible 3SAT instances at the diculty peak,
we chose an instance with medium diculty. We ran Walksat on this instance
until it reached a solution. We then obtained an example of trajectory that we
chopped up into sections of 100 ips. We measured the global dispersion of this
trajectory as well as the dispersion of each section. Figure 1 presents the results.
The same measures, presented in figure 2, were done on a pure random process
that repetitively draws a variable at random among the variables of the formula
and then ips this variable. The random process was run during the same number
of ips as our Walksat example trajectory.

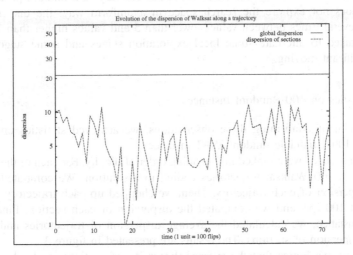

Fig. 1.   Evolution of the dispersion of Walksat along an example trajectory that find a
solution.

The dispersions of the sections (resp. the global dispersion) of the Walksat
trajectory are clearly lower than the dispersion of the sections (resp. the global

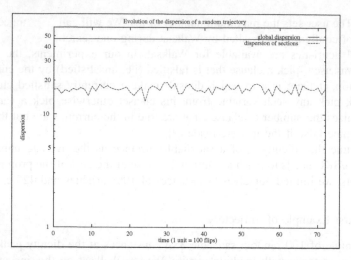

Fig. 2.   Evolution of the dispersion of a random process.

dispersion) of the random process. We also notice that the segment dispersions of Walksat uctuate more than the segment dispersion of the random process along the trajectory. Lastly, in the two trajectories, the section dispersions are much lower than the global dispersions.

Both for sections of 100 ips and for whole trajectories, Walksat disperses itself less than a random process. But above all, contrary to a random process, Walksat does not explore the search space in a uniform way: the dispersion uctuates several times between values lower than 5 and values higher than 10, as if it wanted to alternate some local exploration stages and some stages of more significant moving.

3.3   Statistics on 100 Random Instances

With the aim of confirming these observations, we achieved statistics on our sample of 100 satisfiable random 3SAT instances.

These instances were ranked in ascending order of diculty. For each of them, we produced 100 Walksat trajectories ending to a solution. We computed the global dispersion of each trajectory. Then, we chopped up each trajectory into sections of 100 ips, and we computed the dispersion of each section. Finally, for each instance, we calculated the average dispersion of trajectories and the average dispersion of sections. The results are presented in figure 3.

As above, we notice that the average dispersion of sections is clearly lower than the average dispersion of trajectories. Interestingly, when the diculty of instances increases, the average dispersion of trajectories tends to increase while the average dispersion of sections tends to decrease slightly.

At this stage, we aimed to verify and quantify the uctuation of dispersion along the trajectories. To this end, for each trajectory, we counted the number

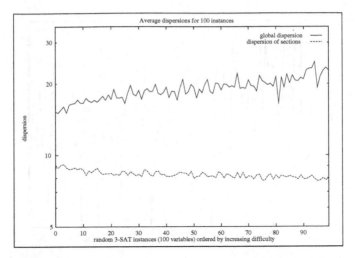

Fig. 3. Average dispersions of trajectories and sections for 100 instances ordered by increasing diculty.

of times the dispersion increases or decreases by a factor of 2 during the search. Each of these variations can stretch over several successive sections. Each section is taken into account in at most one variation. The resulting value was divided by the number of sections in the trajectory. We call   variation ratio this final value. If this variation ratio is close to 1 then the dispersion often increases and decreases by a factor of 2. On the othre hand, if this ratio is close to 0 then most of the section dispersions dier from other ones by a factor of lower than 2. The results are presented in figure 4.

Clearly enough, the section dispersion of the trajectories obtained from our sample of 100 instances (that is to say, 10000 trajectories) typically varies by a factor of 2 on average every 300 to 400 ips. In comparison, the variation ratio of a random process is less than 0.1.

Because the trajectories of Walksat begin from a random configuration and end in a solution, we investigated the possibility that the above results were biased by a specific behavior of Walksat at the beginning and/or at the end of the search. That is why we repeated the previous measures on the 50 central part of each trajectory.

The results obtained from the central parts of the trajectories were very close to the results obtained from the whole trajectories, that is an average global dispersion of about 8 and a variation ratio between 0.25 and 0.35 with the exception of the 10 easiest instances, because the trajectories related to these instances are very short.

3.4   Proximity between Process and Solutions

Because the aim of the Walksat procedure is to find a solution, we think it is interesting to know if the process   focuses on solutions, i.e., if the average

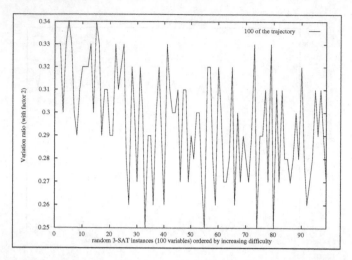

Fig. 4. Average of variation ratio by instance for 100 instances ordered by increasing diculty.

points of the trajectories are close to some solutions. Then, for each section of each trajectory of each instance, we computed the   average solutions , that is the solutions that are closest to the considered sections, according to our proximity measure.

Figure 5 presents the average proximity between the sections and their average solutions as well as the average dispersion of the sections.

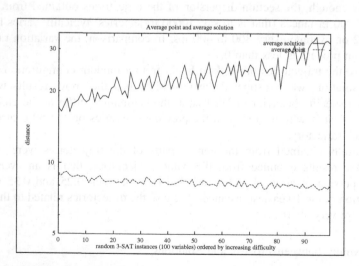

Fig. 5. Average proximity between the sections and their average solutions for 100 instances ordered by increasing diculty.

Thanks to theorem 1, these data allow demonstrate that for the easiest instances, the distance between the solutions and the average points is typically higher than 5. For the hardest instances, this distance is at least 15 [1].

Clearly enough, during the intensification stages, the process focuses on configurations that are not solutions. This oset between the average points and solutions seems to be more pronounced for the most dicult instances.

4   Related Works

As previously mentioned, many studies invoke the notion of population diversity in order to analyze and/or improve the convergence of genetic algorithms. Typically, the population diversity is a measure of spatial dispersion using a metric on the chromosome space.

Some methods of diversity maintenance do not use an explicit definition of a diversity measure. For example, some variants of genetic algorithms favour the repartition of their population into several niches, which are dispersed in the search space. Such dispersion is obtained by reducing the fitness (i.e. objective value) of the chromosomes that are similar to other ones in the population (according to the chosen metric) [6,11]. Other variants avoid creating any chromosome that would be too close to an existing one [8], or favour the mutations that tend to increase the average distance between the mutant chromosomes and the rest of the population [13]. In other works, the diversity of a population is explicitly defined as the average of the distance between any possible pair of chromosomes, that is the average of $\frac{n(n-1)}{2}$ distances for a population of n chromosomes [1].

In the case of stochastic local search algorithms, where the "population" is reduced to a singleton, the notion of spatial dispersion is only relevant when it is defined on a sequence of configurations successively visited by the search process during a given number of iterations. There are few references to such approaches in the literature. However, D.Schuurmans and F.Southey recently introduced the notions of mobility and coverage of a local search process [12]. Mobility is defined as the distance between configurations that are a given number of steps apart in the search sequence. Given a set of points visited by the process, coverage states the maximum distance between any unvisited point in the search space and the nearest visited point.

The study of some aspects of the behavior of the local search processes and genetic algorithms is also related to the works based on the notion of search landscapes [7]. A search landscape is an abstraction that represents the data a local search process needs to work, that is a neighbourhood relation and an objective function. Several studies have contributed to connect some characteristics of the search landscape and the eciency of some evolutionary search processes. In particular, in a study dedicated to the landscapes of random SAT instances,

---

[1] If d is the dispersion of a section of trajectory and   p the proximity between this section and the average solutions, then (p   d) is a lower bound of the distance between any average point of this section and any solution.

J.Frank and al. [5] isolated some regions with specific topological properties, such as plateaus and local extrema, that tend to trap the local search processes. With another approach, A.Sidaner and al. [15] proposed a measure that quantifies the attraction of a search process by a given configuration in this search space. This technique allowed them to show that in the search landscapes of random 3SAT instances, the most attractive configurations are not always solutions.

5   Synthesis and Perspectives

In this paper, we proposed a measure of spatial dispersion that is intuitive and ecient both in terms of time and space complexity, on a space of binary strings with Hamming distance. It was applied to show new aspects of the behavior of Walksat on random SAT instances.

Compared to a pure random process that is not guided by a search strategy, Walksat focuses on certain areas of the search space. But this intensification of the search uctuates significantly over time so much so that on the most dicult instances, Walksat clearly alternates several intensifiation and diversifiation stages. The intensifiation stages correspond to focusing on dierent regions in the search space, while the diversification stages correspond to moving between these regions. Given that such uctuations do not occur with a pure random process, we think that the search space includes some singularities that attract and trap the process, as well as some areas that favour the widening of the search. During the search, the process typically visits several "singularities" before reaching a solution.

We think that the study of the areas the process focuses on, as well as the region where the process disperses itself, is a promising research perspective, that could be compared to some existing studies on search landscapes [5]. We actually suspect that the sporadic search intensifications are due to the local (and global) extrema. We consider that it is necessary to rigorously verify this hypothesis, because we think that a good understanding of the behavior of local search processes, and more generally the evolutionary processes, requires a better comprehension of how the local properties of the search landscapes aect the global behavior of these processes.

Moreover, compared to the dispersion measurements of the process, the measurements of the proximity between the process (i.e. its trajectories and sections of trajectories) and the solutions clearly show that Walksat focuses its search on regions without a solution. The more dicult the instances are, the farther the average points of the process are from the solutions. It seems that the search landscapes of hard instances contain some deceptive regions that attract the process and tend to keep it quite far from the solutions. These observations open new perspectives related to the hypothesis that the search landscapes of some instances include some local sources of diculty. In particular, we plan to propose new local search heuristics that will take into account the spatial dispersion and the average points of the search process. The control of dispersion could allow the adaptation of the intensification of the search process in order to increase its robustness and/or eciency. Locating the average points could allow one to direct the search process in such a way that it does not stay in the same region for too long a time. Such ideas could also be applied to genetic algorithms and lead to new techniques for maintaining the diversity of the population.

References

1. Barker A.L., Martin W.N.: Dynamics of a Distance Based Population Diversity Measure. Congress on Evolutionary Computation, IEEE Press, 2000.
2. Chv'atal V., Szemer' edi E.: Many hard examples for Resolution. In Journal of the ACM, pages 759-768, 1988.
3. Cook S.A.: The complexity of theorem proving procedures. In Conference Records of Third Annual ACM Symposium on Theory of Computing, pages 151-158, 1971.
4. Dubois O., Andre P., Boufkhad Y., Carlier J.: SAT versus UNSAT. In Clique, Coloring and Satisfiability: Second DIMACS implementation challenge, Volume 26 of DIMACS series in Discrete Mathematics and Computer Science, pages 415-136, American Mathematical Society, 1996.
5. Frank J., Cheeseman P.,Stutz, J. (to appear): When gravity fails: local search topology. Gupta and Nau (1992). On the complexity of blocks-world planning. Artificial Intelligence (1997) 56, 139–403
6. Goldberg D.E., Richardson J.J.: Genetic algorithms with sharing for multimodal function optimization. Gen. Algs. and their Apps.: Proc. 2nd Intl. Conf. Gen. Algs., Cambridge, MA, July 1987, 28–31 (Lawrence Erlbaum).
7. Jones T., Forest S.: Fitness Distance Correlation as a Measure of Problem Diculty for Genetic Algorithms. In Proceedings of the 6 [th] International Conference on Genetic Algorithms (1995) 184-192.
8. Mauldin M.L.: Maintaining diversity in genetic search. In Proceedings of the National Conference on AI (1984) 247–250. AAAI.
9. McAllester D., Selman B., Kautz H.: Evidence for Invariants in Local Search. In Proceedings of AAAI-97 (1997) 321-326.
10. Michalewicz Z.: Genetic algorithms + Data Structures = Evolution Programs. Artificial Intelligence, Springer Verlag. New York, 1992.
11. Miller B.L., Shaw M.J.: Genetic Algorithms with Dynamic Niche Sharing for Multimodal Function Optimization. IEEE International Conference on Evolutionary Computation, pp786-791, Piscat   away, NJ: IEEE Press 1995.
12. Schuurmans D.,Southey F.: Local search characteristics of incomplete SAT procedures. In Proceedings of the Seventeenth National Conference on Articial Intelligence (AAAI-2000), pages 297-302, 2000.
13. Sefrioui M., P' eriaux J.: Fast convergence thanks to diversity. In Proc. of the Fifth Annual Conference on Evolutionary Programming (1996) 313-321
14. Selman B., Kautz H., Cohen, B.: Noise Strategies for Local Search. In Proceedings of AAAI-94 (1994) 337-343.
15. Sidanor A., Chabrier J.-J., Bailleux O.: Towards a quantification of attraction in stochastic local search. Proc. of the Workshop on Empirical Methods in Artificial Intelligence at ECAI'00, Berlin, Germany, pp 39-41, August 2000.

The Importance of Selection Mechanisms in Distribution Estimation Algorithms

Andrew Johnson   and Jonathan Shapiro

Computer Science Department, University of Manchester, Manchester, M13 9PL, UK

Abstract. The evolutionary algorithms that use probabilistic graphical models to represent properties of selected solutions are known as Distribution Estimation Algorithms (DEAs). Work on such algorithms has generally focused on the complexity of the models used. Here, the performance of two DEAs is investigated. One takes problem variables to be independent while the other uses pairwise conditional probabilities to generate a chain in which each variable conditions another. Three problems are considered that dier in the extent to which they impose a chain-like structure on variables. The more complex algorithm performs better on a function that exactly matches the structure of its model. However, on other problems, the selection mechanism is seen to be crucial, some previously reported gains for the more complex algorithm are shown to be unfounded and, with comparable mechanisms, the simpler algorithm gives better results. Some preliminary explanations of the dynamics of the algorithms are also oered.

1   Introduction

Instead of maintaining a population of promising solutions as a Genetic Algorithm does, Distribution Estimation Algorithms (DEAs) are evolutionary algorithms that use probabilistic graphical models to represent statistical properties of populations. Each generation a probabilistic model is inferred from selected individuals and is then sampled from to give a new generation. The model represents a probability distribution over the search space. DEAs have attracted much attention in recent years (see [1] for a review) and vary primarily in the complexity of model used. The simplest, such as Population-Based Incremental Learning (PBIL) [2], assume that problem variables are independent. PBIL has a probability vector containing one element per variable initialised to 0.5. Each generation, the vector is shifted towards the selected individuals in proportion to a constant,   . The distribution is given by

$$P(X) = \prod_{i=1}^{l} P(X_i) , \tag{1}$$

The first author is supported by a UK EPSRC Studentship.

P. Collet et al. (Eds.): EA 2001, LNCS 2310, pp. 91–103, 2002.

Fig. 1. A chain model

where X is a string, 1 is its length and X_i is a variable in X. MIMIC [3] diers from PBIL in that it models dependencies between pairs of variables, imposing a chain structure as in Fig. 1 which gives the following distribution over the search space

$$P(X) = P(X_{i_1}) \prod_{i=2}^{1} P(X_{m(i)} X_{m(i\ 1)}),\qquad (2)$$

where i_1 represents the bit position in the string of the variable at the root of the chain, and m(i) gives the bit position in the string of the variable in the ith position in the chain.

The ordering of variables in the chain is determined using information measures. The variable with lowest entropy forms the root of the chain and conditional entropy is used in a greedy algorithm to find the pairs of variables containing the most information. Other DEAs use trees [4] or Bayesian Networks, e.g. [5], to model more sophisticated relationships.

It is generally thought that more complex models give better results than simpler ones and a range of experimental evidence has been produced to support this, by authors proposing DEAs and also in other studies e.g., [6,7]. However, the test problems have often been tailor-made to fit the algorithm so that the interactions between variables are known to fall within the particular DEA's model class.

Whilst previous work has concentrated on model complexity, the other major element of a DEA is the choice of selection method/model update rule used. Various selection methods have until now been tried with no consideration of their eect. Here we investigate two DEAs, PBIL and MIMIC on three problems. We find that indeed on an 'ideal' problem MIMIC obtains better results than PBIL but that on other problems, MIMIC is outperformed by the simpler algorithm. Selection methods are found to be crucial in determining the relative performance of the algorithms.

2 The Algorithms and Selection Mechanisms

Table 1 gives the names that will be used to refer to the five implementations considered in this paper. The table specifies the combination of model and selection method/update rule. Algorithm names are denoted in typewriter font.

Incremental selection decays elements of the probability vector by a factor of 1 for each selected individual and then adds (0.005) to elements corresponding to bits that have a value of 1 in the two selected individuals, X i.e.

Table 1.   Algorithm naming

chain/independent	Selection/update	Name
independent	incremental	PBIL
independent	steadystate	PBIL3
independent	matrix-based	PBILm
chain	matrix-based	chain
chain	steadystate	chain3

$$P^{t+1}(X_i = 1) = (1 \qquad)P^t(X_i = 1) + \quad X_i \ .$$

The chain   implementation is taken from [4] (although we use conditional entropy, as in [3], rather than mutual information) and uses 'matrix-based' selection. In this method, a 'frequency matrix' is used "containing a number $A[X_i = a, X_j = b]$ for every pair of variables X_i and X_j and every combination of binary assignments to   a and b where $A[X_i = a, X_j = b]$" [from [4]]. All elements of the matrix are initialised to 1000. Each generation the elements are decayed by a factor of 0.99 then 1 is added to the elements that correspond to the variable combinations appearing in the four selected individuals. The counts stored in the matrix are used to calculate the entropies needed to infer the probability distribution.

Steadystate selection [8] is an attempt at incorporating the incremental nature of steadystate selection used in GAs, although our version has an important dierence. We maintain two separate populations of equal size. One is the 'steady' population that is used to build the model and the other is the current, sampled population. Every generation the best   = 15 members are taken from current population and inserted into the steady population in place of the worst   members.

3   A Spin-Glass Function

In this section we show that for a function that exactly matches a chain model, MIMIC   algorithms obtain better solutions than   PBIL . A spin-glass function defined as

$$f(X) = \sum_{i=1}^{l-1} J_{i,i+1}(X_i \quad 1/2)(X_{i+1} \quad 1/2) \ ,$$

where the weights are distributed uniformly in the interval [1, 1], has two global optima and, on average, $2^{l/3}$ local optima [9]. This imposes a chain structure as the global optimum can be found by firstly setting the root variable then the other variables sequentially along the chain; to maximise fitness, neighbouring variables should be set to the same value if $J_{i,i+1}$ 0 and to dierent values otherwise. In Table 2 we present results for four implementations of   PBIL   and

MIMIC on spin-glass functions of length 80. The final column presents the percentage of runs on which the algorithms obtain a fitness of greater than 98 of the global optimum. The chain-based implementations clearly outperform their independent model counterparts, consistently getting very close to the global optimum. Steadystate algorithms require 6-12 times fewer generations.

Table 2. Results on spin-glass function, averages over 20 runs

Algorithm	Mean fitness of best individual	Mean generation of best individual	of runs within 98 of global optimum
PBIL3	9.11	227	0
PBILm	9.17	1222	10
chain3	9.57	67	90
chain	9.64	776	100

4   Four-Peaks

The next problem that we consider is the four-peaks function, defined as

$$z(X) = \text{number of leading contiguous zeros}$$
$$o(X) = \text{number of trailing contiguous ones}$$

$$\text{Reward} = \begin{cases} \text{stringLength} + t & \text{if } (o(X) \; t \;) \; (z(X) \; t \;) \\ 0 & \text{otherwise} \end{cases}$$

$$f(X) = \text{MAX}(o(X), z(X)) + \text{Reward} \quad .$$

Four-peaks is so-called because it has four optima, two local optima that are strings of all 0's or all 1's and two global optima that are $t + 1$ leading 0's followed by all 1's, and $t + 1$ trailing 1's preceded by all 0's. The problem is also tunably hard through the parameter t; as t increases so does the size of the basin of attraction of the local optima and hence the diculty of the problem. The string length is taken to be 80, meaning the local optima have a fitness of 80 and the global optima a fitness of 159.

Clearly, four-peaks is a problem of limited real-world interest. However, it has been used by other authors [2,3,4] as a 'proof of concept' for DEAs and, in particular, for chain-model DEAs. In addition, its characteristics are well-known so we use it as the basis of our investigation. Four-peaks is not chainlike; it has weak, long range dependencies ($P(X_i = 1)$ depends on $X_{i\;1}, X_{i\;2}, \dots$). However, if the parameters of the chain were such that the two local optima were highly and equally likely, the global optima would also be likely.   PBIL does not have the potential to model the two local optima simultaneously.

4.1 Results

Previous Implementations.   In [3] results are reported for   MIMIC that emphasise its rapid convergence on an easy instance (t = 8) when a chain-based algorithm is seen to require ten times fewer generations than   PBIL. We see a similar speed-up for chain3 (is set to 15 for this reason). In [4] an improvement is claimed for chain on dierent grounds as it proves more consistent than   PBIL in obtaining the reward. The reward rate is 100 on the easy instance but, on a hard instance (t = 20) falls to 70 for   chain and to 50 for   PBIL. Our findings, shown in Table 3, corroborate these results. Algorithms using steadystate selection are allowed 300 generations, others 1300.

Table 3.   Results on 4-peaks, averages over 30 runs

Algorithm	t	Reward rate	Mean best fitness when reward found	Mean generation of best when reward found	CPU time (secs)
PBIL	8	100	159	965	380
PBILm	8	100	159	1270	1194
PBIL3	8	100	159	153	165
chain	8	100	159	1006	3905
chain3	8	100	159	120	212
PBIL	20	53	144	694	982
PBILm	20	73	148	980	3192
PBIL3	20	57	153	101	164
chain	20	73	153	857	6082
chain3	20	33	144	93	1212

New Implementations.   All previously published implementations dier in their selection mechanisms so in Table 3 we also give results for algorithms where the eect of the selection method has been removed.   chain3 and PBIL3 both use steadystate selection, and   PBILmand chain both use matrix-based selection. It is apparent that those algorithms using matrix-based selection require a greater number of generations to converge, but are most consistent in obtaining the reward. Those with steadystate selection converge quickly but are less robust in obtaining the reward.

The previous claimed superiority of   MIMIC style algorithms turns out to be unfounded. Firstly, it can be seen in Table 3 that   chain3 and PBIL3 require an almost identical number of generations to obtain the best solution. Furthermore, PBIL3 typically finds better solutions and is considerably more reliable in getting the reward. Secondly,   PBILmis as consistent as   chain , both exhibit reward rates of 73.   PBIL3 is an improvement over   PBIL (the original implementation) under all criteria. In the final column, we give the CPU time for our implementations of the algorithms (in Matlab, run on a 450MHz Pentium III PC). For the easy instance, timings are the time needed to find the global optimum and, for the hard instance, are the average time needed to produce a solution with fitness of at least 150. Timings could be greatly reduced through more ecient implementations but nevertheless give an indication of orders of magnitude dierences. PBIL3 is by far the fastest algorithm, requiring around eight times less CPU time than the fastest   MIMIC   implementation. The matrix-based algorithms are very slow, although PBILmis faster than   chain   by a factor of 2.

With implementations of   PBIL   and   MIMIC   utilising equivalent selection methods we now attempt to better understand some aspects of the algorithms' dynamics.

4.2   Measuring the Pairwise Correlations

In this section we describe a way of quantifying the importance of pairwise terms in MIMIC . In the case of PBIL   we can re-write (1) as

$$P(X) = \prod_{i=1}^{l} {}_{i}^{X_i} (1 \quad {}_{i})^{(1 \quad X_i)} , \tag{3}$$

where X is a string, l is its length and $_i$ represents $P(X_i = 1)$. This can in turn be written as

$$P(X) = \exp \left(\sum_{i=1}^{l} w_i X_i + F \right) . \tag{4}$$

Definitions for w_i and F appear in (7). For   MIMIC , (2) is written as

$$P(X) = \prod_{i=2}^{l} {}_{i}^{X_i X_{i\ 1}} (1 \quad {}_{i})^{(1\ X_i)X_{i\ 1}} {}_{i}^{X_i(1\ X_{i\ 1})} (1 \quad {}_{i})^{(1\ X_i)(1\ X_{i\ 1})}$$

$$ {}^{X_1}(1 \quad)^{(1\ X_1)} ,\tag{5}$$

where for i   2, $_i = P(X_i = 1 \mid X_{i\ 1} = 1)$, $_i = P(X_i = 1 \mid X_{i\ 1} = 0)$ and $= P(X_1 = 1)$. Equation (5) is equivalent to

$$\exp \left(\sum_{i=2}^{l} p_i X_i X_{i\ 1} + \sum_{i=1}^{l} s_i X_i + G \right) . \tag{6}$$

With a   0,1   alphabet the weights in (4) and (6) are only relevant when $X_i, X_{i\ 1}$ are equal to 1. For this reason we substitute an alphabet of   -1,1 . In this case, the weights are given by

$$w_i = \ln \frac{\overline{\mu_i}}{1-\mu_i} , \quad F = \sum_{i=1}^{l} \ln \mu_i(1-\mu_i) , \tag{7}$$

$$p_i = 0.25 \, \ln \frac{\mu_i}{1-\mu_i} \frac{1-\mu_i}{\mu_i} , \tag{8}$$

$$s_i = 0.25 \, \ln \frac{\mu_i}{1-\mu_i} \frac{\mu_i}{1-\mu_i} \frac{\mu_{i+1}(1-\mu_{i+1})}{\mu_{i+1}(1-\mu_{i+1})} , \tag{9}$$

$$G = 0.25 \sum_{i=1}^{l} \ln(\mu_i(1-\mu_i)\mu_i(1-\mu_i)) \text{ where } \mu_{l+1}, \mu_{l+1} = 0.5 .$$

The p_i measure the extent to which neighbouring variables in the chain should take the same value (p_i 0), or dierent values (p_i 0). If the p_i are large then pairwise correlations are strong and, conversely, as p_i 0 MIMIC becomes more like PBIL . F and G are normalisation terms that ensure the probabilities sum to 1.

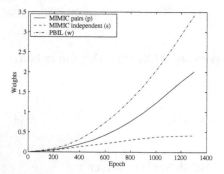

Fig. 2.   Mean absolute weights for matrix-based selection

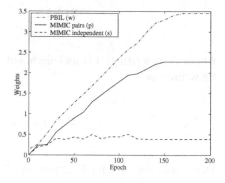

Fig. 3.   Mean absolute weights for steadystate selection

Matrix-based selection.   In Fig. 2 we show the weights (as defined in equations 7,8,9) learnt by matrix-based algorithms on a version of the four-peaks problem using the 1, 1 alphabet. The magnitude of the pairwise weights demonstrates that chain does discover correlations which, after several hundred generations, become more important in determining the probability of a string than the independent variable weights.

Steadystate selection.   The plot of the same equations for steadystate selection in Fig. 3 shows that the strength of the pairwise weights increases much more rapidly than for matrix-based selection and that from the very start of the run, the pairwise weights are considerably more important than the independent weights. This is in keeping with experimental results which show that steadystate algorithms converge more quickly.

4.3   Model Selection

Model selection tests compare how well dierent models fi a given data set which, in our case, is a set of populations. By using one such test we can estimate whether the pairwise correlations, quantified for MIMIC above, result in signi6antly dierent populations from those generated by PBIL. The Likelihood Ratio Test (LRT) determines whether a simpler model M_1 is an adequate fit of a data set relative to a more complex one, M_0, allowing for M_0's greater number of parameters. The LRT test value,   , is given by 2 $\log(L_1/L_0)$, where L_i is the likelihood of the data under model   i. The value   follows a ^{2} distribution with the number of degrees of freedom equal to the dierence in the number of parameters that define the two models. We capture the population every 100 generations during runs on the easy 4-peaks instance, to give 13 samples of 200 individuals. A PBIL style probability vector and a MIMIC chain are fitted to each of the samples. The likelihood of each individual is then defined as the mean of its likelihood under the fitted models.

First we measure how MIMIC might better model a set of populations generated by PBILm The increased modelling power of MIMIC is demonstrated as the PBIL vectors are deemed to be an inadequate fit of PBIL-generated populations compared to the chain models; the critical value at the the 99.9 confidence level for the $^2_{(80)}$ distribution is 124.84 and our test value is 820. Next, we apply the same method to a sample of populations generated by   chain . Importantly, the test value is much higher than before at 2296. Thus we can conclude that the correlations seen in Fig. 2 produce populations with important pairwise interactions that PBIL would be unlikely to generate, since PBIL is an extremely poor fit of these populations.

4.4   Understanding the Dynamics

In this section we aim to explain two performance dierences revealed in section 3. We examine the dierences in terms of the much discussed exploration/exploitation issue in evolutionary computation. The level of exploration performed during search can be approximated by some measure of population diversity, here population fitness variance will be used. A convenient measure of exploitation is obtained for DEAs by making use of the distributions - we calculate the likelihood of a population under the model updated from its selected individuals. If this figure is high then so will be the level of exploitation, since the current population is closely modelled and the next generation is likely to be similar.

Matrix-based Algorithms.   Of the two implementations using matrix-based selection, chain requires around 15 fewer generations than   PBILmto find its

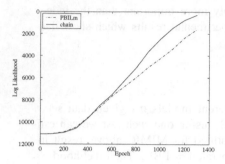

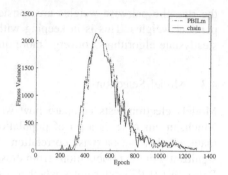

Fig. 4.   Likelihood of the populations of
PBILmand chain   on easy instance

Fig. 5.   Population fitness variance of
PBILmand chain   on easy instance

best solution. Figure 4 illustrates the higher likelihood of populations at any
given generation under   chain   compared to PBILm Therefore, chain   models
populations more accurately than   PBILm In Fig. 5 the variances in fitnesses
of populations are shown. The plot's two curves are strikingly similar. Taken
together, Figs 4 and 5 indicate that, when using matrix-based selection, the
more complex model allows greater exploitation without sacrificing exploration.

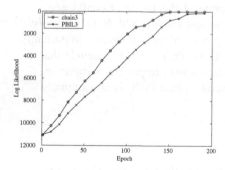

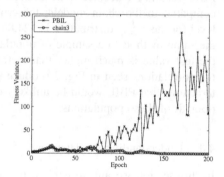

Fig. 6.   Likelihood of current populations of PBIL3 and chain3 , t = 20

Fig. 7.   Population fitness variance of
PBIL3 and chain3 , t = 20

Steadystate Algorithms.   Table 3 shows that   chain3   scores a much lower reward rate than does   PBIL3 on the hard 4-peaks instance. Again this can be understood by looking at the relative levels of exploration and exploitation. As might
be expected from the stronger pairwise correlations discovered by   chain3 , the
gap between the likelihood of the populations under   PBIL3 and chain3 , shown
in Fig. 6, is wider than in Fig. 4 for matrix-based algorithms. The corresponding

plot of population fitness variance in Fig. 7 is revealing; the similarity between the two curves seen for matrix-based algorithms is not present and the variance for chain3 is remarkably low. We conclude that on this problem the stronger correlations learnt by MIMIC, using steadystate selection, result in premature convergence, i.e., insucient exploration.

We have experimented with increased population sizes for   chain (with the number of generations reduced to keep the total number of function evaluations the same) - since it has a larger number of parameters to estimate it may require larger sample sizes - but this has not yielded improved results.

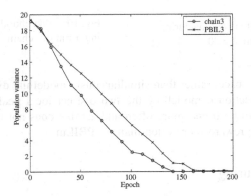

Fig. 8.   Population variance during runs of   PBIL3 and chain3 , t = 20

In Fig. 8 the eect of the fness function is removed and the variance of the population is plotted. This is defined as the sum of the variance of individual variables for PBIL3 and to this quantity two times the covariance between neigh-bouring variables is added for   chain3 (since we are calculating a sum of vari-ances but variables connected in the chain are not independent). By inspection of Fig.s 6 and 8, it is clear that the likelihood of populations and their variance are strongly inversely correlated. Thus DEAs allow the exploration/exploitation trade-o to be visualised.

Points in the Search Space.   So far the likelihood of the population has been considered, but it is also possible to calculate the likelihood of particular points in the space (the extent to which these points are exploited), in order to understand another observation from Table 3, i.e. the failure of   chain to obtain a higher reward rate than   PBILm Figures 9 and 10 plot the likelihoods of the four optima during runs of PBILmand chain . We see that in both graphs the likelihood of two of the optima increases over the run. In the case of   PBILmthe likelihood of the two least favoured optima falls rapidly. This drop is not observed for   chain but, crucially, the likelihood of these two optima does not increase over the run. So, it appears that the greater modelling power of MIMIC is concentrated over a

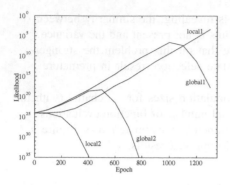

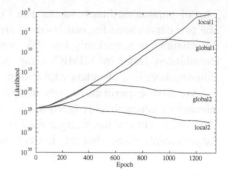

Fig. 9.   Likelihood of the optima during a run of PBILm when t = 20

Fig. 10.   Likelihood of the optima during a run of chain when t = 20

small region of the space, rather than simultaneously modelling distinct regions. As mentioned in section 4 modelling the two distinct local peaks would allow the global optima to be found more often and, as this does not happen,   chain does not obtain the reward more often than   PBILm

5   Graph-Colouring

Lastly, we consider Max   k-colouring, a well-known problem of colouring a graph with k colours such that the number of connected nodes sharing a colour (clashes) is minimised. Our test graphs are of various types; some have randomly connected nodes, others have clusters of connected nodes. The average vertex degree is 22 which clearly constitutes too high an order of variable dependencies to be modelled by a chain. All graphs have 50 nodes and   k is 4, so two bits are required to represent each node. In Fig. 4 we present some results. On the

Table 4.   Graph-colouring results, averages over 30 runs

Algorithm	Colourable with 0 clashes			Uncolourable
	Success rate	Mean epochs	CPU time (secs) to colour with 0 clashes	Mean no. of clashes
PBIL3	64	124	100	81.5
PBIL	92	830	214	82.6
PBILm	93	1227	4769	81.5
chain	100	1123	6529	82
chain3	51	108	410	83

colourable graphs, as with four-peaks, matrix-based variants are slow but very successful. Here, PBILm achieves a slightly lower reward rate than   chain   and the dierence between   PBIL3 and chain3 's success rates is slightly less than on four-peaks. Nevertheless, the we observe the same trends as in four-peaks, including the computational eciency of   PBIL   implementations compared to MIMIC   and the poor performance of   chain3 . Column five shows the results of trials on uncolourable graphs. There is no notion of success or reward, so to account for the increased robustness of matrix-based algorithms we take results for the best half of runs for all algorithms. Here,   PBIL   produces slightly better solutions than   MIMIC .

6   Conclusions

Implementations of   MIMIC   achieve better solutions than   PBIL   on a fitness function ideally suited to their model class, regardless of the selection method used. However, on other problems, whose characteristics match the chain model less closely, some previously reported gains for   MIMIC   are found to be due to dierences in the selection mechanisms. With equivalent mechanisms   PBIL performs comparably or better than the best performing version of   MIMIC , but at a fraction of the computational eort. Some dierences in the dynamics of the algorithms are illustrated, as   MIMIC   is found to model populations more accurately, but can suer from premature convergence.

This study shows the importance of selection mechanisms in DEAs; selection can be the determining factor in the performance of the algorithm, rather than the type of model. Of course, we have only compared the two simplest models and further work should establish whether the conclusions extend to other DEAs. Our set of test problems is relatively small but also varied. This work provides evidence that, rather than concentrating solely on model complexity, investing eort in understanding and optimising the dierent selection mechanisms and update rules in DEAs is justified.

References

1. M. Pelikan, D. Goldberg, and F. Lob. A survey of optimization by building and using probabilistic models. Technical Report 99018, University of Illinois at Urbana Champaign, Illinois Genetic Algorithms Laboratory, 1999.
2. S. Baluja and R. Caruana. Removing the genetics from the standard genetic al-gorithm. In A. Prieditis and S. Russell, editors,   Procee dings of ML-95, Twelfth International Conference on Machine Learning   , pages 38–46, 1995.
3. J. de Bonet, C. Isbell, and P. Viola. Mimic: Finding optima by estimating probability densities. Advances in Neural Information P   rocessing Systems , 9, 1996.
4. S. Baluja and S. Davies. Using optimal dependency-trees for combinatorial opti-mization: Learning the structure of the search space. In D Fisher, editor,   Procee dings of the Fourteenth International Conference on Machine Learning (ICML-97)   , pages 30–38, 1997.
5. M. Pelikan, D. Goldberg, and E. Cant´   u-Paz. Boa: The Bayesian optimization algo-rithm. Technical Report 99003, University of Illinois at Urbana-Champaign, Illinois Genetic Algorithms Laboratory, 1999.

6. P. Bosman and D. Thierens. Linkage information processing in distribution estimation algorithms. In W. Banzhaf, J. Daida, A.E. Eiben, M.H. Garzon, V. Honavar, M. Jakiela, and R.E. Smith, editors, *Procee dings of the GECCO-99 Genetic and Evolutionary Computation Conference*, pages 60–67. Morgan Kaufmann, July 1999.
7. J. Schwarz and J. Ocenasek. Experimental study: hypergraph partitioning based on the simple and advanced algorithms BMDA and BOA. In *Procee dings of the fifth international conference on Soft Computing*, pages 124–130, 1999.
8. G. Syswerda. A study of reproduction in generational and steady state genetic algorithms. In G. Rawlins, editor, *Foundations of Genetic Algorithms*, pages 94–101, 1991.
9. H. Chen and S. Ma. Low-temperature behaviour of a one-dimensional random Ising model. *Journal of Statistical Physics*, 29:717, 1982.

Surrogate Deterministic Mutation: Preliminary Results

K. Abboud and Marc Schoenauer

CMAP, URA CNRS 756,
Ecole Polytechnique,
Palaiseau 91128, France;
kamal,marc @cmapx.polytechnique.fr

Abstract. A new mutation operator based on a surrogate model of the fitness function is introduced. The original features of this approach are 1-the model used to approximate the fitness, namely Support Vector Machines; 2-the adaptive granularity of the approximation, going from space-wide to closely localized around the best-so-far individual of the population; 3-the use of a deterministic optimization method on the surrogate model. The goal is to accelerate the convergence of the evolutionary algorithm, and not to reduce the number of evaluations of the actual fitness by evaluating the surrogate model instead. First results on benchmark functions of high dimensions show the potential improvement that this approach can bring in high-dimensional search spaces, and points out some limitations.

1 Introduction

Hybridizing EAs with some deterministic local search is now a widely applied technique in the context of combinatorial optimization. Such algorithms are called mimetic algorithm [20,17], and are known to give the best results on many OR benchmark problems [15,14,9] However, such hybridization is not so common in the framework of parametric optimization [23,12]. One reason for that is that most local search methods that are know to be ecient for parametric optimization require the use of derivatives of the objective function.

On the other hand, many works in optimization in general, and in evolutionary computation in particular, use surrogate models in lieu of the actual objective function when the latter is very costly. But such approaches generally try to approximate the actual fitness globally, to avoid errors when using the surrogate model for further steps of the optimization process. Nevertheless, as search proceeds, the only interesting part of the fitness landscape is the one surrounding the optimum, and trying to get a good model over the whole search space can be considered a waste.

In an eort to hybridize powerful deterministic optimization algorithms within parametric evolutionary optimization, this paper proposes a new mutation operator that locally builds a surrogate model for the objective function around the selected parent, and deterministically optimizes that surrogate model

P. Collet et al. (Eds.): EA 2001, LNCS 2310, pp. 104–116, 2002.

to generate the ospring. Moreover, by using for the surrogate learning a exible model that allows one to balance between accuracy of the resulting approximation and its global smoothness, it will be possible to control the degree of locality of the resulting operator, going from wide exploration to accurate exploitation.

The paper is organized as follows. Section 2 surveys previous work using surrogate function in the framework of Evolutionary Computation. The Surrogate Deterministic Mutation (SDM) is presented in detail in section 3, together with the approximation model and the deterministic optimization procedure it will based upon. The handles that allow the tuning of the SDM are described in section 4, and some experimental results are given in section 5, obtained on standard benchmark functions of the literature on high dimension spaces. Robustness with respect to some parameters is also discussed, and some to-do list of further work is also given.

2   Surrogate Evolution

2.1   Surrogate Optimization

When tackling a real-world optimization problems, whatever the optimization method, the main source of computational cost generally is the computation of the objective function. Hence many researchers have suggested to replace the objective function(s) with some easy to compute surrogate model [1,3]: Some examples of the objectives are gathered in an initial phase (either purposely randomly generated, or from past experiments). These examples are used to build an approximate function for the actual objective. This approximation are then used in place of the actual objective during the optimization process.

It is well known that the price to pay for the robustness of Evolutionary Algorithms is that a very large number of fitness computations are generally necessary to find a good solution (of the order of several thousands). This makes the use of surrogate models even more critical than for deterministic methods whenever the cost of a single evaluation is more than a few minutes. However, Evolutionary Algorithms also are more exible than standard deterministic optimization algorithms: for instance, it is possible to use both the original objective and the surrogate model in the same run of the algorithm.

Two important issues have to be considered: what mathematical model to choose for the approximation; and what strategy to use during the evolutionary optimization.

With respect to the model for approximation, early work embedding approximation into evolutionary optimization had chosen some simple linear approximation [18,10]. But of course linear models are inappropriate for most real-world (non-linear) problems, and the two most popular models that have emerged recently are Neural Networks [19,13] and krieging [21,8].

The problem of when and how to update the surrogate model is far more dicult. Grefensteteš pioneering work [11,10] used (linear) objective function approximation to tune the parameters of the genetic algorithm, and use the resulting optimal parameter set to run a GA with the actual objective. However, this approach suers from the limitations of random sampling: if the fness is complex, it is very unlikely that the approximate function has the same optima as the original fitness. However, as already mentioned, the most popular approach used in evolutionary surrogate optimization is to mix both the approximation and the actual computation: While the evolution proceeds using the surrogate model of the fitness function, some individuals, at given generations, are evaluated using the actual objective function, and unfortunately, it seems that the best strategy about the choice of when and who to accurately evaluate is problem-dependent, though recent work has starting to investigate some criteria that could be used to actually ensure that the surrogate model does not have false minima (i.e. minima that do not correspond to minima of the original fitness) [8,13].

Nevertheless, while avoiding a too large bias due to the error coming from the approximation, such mixed strategies allow one to gradually refine the approximation as the population is moving toward areas of the search space that have probably not been sampled by the initial training examples, and results can be obtained for problems that were far out of reach for Evolutionary Algorithms otherwise [19].

2.2   Discussion

In all above-mentioned work, the whole optimization process entirely relies on the surrogate model. Hence both   global and local   approximation accuracy are required from the approximation: the well-known exploration vs exploitation dilemma is hence the limiting factor of those approaches.

This has two main consequences. First, globally accurate approximations require many examples, and generally have large computational complexity: the complexity of the computation of the approximation is generally assumed negligible when a small number of examples are considered, though it is obvious that this will not hold for instance for Neural Networks and krieging models if a large number of examples have to be considered [8]. Furthermore, as the approximate model tries to fithe exact objective function, it is as dicult to optimize, trying to fit all local optima for instance.

This paper tries to overcome the above diculties by 1- only partially fitting the exact objective function, starting with a global raw approximation and refining it locally around points that have been identified as possible optima; 2- using Support Vector Machines as approximation tool, as SVMs do scale up nicely with the number of examples; 3- deriving simple enough surrogate models so they can be optimized rather accurately by some fast deterministic method.

3   Surrogate Deterministic Mutation

This section introduces a new mutation operator, termed Surrogate Deterministic Mutation (SDM), to be used within an Evolutionary Algorithm, and based on the a surrogate model of the objective function at hand and a classical deterministic optimization procedure. The main steps of SDM operator are

1. select promising parents   from the population. Promising parents are individuals that are likely to be local optima; in the work presented here, only the best individual of the current population is considered.
2. gather examples of actual objective values   around that parent; depending on how close to the parent those examples are chosen, one can tune the approximation from global to local. Examples are chosen first in the individuals that have been encountered in the past evolution and have already been evaluated, or are drawn anew from the given neighborhood if necessary – and need to be evaluated.
3. build the approximation of the objective function   using those examples; here, Support Vector Machines with Gaussian kernels is used [6].
4. minimize this approximate function   using a local method; in the work presented here, a (fast) standard deterministic method, L-BFGS-B [4] is used.
5. keep the resulting optimum of the surrogate model   in the ospring if it is dierent from the parent (no cloning).

The remaining of this section will now first discuss the choices made for points 3 and 4 above.

3.1 Support Vector Machines

Support Vector Machines (SVMs) have received a lot of attention recently as powerful learning techniques for large-scale classification and regression problems. From the seminal work of Vapnik [22], theoretical foundations make SVMs robust well-understood tools. A detailed introduction can be found in [7].

In this work, SVMs with Gaussian kernels have been chosen, as they lead to one of the most numerically ecient implementation SVM learning [6]. In that case, the approximate function has the form

$$f(x) = {}_0 + \sum_{i=1}^{n} {}_i K(x_i, x)$$
$$\text{with } K(y, z) = \exp(\frac{y\ z^{\ 2}}{2})$$

(1)

The points x_i are some samples from the example set called the   support vectors. Their identifiation, together with the corresponding coecients $_i$ leads to a quadratic constrained optimization problem, whose matrix is very sparse in the case of the Gaussian kernels, making the derivation of a given approximation very fast. Parameter   is user-defined, and its important impact on the resulting model will be detailed in section 4.3.

3.2 Optimization of Surrogate Model: L-BFGS-B

All local optimizations on the surrogates models will be performed using   L-BFGS-B , a fast deterministic optimization procedure [4] for large-scale bound-constrained or unconstrained problems. BFGS is a well-known quasi-Newton method [2] that computes an approximate of the Hessian matrix using only gradient information. The   limited memory version   optimizes the memory usage [5] while bound constrains are handled through a projected gradient method. The implementation used here is based on [24].

4   SDM Parameters

There remain many crucial open issucs in the above procedure: the most important one is probably the choice of the neighborhood, that determines the compromise between exploration and exploitation that SDM will reach; but one also need to pay attention to the number of example points that will be used to generate the surrogate mode, and to the granularity of the SVM (parameter   of section 3.1). Of course, as the long term goal is to provide an operator that works at dierent scales (i.e. starts by global exploration before turning to local fine-tuning), it is unrealistic to believe that the above tunings can be set once at the beginning of the evolutionary process. Moreover, the royal road to reduce parameter dependency is adaptation: next sections will be devoted to introducing adaptive schemes for these crucial parameters of the SDM operator.

4.1   Exploration vs. Exploitation

The degree of exploration of the SDM operator is determined by a real number called   : all examples will be gathered in a ball of radius   centered on the parent. An initial large value will ensure a good exploration, while a decreasing scheme should gradually focus the search close to the parent. In the present work, after a user-defined initialization,   is updated every generation using

$$\sigma_n = \max(2\ d_{n-1},\ \alpha\sigma_{n-1}) \tag{2}$$

where d_{n-1} is the maximum distance between a parent and its ospring obtained using SDM at previous generation, and 0 1 is a user defined parameter to bias the variations of   toward decrease. Note that values of   will increase as soon as previous mutation did lead to ospring lying far from its parent, but that it will never decrease faster than a factor   even in case of unlucky mutations. Tuning the value of   is discussed in section 5.3.

4.2   Examples Gathering

The number of examples N_{ex} comes next as an important parameter: the more the better, as far as approximation is concerned (while respecting the locality defined by   above), except that examples need to be evaluated.

The first idea to avoid the evaluation of many examples for a single application of SDM is to use as many already evaluated points as possible. Hence a memory of past evaluations has to be stored. When some examples are needed, the points in memory are first considered (from most recent to oldest to avoid

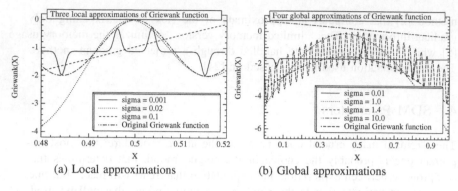

(a) Local approximations   (b) Global approximations

Fig. 1. One-dimensional Griewank function (see equation 4) and its approximation by Gaussian SVMs for dierent values of

using again and again the same points), but only of course if they are close enough from the parent according to the current value of   .

Note that the number of examples N_{ex} is increased by a user-defined factor   whenever SDM fails (see below) to fid an ospring that is better than the parent – or until a (user-defined) maximum number N_{max} is reached.

4.3   Tuning the SVM Learning

The important parameter in Gaussian SVM is the value of   . A too large   results in a at function while a too small   leads to function that is at with some very narrow peaks at each example toward its target value (see Figure 1). However, whereas it seems dicult to   a priori characterize what a good value for   can be, it seems in the global plot (figure 1-b) that a whole range of acceptable values (here, at least [1 .0, 1.4]) will actually lead to improvement.

Hence an a posteriori argument can be used: several values of   are tried using the same example set; each resulting model is then minimized using L-BFGS-B procedure (see section 3.2); the result with minimum fitness is finally retained as overall ospring. More precisely, starting with two user-defied parameters $_{min}$ and $_{max}$ (typically, 30   and .25   ), SDM proceeds as follows:

1. Draw k values of   uniformly in [$_{min}$, $_{max}$]; generate the corresponding SVM model and minimize it using L-BFGS-B;   evaluate   the result;
2. If it a better fitness than that of the parent is found, say for value $_0$, do step 1 again, but on the interval [0 .9 $_0$, 1.1 $_0$]; then go to step 4;
3. If not, add a another set of N_{ex} points to the example set and do step 1 again;
4. Update $_{min}$ and $_{max}$ for next generation

$$_{min} = ((1 \quad) \quad _{min} + \quad) \frac{n}{n \ 1}$$

$$_{max} = ((1 \quad) \quad _{max} + \quad) \frac{n}{n \ 1} \tag{3}$$

where n is the generation number, and $\quad_n$ and $\quad_{n-1}$ are the corresponding values of .

5. Return best individual encountered (either the original parent, or the best ospring found if it is better than the original parent).

This procedure requires, in the worst case, $2 \quad k + (1 + \quad) \quad N_{Ex}$ evaluations: if no known point lies within the ball of radius   , a first set of k examples has to be drawn (and evaluated) anew; step 1 above requires   k evaluations; if no better individual is found, N_{Ex} individuals are drawn (and evaluated), and step 1 is run again. However, in most cases, some points do exist in the ball around the parent, are save some initial evaluations; moreover, the first execution of step 1 above very often leads to an improvement, that ends the procedure.

5   First Results

5.1   Experimental Settings

The SDM specific parameters defined in section 4 have been chosen as follows:

– The initial value for   , tuning the degree of locality of the approximation, seems to be problem dependent. A first choice is the range of the variables (same for all variables in all tests). However, it had to be adjusted for Baluja function (equation 5 below);
– The rate   of decrease of the value of   (equation 2) also seems critical (see section 5.3);
– The number of samples in the example set is set to the size of the population, its increase factor in case of failure of SDM to improve the fitness,   , is set to 0.1, and the maximum number of points in the example set is set to twice the population size (see section 4.3);
– the number k if trials to fid an ecient value of   (see section 4.3) is set to 10, while the update parameter for   (in equation 3) is set to 0.1.

In order to assess the performance of the SDM within an Evolutionary Algorithm, three algorithms have been compared on three functions commonly used in evolutionary parametric optimization. All results are averaged over 10 runs.

– An Evolutionary Algorithm based on SDM, using the following parameters (unless otherwise stated in the text): population size 100, selection by tournament of size 10, generational replacement, arithmetical crossover applied with probability 0.5, uniform mutation applied with probability 0.5 (when SDM is not applied);
– A Hill-Climber that uses SDM as its "move" operator: note that this algorithm is equivalent to a (1,1)-ES that would use SDM as the only operator;
– A standard (30,210)-ES with global crossover and self-adaptive mutations (one standard deviation per object variable).

5.2   Test-Cases

Sphere function: The well-known sphere function (Sphere $(x) = \sum_{0}^{n} x_i^2$ is the simplest test-case for any optimization algorithm: being a quadratic function, it is straightforward to minimize for any gradient-based deterministic method (e.g. BFGS alone). Moreover, exact theoretical convergence results are available for Evolution Strategies on that function - and this makes it a good candidate for benchmarking.

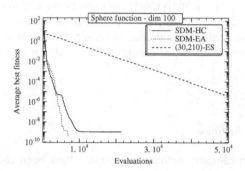

Fig. 2.   Comparative results (averages over 10 runs) for the Sphere function

Figure 2 displays the average results for a problem of dimension 100: it is clear that SDM alone is able to solve that easy problem, i.e. both the evolutionary algorithm and the hill-climber have the same performance, outperforming the standard ES.

It is worth mentioning here that the results of both SDM-based algorithms seem insensitive to the value of   (see section 5.3).

Griewank function: Griewank function is an example of a smooth (i.e. in-fiitely dierentiable) but highly a multi-modal function. However, it shows a global trend toward the global maximum (0) when high frequency oscillations are removed. It is defined on [100, 100]^n by

$$\text{Griewank } (X) = \sum_{i=1}^{n} x_i^2 \qquad \prod_{i=1}^{n} \cos(\frac{x_i}{i}) \tag{4}$$

Here again the two SDM-based algorithms outperform the standard evolution strategy (see figure 3-a): this is due to the large-scale smooth shape of the fitness landscape, the high modality only resulting from some high-frequency perturbations (see figure 1-b for the plot in the 1-dimensional case). Here, however, the generational EA reaches regions of high fitness more quickly than the hill-climber, benefiting from shortcuts in the fitness landscape due to crossover. Also note that the best value of   are dierent (systematic experiments were performed before reaching those values, that are the one used to obtain the pots of figure 3-a): 0.85 for the hill-climber and 0.5 for the generational EA.

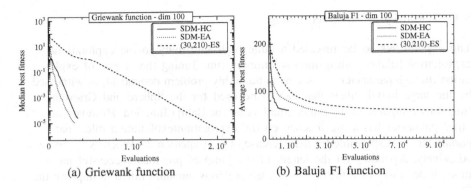

(a) Griewank function   (b) Baluja F1 function

Fig. 3. Comparative results (averages over 10 runs)

Baluja F1 function: Baluja F1 function is a non dierentiable highly multimodal function. The global minimum (0) is very dicult to reach for any optimization algorithm. It is defined on [2.56, 2.56]ⁿ by

$$F_1 = \sum_{i=1}^{n} \left| \sum_{j=1}^{i} y_j \right|$$ (5)

The results (see figure 3-b) are rather dierent from those on Griewank function: here both SDM-based algorithm perform a little better than the standard
ES, but all three algorithms start hovering at a rather high level of fitness –
and none of the final solutions is satisfactory. Moreover, the hill-climber is faster
than the generational EA, but stops abruptly and never reaches so good final
values, even though the   parameter, as for Griewank function, had to be taken
larger (0.85 vs 0.5), meaning that the scope of the approximation decreases more
slowly.

But an important issue was raised by this test case: the initial setting of
had to be manually tuned (and finally set to 0.1), while for both previous cases,
a large initial value was the right choice (anything of the order of magnitude of
the range of variables). This is probably because the fitness landscape does not
seem to have such nice global trend toward the minimum that Griewank (and of
course the Sphere) so nicely exhibits. Note that the behavior of   in all tests was
to start by increasing (up to around 0.5) before decreasing for local refinement.
But an initial guess of 0.5 didn't produce any good result.

The above remark appealed for another baseline experiment: replace the minimization of the surrogate model (both within the generational EA and within
the hill-climber) by that of the original fitness itself (that is cheap to compute
here anyway). However, for both Griewank and Baluja functions, such algorithm
never gave any meaningful result and got stuck in the first local optimum. So it
seems that using a surrogate does indeed produce some sort of smoothing eect
that removes some local minima.

5.3   Sensitivity to Parameter Settings

The experiments so far revealed a quite high sensitivity to the exploration/-exploitation balance during the evolutionary run. Tuning the degree of exploration through parameter   seems to be highly problem-dependent, as witnessed by the large initial values that had to be used for the Sphere and Griewank functions compared to the rather small value for Baluja function. However, another parameter has a big inuence on the exploration/exploitation dilemma: parameter   , responsible for the decrease of   (equation 2). Whereas   increases adaptively, depending on the length of the jump of previous successful mutations, it decreases by a factor   in case previous mutation failed to improve the fitness.

Experiments on Griewank function, both for the SDM-based Evolutionary algorithm and the SDM-based Hill-Climber have been performed with dierent values of   . Results are displayed in Figure 4.

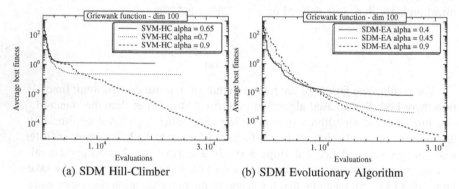

(a) SDM Hill-Climber   (b) SDM Evolutionary Algorithm

Fig. 4.   Sensitivity with respect to parameter   , responsible for the decrease of the amount of exploration

The general trends already mentioned are enforced. The range of useful values for the hill-climber is rater small, and involves values close to 1: SDM alone requires a rapid focus of the approximation. on the other hand, when an evolutionary paradigm is added, a slower decrease of the focus probably allows more exploration, resulting in o-line better results.

5.4   Further Work

The experiments presented in this paper are the very first ones obtained using the idea of minimizing a surrogate model inside an evolutionary algorithm, and hence many issues require further attention.

The gathering of examples should be made more similar to the way ES generate children: at the moment, the variation of   are totally independent of other parameters, while it could be somehow related to the way the standard deviations of some self-adaptive Gaussian mutation evolve (at teh moment, only raw uniform mutation was used together with SDM). A possibility is even to use only the children of the best   individuals (plus all known individuals inside that region?), thus removing totally the need for the exogenous   parameter.

In the present work, SDM is applied systematically to the best individual in the population. There should be some ways to apply it to dierent parents (e.g. after some possible multi-modality has been identified), or not to apply it at all (either deterministically, when detecting that is will most probably be useless, or stochastically).

More generally, SDM introduces many user-defined parameters: parameter tuning is already the responsible for the most time-consuming part of Evolutionary Optimization. Adaptability or self-adaptability can be a way to circumvent this diculty, but only if the sensitivity of the algorithm with respect to the additional parameters that adaptability requires is actually small compared to the sensitivity to the original parameter: this does not seem to be the case at the moment for the exploration/exploitation tuning (section 5.3).

6   Conclusion

Evolutionary Algorithms should not ignore the recent progresses made in learning algorithms, where SVMs now allow one to accurately and quickly solve large-scale approximation problems. Indeed, an Evolution Algorithm visits many points of the search space during its life-time, thus gathering many examples of the fitness function at hand. The Surrogate Deterministic Mutation proposes a way to use this knowledge inside an evolutionary algorithm, and preliminary results are encouraging. Of course, many issues need to be further addressed, and many questions remain to be answered: when and where to apply SDM (e.g. coupled with some niching mechanism); would other types of selection/replacement give better results (e.g. SSGA or ES or EP or   ...); how to decrease the number of user-defined additional parameters.

However, the most important issue, as far as the Evolutionary Computation community is concerned, is the relative contribution of Evolutionary ideas inside the resulting algorithm, as the simple hill-climber using SDM seems to perform almost as good as the generational EA using SDM once per generation, on its best individual. Though we are convinced that indeed evolution does add some global point of view to the resulting algorithm, only careful further studies can answer that question.

References

1. J-F.M. Barthelemy and R.T. Haftka. Approximation concepts for optimum structural desig n - a r eview. Structural Optimization   , 5:129–144, 1993.
2. F. Bonnans, J.C. Gilbert, C. Lemarechal, and C. Sagastiz´   abal. Optimisation Num´erique, aspects th´ eoriques et pratiques , volume 23 of Math´ematiques Applications . Springer Verlag, 1997.

3. A.J. Booker, J.E. Dennis Jr., P.D. Frank, D.B. Serafini, V. Torczon, and M.W. Trosset. A rigorous framework for optimization of expensive functions by surrogates. Structural Optimization , 17(1):1–13, 1999.

4. R. H. Byrd, P. Lu, and J. Noceda. A limited memory algorithm for bound constrained optimization. SIAM Journal on Scientific and Statistical Computing , 16(5):1190–1208, 1995.

5. R. H. Byrd, J. Nocedal, and R. B. Schnabel. Representation of quasi-newton matrices and their use in limited memory methods. Mathematical Programming , 63(4):129–156, 1994.

6. R. Collobert and S. Bengio. Support Vector Machines for Large-Scale Regression Problems . IDIAP-RR-00-17, 2000.

7. N. Cristianini and J. Shawe-Taylor. An introduction to Support Vector Machines . Cambridge University Press, 2000.

8. M.A. El-Beltagy, P.B. Nair, and A.J. Keane. Metamodeling techniques for evolutionary optimization of computationally expensive problems: Promises and limitations. In D.E. Goldberg al., editor, GECCO'99 , pages 196–203. Morgan Kaufmann, 1999.

9. P. Galinier and J. Hao. Hybrid evolutionary algorithms for graph coloring. Journal of Combinatorial Optimization , 3(4):379–397, 1999.

10. J. J. Grefenstette. Predictive models using fitness distributions of genetic operators. In L. D. Whitley and M. D. Vose, editors, FOGA 3 , pages 139–161. Morgan Kaufmann, 1995.

11. J. J. Grefenstette and J. M. Fitzpatrick. Genetic search and approximate function evaluation. In J. J. Grefenstette, editor, Procee dings of ICGA , pages 160–168. Laurence Erlbaum Associates, 1985.

12. P. Husbands, G. Jermy, M. McIlhagga, and R. Ives. Two applications of genetic algorithms to component design. In T. Fogarty, editor, AISB Workshop on Evolutionary Computing , pages 50–61. Springer Verlag, 1996.

13. Y. Jin, M. Olhofer, and B. Sendho. On evolutionary optimisation with approximate fitness functions. In D. Whitley al., editor, GECCO'2000 , pages 786–793, 2000.

14. P. Merz and B. Freisleben. A genetic local search approach for the AP. In Th. Back, editor, Procee dings of ICGA'97 , pages 465–470. Morgan Kaufmann, 1997.

15. P. Merz and B. Freisleben. Genetic local search for the TSP: New results. In Procee dings of ICEC'97 , pages 159–164. IEEE Press, 1997.

16. K.-H. Liang, X. Yao, and C. Newton. Combining landscape approximation and local search in global optimization. In Proceedings of the CEC'99, pages 1514–1520, Piscat away, NJ, 1999. IEEE Press.

17. P. Merz and B. Freisleben. Fitness landscapes and memetic algorithm design. In D. Corne, M. Dorigo, and F. Glover, editors, New Ideas in Optimization , pages 245–260. McGraw-Hill, London, 1999.

18. G. Mosetti and C. Poloni. Aerodynamic shape optimization by means of hybrid genetic algorithms. In 3rd International Congress on Industrial and APplied Mathematics , 1995.

19. C. Poloni and V. Pediroda. GA coupled with computationaly expensive simulations: tools to improve eciency. In Genetic Algorithms and Evolution Strategies in Engineering and Computer Sciences , pages 267–288. John Wiley, 1997.

20. N. J. Radclie and P. D. Surry. Formal memetic algorithms. In T.C. Fogarthy, editor, Evolutionary Computing , pages 1–16. Springer Verlag LNCS 865, 1994.

21. A. Ratle. Accelerating the convergence of evolutionary algorithms by fitness landscape approximation. In Th. B ack et al. editors, Procee dings of PPSN V , pages 87–96. Springer-Verlag, LNCS 1498, 1998.
22. V. N. Vapnik. The Nature of Statistical Learning . Springer Verlag, 1995.
23. D. Waagen, P. Diercks, and J. McDonnell. The stochastic direction set algorithm: A hybrid technique for finding function extrema. In D. B. Fogel and W. Atmar, editors, Procee dings of EP'92 , pages 35–42, 1992.
24. C. Zhu, R. H. Byrd, and J. Nocedal. L-BFGS-B, Fortran routines for large scale bound constrained optimization. ACM Transactions on Mathematical Software , 23(4):550–560, 1997.

The Eects of Partial Restarts in Evolutionary Search

Ingo la Tendresse [1], Jens Gottlieb [2], and Odej Kao [1]

[1] Department of Computer Science, Technical University of Clausthal
Julius-Albert-Str. 4, 38678 Clausthal-Zellerfeld, Germany
tendresse,okao   @informatik.tu-clausthal.de
[2] SAP AG
Neurottstr. 16, 69190 Walldorf, Germany
jens.gottlieb@sap.com

Abstract.   A stagnation of evolutionary search is frequently associated
with missing population diversity. The resulting degradation of the overall performance can be avoided by applying methods for diversity management. This paper introduces a conceptually simple approach to maintain diversity called   partial restart   . The basic idea is to re-initialize parts
of the population after certain time intervals, thereby raising the probability of escaping from local optima that have dominated the recent
search progress. The usefulness of the proposed technique is evaluated
empirically in two characteristic problem domains, represented by the
satisfiability problem and the onemax problem. The main goal is to identify problem structures where partial restarts are promising, and to gain
a better understanding of the relations between dierent variants of partial restarts.

1   Introduction

Premature convergence is a serious issue in evolutionary algorithms since it might
significantly degrade the overall performance. It is usually perceived as a stagnation of the search process and is often associated with low population diversity.
For that reason several diversity management techniques like crowding [Mah92],
sharing [Gol89], incest prevention [Esh91,ES91] or duplicate elimination [RG99,
Esh91] have been proposed. This paper is concerned about a conceptually very
simple approach to maintain diversity, which we call   partial restart   . Its basic idea
is to re-initialize parts of the population after certain time intervals. This idea
was already investigated by Maresky et. al., who reported good results for selectively destructive restarts on numerical problems [MDG+95]. Their approach is
based on re-initialization at gene level, which is performed as soon as a certain
convergence level is reached. Eshelman originally proposed this idea [Esh91], using restarts based on applying highly disruptive gene-level mutation operators
to the best individual of the current population. In contrast to that, we propose
to re-initialize the population at individual level rather than at gene level. Our
approach maintains the best solutions, whereas Maresky et. al.'s approach tends

P. Collet et al. (Eds.): EA 2001, LNCS 2310, pp. 117–127, 2002.

to change all individuals and hence destroys the best solution with high probability, due to the strong interaction of genes. In order to focus on partial restarts' basic eects and to prevent positive or negative side eects of adaptive schemes to determine the schedule for restarting, we rely on a fixed restart schedule at predetermined generation numbers.

Evolutionary search is strongly aected by local optima of the search space. Local optima with large basins of attraction frequently cause premature convergence, in particular if the population diversity is low. Partial restarts oer a simple way to increase population diversity, which raises the probability of escaping from local optima that have dominated the recent search progress.

This paper evaluates partial restarts concerning their usefulness in dierent problem domains. Our goals are to identify problem structures where partial restarts are promising, and to gain a better understanding of the relations between dierent variants of partial restarts covered by our general framework. The most prominent variants are classical evolutionary search and complete restarts, which correspond to a single run and several independent single runs, respectively. Therefore, we selected two bit string based problems with dierent characteristic properties: The satisfiability problem and the onemax problem. While the former is dicult to solve –it is NP-complete, its fitness landscape contains many local optima, and classical genetic algorithms fail – the latter is trivially solvable for the Hamming neighborhood.

We proceed with an introduction of partial restarts in Section 2. The satisfiability problem and the onemax problem are discussed in Sections 3 and 4, respectively. Empirical results for partial restarts and both problems are presented in Section 5. Finally, conclusions are given in Section 6.

2   Partial Restarts

The partial restart mechanism is intended to increase population diversity after certain stages of the search. Its behavior is parameterized by the   partial restart cycle prcycle   IN and the partial restart rate   prrate   [0, 1] as follows:

- a partial restart is invoked after   prcycle   subsequent generations without restarting, and
- during each restart, the worst   prrate   popSize   individuals are replaced by newly initialized individuals.

The partial restart mechanism can easily be integrated into existing evolutionary algorithms, as shown in Figure 1 for a steady-state replacement scheme, in which one ospring is produced in each generation. The computational costs of a partial restart depend on sorting the population according to the fitness and initializing the desired fraction of the population. These costs are negligible except for problems with complex initialization routines.

This definition of partial restarts covers several important cases that are commonly investigated:

```
procedure     evolution
begin
   initialize(population[1   ... popSize ])
   set t = 1
   while   not termination condition     do
      begin
         sortSmallestFitnessFirst(population)
         if t    0 (mod  prcycle ) then
            begin
               initialize(population[1   ... popSize    prrate   ])
            end
         else
            begin
               produce an ospring by mutating a selected individual
               replace population[1] by the ospring
            end
         set t = t + 1
      end
end
```

Fig. 1.   Pseudo code of steady-state evolutionary algorithm with partial restarts

- prrate   = 0 (or prcycle =   ) represent a classical evolutionary algorithm
  without restart, which resembles a single run,
- prrate   = 1 yields complete restarts, i.e. the search consists of several independent runs, and
- prrate   = 1 and prcycle   = 1 represent random search.

Partial restarts can be perceived as loosely coupled single runs, where above
parameters specify the lengths of the single runs and the degree of dependence
between the single runs.

In general, the following eects of the parameters   prcycle   and prrate   can
be expected:

- Small prcycle   and large prrate   values cause a frequent re-initialization of
  large parts of the population. This kind of search behavior resembles random
  search and hence poor results are expected.
- The case prrate   = 1 corresponds to complete restarts [BFM97]. Using too
  small values for prcycle   make the evolutionary algorithm act like random
  search. Extremely high values for   prcycle   represent classical evolutionary
  search, i.e. a single long run. It is interesting to check whether some intermediate value for prcycle   is beneficial here, since it could lead to a better
  exploitation of the overall time available.
- Small values for prcycle   limit the progress possibilities of the solution candidates in the current generation, and thus these could be replaced before
  having positive eects on the search. The short time interval given for the

progress might not be sucient for a clear separation between promising and
poor individuals.

- The higher the value for prcycle , the more rarely a re-initialization is performed and thus a significant change in performance may not be expected
  compared to classical evolutionary search.
- Small values for prrate   restrict the eects of re-initialization and hence its
  impact on the search process.

Our study analyzes partial restarts as well as related issues, like the relation
between a single long run and several independent short runs, and considers
two structurally completely dierent problems, which are presented in the next
sections.

3   The Satisfiability Problem

The satisfiability problem (SAT) is based on a set of Boolean variables $x_1, \ldots, x_n$
and a Boolean formula $f : \mathbb{B}^n \quad \mathbb{B}, \mathbb{B} = \quad 0, 1$. The question is whether
a variable assignment $x = (x_1, \ldots, x_n) \quad \mathbb{B}^n$ exists such that $f(x) = 1$.
A SAT instance is called satisfiable if such x exists, and unsatisfiable otherwise. In general f is assumed to be in conjunctive normal form (CNF), that is
$f(x) = c_1(x) \quad c_m(x)$ where each clause c_i is a disjunction of literals, and a
literal is a variable or its negation. The class k-SAT contains all SAT instances
in CNF such that each clause contains exactly k distinct literals. While 2-SAT
is solvable in polynomial time, k-SAT is NP-complete for $k \quad 3$ [GJ79]. Due to
its complexity and practical relevance, 3-SAT has been tackled by several evolutionary algorithms, but only those approaches incorporating problem-specific
knowledge in the variation operators or the fitness function were successful.
Adapting the fitness landscape during the evolution process is a promising option [BEV98,GV00], as well as using local optimization and the MAXSAT fitness
function that counts the number of satisfied clauses [MR99].

Our experiments are based on 12 satisfiable 3-SAT instances   [1] that were introduced in [BEV98], ranging from $n = 30$ to $n = 100$ variables. The instances
lie in the phase transition, that is $m = 4.3 \quad n$, which is known to contain dicult instances [MSL92,SGS00]. Although adaptive fitness functions are appealing, we focus on the MAXSAT fitness since this allows analyzing the partial
restart mechanism without noise of adaptation. Classical variation operators
are known to fail for this fitness function, due to misleading low order schema
information and similar schema fitness averages [RW98]. Therefore we use the
problem-specific mutation operator proposed in [GV00], which selects an unsatisfied clause and ips one variable contained in that clause.

Evolutionary algorithms could benefit from partial restarts here, since the fitness landscape of SAT is extremely dicult to solve. There are many solutions
that satisfy most clauses, and hence many local optima exist. Figure 2 exem-
plarily shows the distribution of the fitness values for the 9th instance, which

---

[1] available at   http://www.in.tu-clausthal.de/   gottlieb/benchmarks/3sat

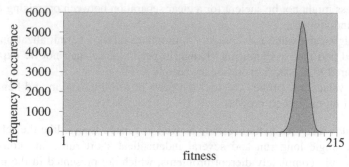

Fig. 2.   An example for the fitness distribution of the 3-SAT instance 9 (n = 50)

consists of n = 50 variables. The distribution has been obtained by randomly initializing 65 000 solution candidates, and it indicates a very high average fitness that is close to the maximum fitness 215. The large number of local optima causes fast convergence: Evolutionary search usually finds very good solutions quite quickly, which is shown in Figure 3 for a representative run on the 10th instance (n = 100). Once such fness level is reached, it is dicult for the evolutionary algorithm to obtain the global optimum since most solutions have a comparable fitness and consequently the fitness does not provide enough guidance.[2] There is the risk of premature convergence, which can make the search ineective. Here, a partial restart could help to re-focus the search in order to prevent premature convergence.

4   The Onemax Problem

The onemax problem (MAX1) is a toy problem commonly used to study the behavior of hill-climbing algorithms and evolutionary search [Ree00]. A solution candidate is represented by a bit string of length   n, and its fitness is the sum of all bits, that is its Hamming weight. Assuming the mutation operator ipping a randomly chosen bit, the fitness landscape is trivial for several reasons. There is one local optimum only – the global optimum – with the complete search space as its basin of attraction. Thus, there are no local optima with surrounding valleys that might prevent hill-climbers from reaching the global optimum. The fitness is perfectly correlated with the number of bit ips needed to reach the global optimum, and it therefore yields perfect guidance towards global optimality.

In contrast to SAT, the onemax problem is expected to be solved by evolutionary search in a straightforward manner. No local optima hinder the search

---

[2] Local search algorithms have also been reported to find solution candidates satisfying all but a few clauses quickly [Hoo98,SGS00]. The same situation occurs for evolutionary algorithms with standard bit mutation, although they need much more time to reach such fitness level.

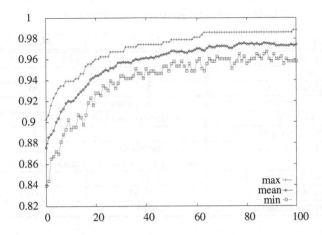

Fig. 3.   Dynamics of the maximal, mean, and minimal fitness for the 10th instance (n = 100) during the first 100 generations

and hence, premature convergence is not an issue. It is questionable whether partial restarts are useful in this domain. Suppose random initialization at Hamming weight level n/2, partial restarts and in particular complete restarts are expected to slow down the progress of evolutionary search: The closer the current population to the global optimum, the more progress already made will be destroyed by re-initializing parts of the population. Anyway, we have selected this problem in order to examine the eects of partial restarts, and to show that restarts may slow down search processes in specific problem domains.

5   Empirical Analysis

The eects of partial restarts within evolutionary search are examined empirically by varying the parameters   prrate   and prcycle   in a systematical way. For each parameter configuration we report results for two commonly used performance measures:

- Success rate (SR) is the rate of successful runs – i.e. in which a global optimum was found – averaged over several single runs for an instance, and
- Average number of evaluations to solution   (AES) measures how many evaluations are necessary to find the global optimum in successful runs for an instance. If no runs were successful, we define AES as 0.

Throughout our experiments we use the general setup shown in Table 1.

5.1   Results for 3-SAT

Intuitively, partial restarts yield the largest gain in cases with lots of sub optimal local optima and just a few global optima. As the problem structure of

Table 1.   General setup for the experiments

setup	3-SAT	MAX1
prcycle	2, . . . , 120	
prrate	[0, 1]	
population size	50	50
parent selection	tournaments of size 2	
replacement	steady-state (kill the worst)	
ospring per generation	1	1
mutation operator	ips one bit in one unsatisfied clause	ips one bit (randomly chosen)
crossover probability	0	0
evaluation limit	30 000	30 000
runs per instance	200	200

3-SAT matches these conditions, we may expect partial restarts to be useful in this problem domain. In order to check the hypotheses stated in Section 2, we performed runs on each 3-SAT benchmark instance and dierent parameter configurations for prcycle and prrate .

Figure 4(a) and Figure 4(b) visualize the obtained SR and the AES for instance 11 (n = 100). In comparison to the standard evolution represented by prrate = 0, partial restarts exhibit a significant increase of the success rate for high values of prrate , like e.g. prrate   0.7. Interestingly, small but positive values of prrate cause a slight SR decrease compared to classical evolutionary search. The best performance, however, is achieved by   prrate = 1 and sucient high values of prcycle . Therefore, complete restarts seem to be a highly eective way to cope with premature convergence. But care must be taken when setting the parameters appropriately. Complete restarts that are performed too often, i.e. for small values of prcycle , represent some kind of random search which deteriorates the performance. In this case, the evolutionary search between two restarts is interrupted too early. In the other case of extremely high   prcycle values the search process resembles the classical evolutionary search, which suffers from premature convergence, resulting in a waste of time that could better be spent into new complete restarts. Thus, there is a trade-o between values of prcycle   that are too low and too high, respectively.

The AES clearly grows for higher values of prrate , and a very low AES is exhibited by those parameter combination with a low success rate. This indicates premature convergence for two reasons. On the one hand, if a solution is actually found, it is found in early stages of the search process. This means that only very few solutions occur in the remaining part of the search, which represents a waste of time. On the other hand, the partial restarts and in particular complete restarts allow to determine solutions in later stages of the search, too. The AES values are higher, but they come together with a significant higher success rate, given the same time limit. Thus, the restarting mechanism prevents a waste

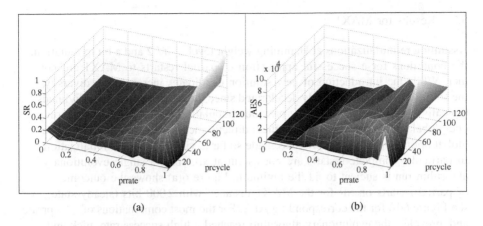

(a)   (b)

Fig. 4. Empirical results for the 3-SAT instance 11: (a) SR; (b) AES

of time due to premature convergence, and it consequently results in a better overall performance.

The effects of partial restarts on instances that are efficiently solvable by classical evolutionary search are rather small. However, we observe slight improvements even for those instances, as shown in Figure 5 for the 3-SAT instance 7 (n = 50). Here the classical evolutionary algorithm performs quite well, reaching a success rate of 0.9. Again, two observations confirm what we also reported for instance 11 before: (i) Complete restarts seem to be the best choice since the highest success rate is achieved for prrate = 1, and (ii) bad results in terms of both, SR and AES, are obtained for low values of prcycle .

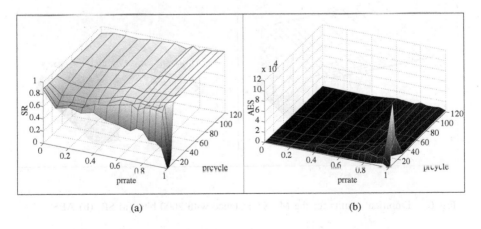

(a)   (b)

Fig. 5. Empirical results for the 3-SAT instance 7: (a) SR; (b) AES

5.2   Results for MAX1

Assuming re-initializations at Hamming weight level   n/ 2 and a bit ip mutation, the conditions for a successful application of partial restarts to MAX1 are not met. While the introduction of new solution candidates increases the diversity, the solution progress is severely decelerated since the newly initialized individuals typically have a fitness below the previous population's average fitness. As only one global optimum exists and the fitness directly measures the distance from the global optimum, the search progress made so far is violated by a re-initialization. In particular complete restarts are not useful at all, since a single evolutionary algorithm run is sucient to fid the optimum. Figure 6(a) shows the outcome of partial restarts on SR for the MAX1 instance with 2 000 bits (steady-state); see Figure 6(b) for the corresponding AES. For the most combinations of   prrate and prcycle   the evolutionary algorithm reached a high success rate. uick and strong partial restarts yield a significantly worse SR, which coincides with a high AES that nearly reaches the limit of 30 000 evaluations. Note that a smaller evaluation limit would have caused stronger eects on the success rate.

The behavior is exactly what could have been expected. The uniform initialization of new solutions candidates yields an average fitness of   n/ 2, which is significantly smaller than the average fitness of an active population, supposed a sucient long time since the last re-initialization. The replacement by newly initialized candidates degrades the overall fitness, and even worse results are obtained for frequent applications of re-initializations. Varying the time intervals between restarts does not have a decisive impact on the success rate, except for extremely small values of   prcycle   which cause the evolutionary algorithm to behave like random search. Obviously, the search could only benefit from partial restarts if the newly initialized individuals have an acceptable fitness.

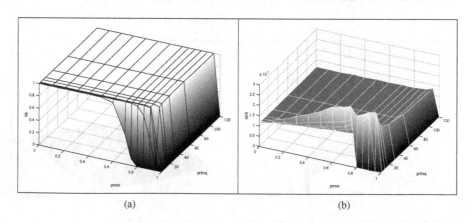

(a)   (b)

Fig. 6.   Empirical results for the MAX1 instance with 2000 bits: (a) SR; (b) AES

6   Conclusion

Partial restarts have been proposed to prevent premature convergence by significantly increasing the population diversity. Our partial restart framework also covers special cases like classical evolutionary search (no restarts at all) and complete restarts. We evaluated several partial restart variants in two dierent problem domains, namely the satisfiability problem and the onemax problem. Our results show that a better performance can be reached by partial restarts for the satisfiability problem, and that particularly complete restarts are promising.

The overall eectivity of evolutionary search for a given time or evaluation limit strongly depends on premature convergence. As soon as evolutionary search gets stuck in some dominating local optima, the remaining search eorts are wasted since usually no further improvements are found. Here, a complete restart oers a simple but eective method of making better use of the time available for the search. For satisfiability problems which can be solved robustly by classical evolutionary algorithms, partial restarts cannot improve the overall outcome. However, partial and in particular complete restarts are promising if classical evolutionary search exhibits a poor performance, like for some of the satisfiability instances.

The onemax problem is easily solved by classical evolutionary search and therefore partial restarts hinder the search progress. This example shows that partial and complete restarts can be beneficial only in problem domains where evolutionary search severely suers from premature convergence at local optima.

The partial restart mechanism is conceptually very simple and can thus be easily incorporated into existing evolutionary algorithms. However, the choice of adequate parameters is a dicult task. In our experiments, we used constant values for both parameters through the complete run. Although even this parameter setup significantly improved the performance for some instances, we expect even better results for an adaptive control of these parameters. We identified premature convergence as important problem that partial restarts cope with, and hence we believe that partial restarts should be performed if premature convergence is recognized. One option would be to re-initialize parts of the population if no fitness improvement was found for several generations, or if all individuals have a high similarity, which both indicate premature convergence.

References

[BEV98]   T. B ack, A. E. Eiben, and M. E. Vink. A Superior Evolutionary Algorithm for 3-SAT. In   Procee dings of the 7th Annual Conference on Evolutionary Programming   , Lecture Notes in Computer Science, Volume 1477, 125 – 136, Springer, 1998

[BFM97]   T. B ack, D.B. Fogel, and Z. Michalewicz (eds.):   Handbook of Evolutionary Computation , Oxford University Press, 1997

[ES91]   L.J. Eshelman, J.D. Schaer: Preventing Premature Convergence in Genetic Algorithms by Preventing Incest   , Proceedings of the Fourth International Conference on Genetic Algorithms, 115 – 122, 1991

[Esh91]   L. Eshelman,   The CHC Adaptive Search Algorithm: How to Have Safe
             Search When Engaging in Nontraditional Genetic Recombination   , In G.
             Rawlins, editor, FOGA -1, pages 265–283. Morgan Kaufmann, 1991
[GJ79]   M. R. Garey and D. S. Johnson.   Computers and Intractability: A Guide
             to the Theory of NP-Completeness   . W. H. Freeman, San Francisco, CA,
             1979
[Gol89]   D. E. Goldberg.   Genetic Algorithms in Search, Optimization, and Ma-
             chine Learning . Addison-Wesley, Reading, MA, 1989
[GV00]   J. Gottlieb and N. Voss. Adaptive Fitness Functions for the Satisfiability
             Problem. In   Procee dings of the 6th International Conference on Paral-
             lel Problem Solving from Nature , Lecture Notes in Computer Science,
             Volume 1917, 621 – 630, Springer, 2000
[Hoo98]   H. H. Hoos. Stochastic Local Search – Methods, Models, Applications .
             Dissertation, Darmstadt University of Technology, 1998
[Mah92]   S. W. Mahfoud. Crowding and Preselection Revisited. In   Procee dings of
             the 2nd Conference on Parallel Problem Solving From Nature   , North-
             Holland, 27 – 36, 1992
[MDG+95]   J. Maresky, Y. Davidor, D. Gitler, G. Aharoni, A. Barak. Selectively
             Destructive Re-start. In   Procee dings of the Sixth International Conference
             on Genetic Algorithms , Morgan Kaufmann, San Francisco, CA, 144–150,
             1995
[MR99]   E. Marchiori and C. Rossi. A Flipping Genetic Algorithm for Hard 3-SAT
             Problems. In   Procee dings of the Genetic and Evolutionary Computation
             Conference , 393 – 400, Morgan Kaufmann, 1999
[MSL92]   D. Mitchell, B. Selman, and H. Levesque. Hard and Easy Distributions
             of SAT Problems. In   Procee dings of the 10th National Conference on
             Artificial Intelligence , 459 – 465, 1992
[Ree00]   C. R. Reeves. Fitness Landscapes and Evolutionary Algorithms. In   Pro-
             ceedings of Artificial Evolution   , Lecture Notes in Computer Science, Vol-
             ume 1829, 3 – 20, Springer, 2000
[RG99]   G. R. Raidl and J. Gottlieb. On the Importance of Phenotypic Dupli-
             cate Elimination in Decoder-Based Evolutionary Algorithms. In   Late-
             Breaking Papers at the Genetic and Evolutionary Computation Confer-
             ence , 204 - 211, 1999
[RW98]   S. Rana and D. Whitley. Genetic Algorithm Behavior in the MAXSAT
             Domain. In   Procee dings of the 5th International Conference on Paral-
             lel Problem Solving from Nature , Lecture Notes in Computer Science,
             Volume 1498, 785 – 794, Springer, 1998
[SGS00]   J. Singer, I. P. Gent, and A. Smaill. Backbone Fragility and the Local
             Search Cost Peak.   Journal of Artificial Intelligence Research   . Volume 12,
             235 – 270, 2000

History and Immortality in Evolutionary Computation

Benoit Leblanc [1], Evelyne Lutton [1], Bertrand Brauns chweig[2], and
Hervé Toulhoat [2]

[1] INRIA, projet FRACTALES 78150 Le Chesnay - France
Benoit.Leblanc, Evelyne.Lutton   @inria.fr
[2] Institut Fran cais du P´etrole   1 et 4, avenue de Bois-Pr´ eau
BP 311 - 92852 Rueil-Malmaison Cedex - France
Bertrand.Braunschweig, Herve.Toulhoat   @ifp.fr

Abstract.   When considering noisy fitness functions for some CPU-time
consuming applications, a trade-o problem arise: how to reduce the
inuence of the noise while not increasing too much computation time.
In this paper, we propose and experiment some new strategies based on
an exploitation of historical information on the algorithm evolution, and
a non-generational evolutionary algorithm.

1 Introduction

Handling noise in Evolution Algorithms has already been studied for the reason
that most real-world problems present some noisy behavior, with many possible
origins. This diculty has often been successfully overcome by raising the population size [3] or by making multiple evaluations of the same individual ([5], [9]),
using an average as fitness score.

We address here problems where fitness is noisy and in the same time computationally expensive, reducing the applicability of the previous solutions. Considering that each fitness evaluation bears important information that we do not
want to lose, an exploitation of the   history of evaluations is a solution to reduce
the misleading noise.

Such a technique has already been experimented by Sano and Kita in [10]
for noisy functions, by Corn and al [2] and Zitzler and al [11] for multiobjective
optimisation. In Section 2, we propose a similar system of history-based fitness
scoring, relying on a   genetic database . Then in Section 3, it is shown that this
genetic database may also be used to produce ospring. A sharing technique
complements this scheme, it is described in Section 4. Finally experiments on
two multimodal test functions are presented.

2 Historical Information

2.1 Motivation

Inspired by the principles of Darwinian evolution, evolutionary algorithms (EA)
are based on the concept of evolving population. The important size of population guarantees the redundancy of information (genes and their expression) and

P. Collet et al. (Eds.): EA 2001, LNCS 2310, pp. 128–139, 2002.

its diversity, so the "death" of old individuals is not a problem, but is rather seen as an important evolution mechanism.

Here we deal with the class of problems where the total number of individuals created during the evolution is limited. This constraint arises for example when the fitness evaluation takes a long time. Moreover if the evaluation is subject to noise, the problem of accuracy of information becomes crucial. As stated before, we cannot aord raising too much the population size or the number of evaluations for the same individual.

To reduce the eect of noise, we therefore propose to use similarities between individuals (many instances of a single individual frequently coexists inside a population). Going further in this direction, we may also consider the whole information produced along the evolution: it often happens that an individual is a copy – or a slightly disturbed copy – of a "dead" ancestor. As we will see below, keeping track of all evaluations performed along the evolution provide another way to reduce the noise of the fitness function.

Moreover, if we can use a metric on the search space that makes sense (i.e. on which we can define a regularity property such as: two individuals that are similar with respect to this metric have similar fitness values), the previous idea may be extended. This implies that we assume some regularity properties of the underlying signal. This is a common hypothesis for many "denoising" techniques in signal analysis [6]. Fitness evaluations may be then averaged for individuals that lie in a given neighbourhood (with appropriate weights, related to the fitness regularity assumption). The resulting computation time overload for the EA remains negligible in the case of time consuming fitness.

2.2   An Implementation for Real-Coded Genomes

Sano and Kita [10] proposed to use the history of search to refine the estimated fitness values of an individual, using the fitness evaluations of individuals similar to it. Their approach is based on a stochastic model of the fitness function that allows to use a maximum likelihood technique for the estimation of the underlying fitness function.

Here we make the assumption that the underlying fitness function is regular with respect to the search space metric. Let us first define:

– The search domain:

$$S = \prod_{i=1}^{m} [a_i, b_i], \text{ with } i \quad 1, ..., m \quad (a_i, b_i) \quad \mathbb{R}^2 \text{ and } a_i \quad b_i \quad (1)$$

– A max distance on S:

$$x, y \quad S, \quad d \quad (x, y) = \max_{i \quad 1, ..., m} \frac{x_i \quad y_i}{b_i \quad a_i} \quad (2)$$

The divider ($b_i \quad a_i$) ensures that each component of a vector has the same weight in the distance regardless of the extent of its domain.

– An euclidian distance on S:

$$\forall x, y \in S, \quad d_2(x, y) = \sqrt{\sum_{i=1}^{m} \left(\frac{x_i}{b_i} - \frac{y_i}{a_i} \right)^2} \tag{3}$$

– The neighbourhood of a point is defined using the max distance:

$$\forall x \in S, \forall \beta \in \mathbb{R}_+, \quad B_\beta(x) = \{ y \in S_g ; d_\infty(x, y) \le \beta \} \tag{4}$$

We now define the regularity of a fitness as the fact that the fitness values of individuals belonging to the neighbourhood of an individual x (being in $B_\beta(x)$), are also close to the fitness value $f(x)$. Holder regularity is a well-fitted tool for this purpose:

Definition 1 (Hölder function of exponent h).
   Let (X, d_X) and (Y, d_Y) two metric spaces. A function $F : X \to Y$ is called a Holder function of exponent $h \ge 0$ and constant k, if for each $x, y \in X$ such that $d_X(x, y) \le 1$, we have:

$$d_Y(F(x), F(y)) \le k \cdot d_X(x, y)^h \tag{5}$$

Although a Hölder function is always continuous, it needs not be differentiable. Intuitively, a Hölder function with a low value of h looks much more irregular than a Hölder function with a high value of h (in fact this statement only make sense if we consider the highest value of h for which (5) holds). The majority of fitness function on real search space is Hölder.

We now want to keep track of the points of S (y_i) that have been evaluated at least one time. Of course the same point may have been evaluated more than one time, so we have to consider the number of evaluation ($\text{inst}(y_i)$) and the average of these evaluations ($\bar{f}(y_i)$). We can then define the following set:

$$\Phi_t = \{ y_i, \text{inst}(y_i), \bar{f}(y_i) \}, i \in \{ 1, ... n \}$$

The index t denotes the number of fitness evaluations that have been taken into account for the construction of Φ_t. It just emphasises that Φ_t can be considered as a genetic database , that is continuously updated along the evolution, i.e. when pairs (individual, fitness evaluation) are computed. However, for clarity we will later drop the t subscript.

Using Φ we can now define a weighted fitness function:

$$\forall x \in S, \ g(x) = \frac{\sum_{y \in B_\beta(x)} w(x, y) \cdot \text{inst}(y) \cdot \bar{f}(y)}{\sum_{y \in B_\beta(x)} w(x, y) \cdot \text{inst}(y)}$$

Where the weight $w(x, y)$ is defined according to the euclidian distance:

$$w(x, y) = 1 - \frac{d_2(x, y)}{\sqrt{m}}$$

We have $w(x, x) = 1$, and as :

$$\max (d_2(x, y), y \quad B \quad (x)) = \quad \overline{m} \qquad \text{and} \quad d \quad (., .) \quad d_2(., .)$$

$w(x, y)$ is always non negative: $\quad x \quad S, \quad y \quad B \quad (x), \quad w(x, y) \quad 0$

can now be used in the following way. Each time that an individual $\quad x$ has been evaluated, its "raw" (not yet averaged) fitness score is used to update the database. The weighted fitness score can be returned with the computation of $g \quad (x)$. The accuracy of the weighted fitness $\quad g \quad$ depends greatly on the regularity assumption on the fitness function. The parameter $\qquad$ is directly related to the regularity of the underlying fitness function (i.e. to $\quad k$ and h), and in the case of an extremely irregular function (i.e. having discontinuities or $\quad h$ near 0), we have to set $\quad = 0$.

3   Classical Fixed Size Population versus Growing Population?

The idea of using historical information has also been developed for multiobjective optimisation by Corn and al [2] and Zitzler and al [11]. Their approach consists in building an "archive" of non-dominated individuals to maintain diversity, that is updated at each generation.

We propose to build a genetic database, as a simple cumulation of all produced individuals. It can be used directly in a real-coded GA, for example, with the following procedure:

1 : Evaluate each individual of the current population.
2 : Add each individual with its raw fitness score to the database.
3 : Compute weighted fitness scores of all individuals with the help of the database.
4 : Apply your favorites selection schemes, genetic operators, replacement schemes, and loop on step 1 until termination.

Moreover this structure may be used to modify the classical birth and death cycle of an EA. More precisely the individuals to be reproduced can be directly selected in this genetic database. This can be seen as a   growing population of immortal individuals. To maintain diversity, a simple tournament selection seems then appropriate: choose randomly n_t (if n_t is the size of the tournament) individuals in   and keep the one having the best weighted fitness. Any individual of the genetic database may thus have ospring at any time. Thereby the information of the whole evolution is not only used to produce more accurate fitness evaluations but oers a simple way to maintain diversity. We should also emphasize the asynchronous aspect of this algorithm, i.e. we do not have to wait the whole current population to be evaluated in order to perform selection, but at any time we are able to choose from all already evaluated individuals. It is adapted to distributed implementation, for example with a client-server model: a genetic server deserves clients that perform the fitness evaluations. The server can manage the database with the following principles:

- A pool of random osprings is initially created.
- For any client request, the server supplies an ospring from its pool until this one is empty.
- As soon as a client has finished the evaluation of its current individual, it is returned to the server that adds the information to the database.
- When the ospring pool is empty the server creates new individuals to fl it again. This creation is made by selecting parents from the database (with a tournament for example) and applying genetic operators.
- In order to have a minimum initial diversity, we impose that when the server creates new individuals a minimum number of individuals (call it min_{par}) has to be present in the database before selection can be applied. If this condition is not fullfled, osprings are generated randomly until the database is suciently large.

4   Sharing

In order to maintain diversity it also seems convenient to use a sharing procedure [4]. We propose the following one, linked to the weighted averaging procedure in a simple way: each time we compute $g(x)$ the following quantity can be computed with few extra computation:

$$x \quad S, \quad W(x) = \sum_{y \quad B(x)} w(x, y) \quad inst(y) \tag{6}$$

This can be seen as a neighbour count which is used in the shared fitness function:

$$x \quad S, \text{ such as } W(x) = 0, \quad h(x) = g(x) \quad 1 + \frac{1}{W(x)} \tag{7}$$

As for each evaluated point, we have $W(x)$ 1, a tournament based on $h(x)$ can be used. The eect of this sharing will be that for an individual without neighbours and evaluated once, we will have $W(x) = 1$ and therefore $h(x) = g(x)$ 2. On the contrary for an individual having many neighbours we will have $h(x)$ $g(x)$. As a consequence, isolated individuals will be given a higher probability to be selected than surrounded ones.

5   Experimental Procedure

5.1   Algorithms and Genetic Operators

The following EA are compared:

- a GA without use of the fitness weighted averaging.
- a GA with fitness weighted averaging (further denoted GAW).
- our immortal evolutionary algorithm (IEA), with tournament.
- IEA + sharing   .

Individuals are encoded as real vectors, the search is represented with equation (1). The euclidian distance is used as a metric on this space.

The classical generational GA will use the stochastic universal selection (SUS, see [1]) with a full replacement of population.

The classical gaussian mutation   will be used for each component with a fixed variance $_i = 0.1$ $(b_i   a_i)$. We did not experiment here the adaptive method, which is commonly used in Evolutionary Strategies.

Regarding crossover , we will test 3 configurations:

1 : Classical arithmetic crossover. If we denote (x, y) the parents, (x , y) the ospring, and   a random uniform number from $[0 , 1]$:

$$x = x + (1)y$$
$$y = (1)x + y \tag{8}$$

2 : Arithmetic crossover with mating restriction (that will be further called mating crossover). Only parents that are close enough are allowed to mate. Practically, if m denotes the dimension of the search space, an euclidian distance of $(0 .1   \overline{m})$ will be the threshold.

3 : No crossover.

5.2   Common Characteristics of Experiments

- All measurements are averaged on   25 runs , for a given configuration.
- A limited number of 3200 fitness evaluations is fixed, i.e. for GA, runs are a 100 generations with population of size   32.
- The unperturbed fness scores are also kept o-line in order to evaluate the accuracy of the algorithms.
- IEA parameters:

  tournament size $n_t = 4$
  number of initial random individual $\min_{par} = 32$.

- Each time that measures are computed on the population of a GA (average fitness scores for example), similar measures are taken for the IEA by grouping new individuals in sets of the same size as the population size. Note that this is done only for measurement purpose.
- When crossover is used, the following probabilities are tested:

  crossover probability $p_c = 0.2, 0.5, 0.8$.
  mutation probability $p_m = 0, 0.01, 0.1$.
  In the case of mutation alone   we set $p_m = 0.02, 0.05, 0.1, 0.2$.

We must outline that all measures of all run are not reported in this paper, as it would require a large amount of figures. We therefore tried to choose the most significant ones to be discussed here, a complete report of these tests is available in [8].

6   f₁: A Multimodal Function

6.1   Definition

We consider the following function:

$$F_1 : [0,1]^3 : \qquad\qquad \mathbb{R}^+ \tag{9}$$
$$(x_0, x_1, x_3) \qquad\qquad \sum_{i=0}^{2} t(x_i)$$

A gaussian noise is added to obtain the noisy fitness function:

$$f_1 = F_1 \quad (1 + N_{(0,0.5)}) \tag{10}$$

With t (see figure 1, left):

$$[0,1] : \qquad [0,1]$$
$$x \qquad \begin{cases} (4 \quad x) & \text{if } x \quad [0, 0.25[\\ (2 \quad 4 \quad x) & \text{if } x \quad [0.25, 0.5[\\ (4 \quad x \quad 2) & \text{if } x \quad [0.5, 0.75[\\ (4 \quad 4 \quad x) & \text{if } x \quad [0.75, 1] \end{cases} \tag{11}$$

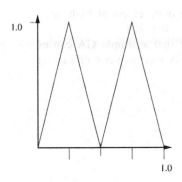

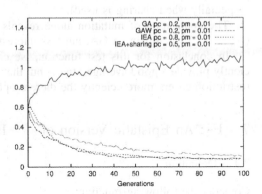

Fig. 1.   Left: Function   t. Right: average noise on   f₁ (on 25 runs).

As t has two optima (at $\frac{1}{4}$ and $\frac{3}{4}$) of the same height, F_1 will have 8 optima of the same height. In order to measure the ability of algorithms to locate many optima, we will count the number of individuals that fall in their vicinity. Precisely, for each optimum ($o_{i,i}$ $_{0,\dots,7}$), we count the number of individuals that verify   d (o_i, x) 0.1, obtaining the list of optimum neighbouring counts ($no_{i,i}$ $_{0,\dots,7}$). If we now sort them in decreasing order, it becomes possible to compute averages over dierent runs.

Finally, as its clear that F_1 is regular (h = 1) we set   = 0.05.

6.2   Results on F_1

Figure 1(right) shows the average noise of fitness evaluations. For the GA without fitness weighted averaging, the quantity plotted is simply $\sum_{x \in \text{population}} f_1(x)$ $F_1(x)$, and for the other algorithms $\sum_{x \in \text{population}} g(x) \cdot F_1(x)$. We remind that in the case of the IEA, the term population denotes only the grouping of the last new individuals for comparison purpose. We clearly see that, in absence of weighted averaging the noise increases as the average fitness increases because of the multiplicative nature of noise. When using the weighted averaging procedure noise decreases significantly.

Figures 2 to 4 show the performance in terms of Average Denoised Fitness (further called   ADF), corresponding to F_1 . The ability to locate optima is measured by the optimum neighbouring counts ($no_{i,i \in 0,...,7}$)[1] , for the three configurations (arithmetic crossover, mating crossover and mutation alone).

Note that in the case of arithmetic crossover, the classical GA with a low crossover probability ($p_c = 0.2$) leads to the highest ADF scores, but concentrate rather on a single optimum. The eect of weighted averaging does not change much the results. The IEA obtains lower performances in terms of ADF, but obtains a better diversity of optima when used with sharing.

In the case of mating crossover, the GA and the GAW performs better when p_c is set to 0.8. It must be outlined that the eective crossover ratio is lower, due to the mating restriction rule. But we see on the IEA runs that it provides good results in terms of ADF and in the same time in terms of optima diversity (especially when sharing is used).

The application of mutation alone reveals to be quite ecient at high rate ($p_m = 0.2$) when using a GA, but less interesting for the IEA.

In conclusion for this test function, we can see that a simple GA can e-ciently provide a good average fitness, but that the IEA + crossover with mating restriction covers more eciently the dierent optima.

7 F_2: An Epistatic Version of F_1

7.1   Definition

Consider the following function:

$$F_2 : [0, 1]^3 : \qquad \mathbb{R}^+$$
$$(x_0, x_1, x_3) \qquad t_2(x_0, x_1) + t_2(x_1, x_2) + t_2(x_2, x_0) \qquad (12)$$

with t_2 being defined with the help of function t (see section 6):

$$t_2(x, y) = \begin{array}{ll} t(x) & \text{if } (x \quad 0.5) \quad (y \quad 0.5) \quad 0 \\ 0.5 \quad t(x) & \text{if } (x \quad 0.5) \quad (y \quad 0.5) \quad 0 \end{array} \qquad (13)$$

---

[1] As all graphs do not have exactly the same ordinate scale, we have drawn a line at (y = 10) for visual comparisons.

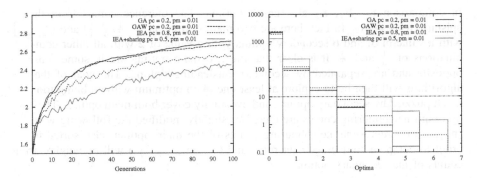

Fig. 2. F_1: Arithmetic crossover. Left: average denoised fitness values. Right: optimum neighbouring counts.

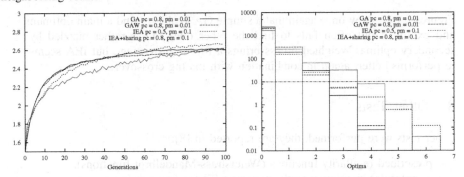

Fig. 3. F_1: Mating crossover. Left: average denoised fitness values. Right: optimum neighbouring counts.

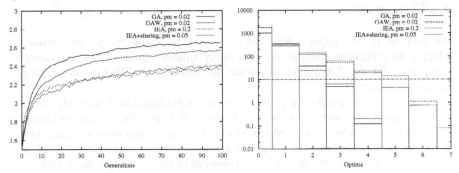

Fig. 4. F_1: mutation alone. Left: average denoised fitness values. Right: optimum neighbouring counts.

The noisy fitness is:

$$f_2 = F_2 \ (1 + N_{(0,0.5)}) \qquad (14)$$

As F_2 is also a regular function ($h = 1$), we set $= 0.05$.

In comparison to F_1, F_2 also has 8 optima. The dierence comes from the epistatic form of t_2. In fact there are two main optima at ($\frac{1}{4}, \frac{1}{4}, \frac{1}{4}$) and ($\frac{3}{4}, \frac{3}{4}, \frac{3}{4}$) with a value of 3, and 6 secondary optima with a value of 2 with all other combinations of $\frac{1}{4}$ and $\frac{3}{4}$. It is clear that the two main optima are in some sense opposite and are separated by secondary misleading optima. The goal of the algorithms will then be to explore at least one main optimum without being too much puzzled by secondary optima, and eventually cover both main optima. The optimum neighbouring counts are therefore slightly modified the following way: ($no_{i,i}$ $_{0,1}$) represents neighbouring counts of the main optima (also sorted in order to compute an average over runs) and ($no_{i,i}$ $_{2,...,7}$) will be neighbouring counts of the secondary optima.

7.2   Results on F_2

Figures 5 to 7 show once again that a simple GA is able to find a main optimum and exploit it but often fails to find the other one, and is rather puzzled by secondary optima. Weighted fitness brings a first improvement, but IEA seems to performs better, again in combination with mating crossover.

8   Other Tests

Other tests were performed, they are reported in [8]:

- prescribed regularity functions (Weierstrass-Mandelbrot functions).
- a molecular simulation application (see [7] for first results).

9   Conclusion

We propose in this paper a new use of history in evolutionary computation, adapted to computationally heavy and noisy fitness functions. An increased complexity for the EA allows to reduce the number of CPU-time consuming fitness evaluations.

Moreover we experimentally show that, when the function is suciently regular in respect to a metric on the search space, it is possible to use similarity of individuals to reduce noise successfully without additional fitness evaluations. We propose a new algorithm using the whole history of evolution to generate new ospring. Experiments with a limited number of fness evaluations on realcoded test functions show that, if GA can overcome the eect of noise in fiding good regions, our immortal evolutionary algorithm (IEA) maintains a better diversity. Of course, we cannot draw firm conclusions on the basis of two testfunctions, but these experiments show in which way we can tune the balance between exploration (discovery of many optima) and exploitation (good average fitness of the population), by balancing crossover, mutation and sharing methods. The problem of an automatic adaptation of   along the evolution will be considered as future work. For that purpose an on-the-y estimation of the regularity of the fitness function could be used.

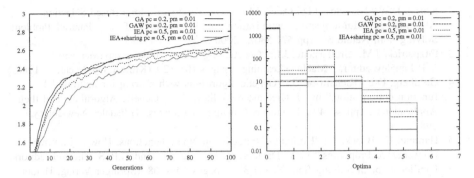

Fig. 5. F_2: Arithmetic crossover. Left: average denoised fitness values. Right: optimum neighbouring counts.

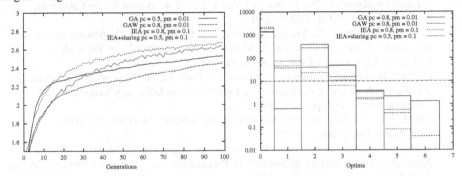

Fig. 6. F_2: Mating crossover. Left: average denoised fitness values. Right: optimum neighbouring counts.

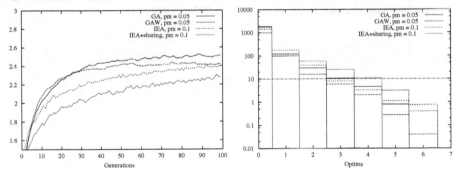

Fig. 7. F_2: mutation alone. Left: average denoised fitness values. Right: optimum neighbouring counts.

References

1. Baker, J.E.: Reducing bias and ineciency in the selection algorithm" in Genetic Algorithms and their applications: Proceedings of the Second International Conference on Genetic Algorithms, 14-21,   1987 .

2. Korne, D.W., Knowles, J.D., Oates, M.J.: "The Pareto Envelope-Based Selection Algorithm for Multiobjective Optimization", in Proceedings of   Parallel Problem Solving from Nature 6 , (pp 571-580), 2000 .
3. Fitzpatrick, J.M., Grefenstette, J.J.: "Genetic Algorithms in noisy environments." in P. Langley, editor,   Machine Learning , pages 101-120, (Kluwer, Dordrecht,   1988).
4. Goldberg, D.E., Richardson, J.: "Genetic algorithms with sharing for multimodal function optimization." in J.J. Grefenstette, editor,   Genetic Algorithms and their Applications , (pp 41-49), Lawrence Erlbaum Associates, Hillsdale, New-Jersey, 1987 .
5. Hammel U., B ack, T.: "Evolution Strategies on Noisy Functions. How to improve Convergence Properties." in Y. Davidor, R. M   anner, and H.P. S chwefel, editors, Parallel Problem Solving from Nature 3   , pages 159-168, (Springer Verlag, Heidelberg, 1994).
6. Lévy V éhel, J. and Lutton, E.: "Evolutionary signal enhancement based on H   older regularity analysis", in Proceedings of EVOIASP2001 Workshop, Como Lake, Italy, Springer Verlag, LNCS 2038,   2001 .
7. Leblanc, B., Lutton, E., Brauns   chweig, B., Toulhoat, H.: "Improving molecular simulation: a meta optimisation of Monte Carlo parameters", in Proceeding of CEC2001, Congress on Evolutionary Computation   , 2001 .
8. Leblanc, B., Lutton, E., Brauns   chweig, B., Toulhoat, H.: "History and never ending life in evolutionary computation: a molecular simulation application", INRIA Research Report, to appear,   2001 .
9. Miller, B.L.: "Noise, sampling, and genetic algorithms", Doctoral dissertation, Illigal Report No. 97001,   1997 .
10. Sano, Y., Kita, H.: "Optimization of Noisy Fitness Functions by Means of Genetic Algorithms Using History of Search", in Proceedings of   Parallel Problem Solving from Nature 6 , (pp 571-580), 2000 .
11. Zitzler, E. and Thiele L.: "Multiobjective Evolutionary Algorithm: A Comparative Case Study and the Strength Pareto Approach ", in   IEEE Transactions on Evolutionary Computation, 2(4)   , (pp 257-272), 1999 .

Origins and Learnability of Syllable Systems: A Cultural Evolutionary Model

Pierre-Yves Oudeyer

Sony Computer Science Lab, Paris
py@csl.sony.fr

Abstract. This paper presents a model of the origins of syllable systems that brings plausibility to the theory which claims that language learning, and in particular phonological acquisition, needs not innate linguistically specific information, as believed by many researchers of the Chomskyan school, but is rather made possible by the interaction between general motor, perceptual, cognitive and social constraints through a self-organizing process. The strategy is to replace the question of acquisition in a larger and evolutionary (cultural) framework: the model addresses the question of the origins of syllable systems (syllables are the major phonological units in speech). It is based on the artificial life methodology of building a society of agents, endowed with motor, perceptual and cognitive apparati that are generic and realistic. We show that agents eectively build sound systems and how these sound systems relate to existing human sound systems. Results concerning the learnability of the produced sound systems by fresh/baby agents are detailed: the critical period eect and the artiﬁial language eect can eectively be predicted by our model. The ability of children to learn sound systems is explained by the evolutionary history of these sound systems, which were precisely shaped so as to fit the ecological niche formed by the brains and bodies of these children, and not the other way around (as advocated by Chomskyan approaches to language).

1 Introduction

Children learn language, and in particular sound systems, incredibly easily and fast, in spite of its apparent idiosyncratic complexity and noisy learning conditions. Many researchers, especially those in the Chomskyan school, believe this can not be possible without a substantial genetically linguistically specific endowment. In fact the role of learning in language development is thought to be very minor ([20]) and reduced to the setting of a few parameters like in the Principles and Parameters theory ([5]) or in Optimality Theory ([1]). Yet, a growing number of researchers (but still the minority among language researchers) have challenged this view, and think that no linguistically specific innate neural device is necessary to account for the oddities of language learning (and structure): rather, they propose that these result from the complex interactions between a number of general motor, perceptual, cognitive, social and functional constraints,

P. Collet et al. (Eds.): EA 2001, LNCS 2310, pp. 143–155, 2002.

and this in a mainly cultural manner ([24]). The word "constraint" is used in its most general meaning: it can be "obstacle" or "opportunity". According to this view, language emerged and evolved so as to fit the ecological niche of initially non-speaking human brains and bodies. In brief, the languages that humans speak were selected so as to be learnable (and not the other way around as suggested in ([19]).

As a consequence, if we take that view as we do here, it seems natural to put oneself in a cultural evolutionary framework: if one wants to understand the principles of language learning, one has to understand the principles of language emergence and evolution, i.e. language epigenesis. This paper follows this idea and illustrates the theory with the case-study of the origins and learnability of syllable systems, which are thought to be a fundamental unit of the complex phonological systems of human languages ([18]). We present a computational model in the spirit of past work on the origins of language ([24], [12]). Among related existing models of the origins of sound systems, there exists two models of the origins of vowel systems: Lindblom [14] showed that the optimization of a number of analytically defined perceptual constraints could predict the most frequent vowel systems, whereas de Boer ([4]) developed an operational multi-agent based model of how vowel systems could have been built culturally. Also, Redford ([21]) made a model similar to Lindblom's concerning syllable systems. Yet, this work is focused uniquely on the properties of sound systems, but does not give cues of how it could actually have been built and how it relates to the cognitive abilities of speakers. The model presented here is inspired from the work of de Boer, in particular for the evolutionary architecture (the imitation game). The dierence is that fist we are dealing with syllables here, and secondly we tried to model constraints in a more embodied and situated manner. Indeed, previous models have shown how important constraints are to the shape of sound systems: when dealing with too abstract constraints, there is a danger to find wrong explanations. Furthermore, Redford showed that certain phenomena can be understood only by considering the interactions between constraints, so models should try to incorporate most of them. The present model builds on a first very simple model detailed in ([27]). It is much more realistic and less arbitrarily biased at both morphological and cognitive levels, and while only studies of eciency were performed with the previous model, structural properties and learnability of the produced sound systems are here presented. Due to space limitations, this paper focuses on the learnability aspects of the behavior of the model and its implications on theories of human sound systems. The fine details of the architecture will be described in a longer paper, and the structural properties are detailed in a companion paper ([25]).

The next section presents an overview of the model with its dierent modules. Then we summarize the eciency of the system as well as the structural properties of the produced syllable systems. Finally, we explore in details their learnability and the implications on theories of language.

2   The Model

2.1   The Imitation Game

Central to the model is the way agents interact. We use here the concept of game, operationally used in a number of computational models of the origins of language ([24], [27]). A game is a sort of protocol that describes the outline of a conversation, allowing agents to coordinate by knowing who should try to say what kind of things at a particular moment. Here we use the "imitation game" developed by de Boer for his experiments on the emergence of   vowel systems.

A round of a game involves two agents, one being called the speaker, and the other the hearer. Here we just retain from their internal architecture that they possess a repertoire of items/syllables, with a score associated to each of them (this is the categorical memory described below). The speaker initiates the conversation by picking up one item in its repertoire and utters it. Then the hearer tries to imitate this sound by producing the item in its repertoire that matches best with what he heard. The speaker then evaluates whether the imitation was good or not by checking whether the best match to this imitation in his repertoire corresponds to the item he uttered initially. He then gives a feedback signal to the hearer in a non-linguistic manner. Finally, each agent updates its repertoire. If the imitation succeeded, the scores of involved items increase. Otherwise, the score of the association used by the speaker decreases and there are 2 possibilities for the hearer: either the score of the association he used was below a certain threshold, and this item is modified by the agent who tries to find a better one; or the score was above this threshold, which means that it may not be a good idea to change this item, and a new item is created, as close to the utterance of the speaker as the agent can do given its constraints and knowledge at this time of its life. Regularly the repertoire is cleant by removing the items that have a score too low. Initially, the repertoires of agents are empty. New items arc added either by invention, which takes place regularly in response to the need of growing the repertoire, or by learning from others.

2.2   The Production Module

Vocal tract.   A physical model of the vocal tract is used, based on an implementation of Cook's model ([6]). It consists in modeling the vocal tract together with the nasal tract as a set of tubes that act as filters, into which are sent acoustic waves pro duced by a model of the glottis and a noise source. There are 8 control parameters for the shape of the vocal tract, used for the production of syllables. Finally, articulators have a certain stiness and inertia.

Control system.   The control system is responsible for driving the vocal tract shape parameters given an articulatory program, which is the articulatory specification of the syllable. Here we consider the syllable from the point of view of the frame-content theory ([18]) which defines it as an oscillation of the jaw (the frame) modulated by intermediary specific articulatory configurations, which represent a segmental content (the content) corresponding to what one may call phonemes. A very important aspect of syllables is that they are not a mere sequencing of segments by juxtaposition: co-articulation takes place, which means that each segment is inuenced by its neighbors. This is crucial because it determines which syllables are dicult to pronounce and imitate. We model here co-articulation in a way very similar to what is described in ([17]), where segments are targets in a number of articulatory dimensions. The dierence is that we provide a biologically plausible implementation inspired from a number of neuroscientific findings ([3]) and that uses techniques developed in the field of behavior-based robotics ([2]). This will be detailed in a forthcoming longer paper. The constraint of jaw oscillation is modeled by a force that pulls in the direction of the position the articulators would have if the syllable was a pure frame, which means an oscillation without intermediary targets. This can be viewed as an elastic whose rest position at each time step is the pure frame configuration at this time step. Finally, and crucially, we introduce a notion of articulatory cost, which consists in measuring on the one hand the eort necessary to achieve an articulatory program and on the other hand the diculty of this articulatory program (how well targets are reached given all the constraints). This cost is used to model the principle of least eort explained in ([14]): easy articulatory programs/syllables tend to be remembered more easily than others. Agents are given initially a set of pre-defined targets that can be thought to come from an imitation game on simple sounds (which means they do not involve movements of the articulators) as described in ([4]). Although the degrees of freedom that we can control here do not correspond exactly to the degrees that are used to define human phonemes, we chose values that allow them to be good metaphors of vowels (V), liquids (C1) and plosives (C2), which mean sonorant, less sonorant, and even less sonorant phonemes (sonority is directly related to the degree of obstruction of the air ow, which mean the more articulators are opened, the more they contribute to a high sonority of the phoneme).

2.3   The Perception Module

The ear of agents consists of a model of the cochlea, and in particular the basilar membrane, as described in ([16]). It provides the successive excitation of this membrane over time. Each excitation trajectory is discretized both over time and frequency: 20 frequency bins are used and a sample is extracted every 10 ms. Next the trajectory is time normalized so as to be of length 25. As a measure of similarity between two perceptual trajectories, we used a technique well-known in the field of speech recognition, dynamic time warping ([22]). Agents use this measure to compute which item in their memory is closest. No segmentation into "phonemes" is done in the recognition process: the recognition is done over the complete unsegmented sound. Agents discover what phonemes compose the syllable only after recognition of the syllable and by looking at the articulatory program associated to the matched perceptual trajectory in the exemplar. This follows a view defended by a number of researchers ([23]) who showed with psychological experiments that the syllable was the primary unit of recognition, and that phoneme recognition came only after.

2.4   The Brain Module

The knowledge management module of our agents consists of 2 memories of exemplars and a mechanism to shape and use them. A first memory (the "inverse mapping" memory) consists of a set, limited in size, of exemplars that serve in the imitation process: they represent the skills of agents for this task. Exemplars consists in associations between articulatory programs and corresponding perceptual trajectories. The second memory (the categorical memory) is in fact a subset of the inverse-mapping memory, to which is added to each exemplar a score. Categorical memory is used to represent the particular sounds that count as categories in the sound system being collectively built by agents (corresponding exemplars are prototypes for categories). It corresponds to the memory of prototypes classically used in the imitation game ([4]).

Initially, the inverse mapping memory is built through babbling. Agents generate random articulatory programs, execute them with the control module and perceive the produced sound. They store each trial with a probability inverse to the articulatory cost involved (prob=1-cost). The number of exemplars that can be stored in this memory is typically quite limited (in the experiments presented below, there are 100 exemplars whereas the total number of possible syllables is slightly above 12000). So initially the inverse mapping memory is composed of exemplars which tends to be more numerous in zones where the cost is low than in zones where the cost is higher. As far as the categorical memory is concerned, it is initially empty, and will grow through learning and invention.

When an agent hears a sound and wants to imitate it, he first looks up in its categorical memory (if it is not empty) and find the item whose perceptual trajectory is most similar to the one he just heard. Then he executes the associated articulatory program. Now, after the interaction is finished, in any case (either it succeeded or failed), it will try to improve its imitation. To do that, it finds in its inverse mapping memory the item (it) whose perceptual trajectory matches best (it may not be the same as the categorical item). Then it tries through babbling a small number of articulatory variations of this item that do not belong to the memory: each articulatory trial item is a mutated version of it, i.e. one target has been changed or added or deleted. This can be thought of as the agent hearing at a point "ble", and having in its memory the closest item being "e". Then it may try "vle", "i", or even "ble" if the chance decides so (indeed, not all possible mutations are tried, which models a time constraints: here they typically try 10 mutations). The important point is that these mutation trials are not forgotten for the future (some of them may be useless now, but very useful in the future): each of them is remembered with a probability inverse to its articulatory cost. Of course, as we have memory limitation, when new items are added to the inverse mapping memory, some others have to be pruned. The strategy chosen here is the least biased: for each new item, a randomly chosen item is also deleted (only the items that belong to categorical memory can not be deleted).

The evolution of inverse mapping memory implied by this mechanism is as follows. Whereas at the beginning items are spread uniformly across "iso-cost" regions, which means skills are both general and imprecise (they have some capacity of imitation of many kind of sounds, but not very precise), at the end items are clustered in certain zones corresponding to the particular sound system of the society of agents, which means skills are both specialized and precise. This is due to the fact that exemplars closest to sound produced by other agents are dierentiated and lead to an increase of exemplars in their local region at the cost of a decrease elsewhere.

It is interesting to remark that what goes on in the head of each agent is very similar to what happens in genetic evolution. One can view the set of exemplars that an agent possess as a population of individuals/genomes, each defined by the sequence of articulatory goals. The fitness function of each individual/syllable is defines by how often it leads to successful imitation when it is used, in both speaker and hearer roles. This population of individuals evolve through a generate and select process, generation being performed through a combination of completely random inventions and mutations of syllables (= one changes one articulatory goal), and selection using the scores of each syllable. The original thing here as compared to many simulations modeling either genetic or cultural evolution, is that the fitness function is not fixed but evolves with time: indeed the fitness of one syllable depends on the population of syllables in the heads of other agents whose fitness itself depends on this syllable. So we have a case of coupled dynamic fitness landscapes. As we will see, what happens is that those fitness landscapes synchronize at some point, they become very similar and stable. Also, the fitness of one syllable depends of the other syllables/exemplars in the memory of the agent: indeed, if a syllable is alone in its part of the space, for example, then few syllables of this area will be produced and other agents will have less opportunity to be practice imitation of this kind of syllable, and so there is a high probability that the syllable will be pruned. The consequence of this is that groups selection also happens.

3   Eciency

The first thing one wants to know is simply whether populations of agents manage to develop a sound system of reasonable size and that allows them to communicate (imitations are successful). Figure 1 and 2 show an example of experiment involving 15 agents, with a memory limit on inverse-mapping memory of 100 exemplars, with vocalizations comprising between 2 and 4 targets included among 10 possible ones (which means that at a given moment, one agent never knows more than about 0.8 percent of the syllable space). In figure 1, each point represents the average success in the last 100 games, and on figure 2, each point represents the average size of categorical memory in the population (i.e. the mean number of syllables in agents' repertoires). We see that of course the success is very high right from the start: this is normal since at the beginning agents have basically one or two syllables in their repertoire, which implies that even if an imitation is quite bad in the absolute, it will still get well matched. The challenge is actually to remain at a high success rate while increasing the size of the repertoires. The 2 graphs shows that it is the case. To make these results convincing, the experiments was repeated 20 times (doing it more is rather infeasible since each experiment basically lasts about 2 days), and the average number of syllables and success was measured in the last 1000 games (over a

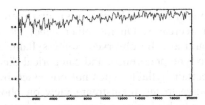

Fig. 1. Example of the evolution of success in interactions for a society of agents who build a sound system from scratch

Fig. 2. Corresponding evolution of mean number of items/categories in repertoires of agents along with time

total of 20000 games): 96.9 percent is the mean success and 79.1 is the mean number of categories/syllables.

The fact that the success remains high as the size of repertoire increases can be explained. At the beginning, agents have very few items in their repertoires, so even if their imitations are bad in the absolute, they will be successfully recognized since recognition is done by nearest-neighbours (for example, when 2 agents have only 1 item, no confusion is possible since there is only 1 category). As time goes on, while their repertoires become larger, their imitation skills are also increasing: indeed, agents explore the articulatory/acoustic mapping locally in areas where they hear other utter sounds, and the new sounds they create are hence also in these areas. The consequence is a positive feed-back loop which makes that agents who knew very dierent parts of the mapping initially tend to synchronize their knowledge and become expert in the same (small) area (whereas at the beginning they have skills to imitate very dierent kinds of sounds, but are poor when it becomes to make subtle distinctions in small areas).

4   Structural Properties

The properties summarized here are detailed in ([25]). The produced syllable systems have structures very similar to what we observe in human languages. On the one hand, a number of universal tendencies were found, like the ranking of syllable types along their frequency (CV   CVC   CCV   CVVC/CCVC/CVCC); Also the model predicts the preference for syllables respecting the sonority hierarchy principle, which states that within a syllable, the sonority (or degree of obstruction of the air ow in the vocal apparatus), first increases until a peak (the nucleus) and then decreases. On the other hand, the diversity observed in human languages could also be observed: some syllable systems did not follow the trend in syllable type preference, and categorical dierences exist (some syllable systems have certain syllable types not possessed by others). This constitutes a viable alternative to the mainstream view on phonological systems, optimality theory ([1]), which require the presence of innate linguistically specific constraints in the genome to account for universal tendencies (an example of constraint is the COMPLEX constraint which states that syllables can have at most one consonant at an edge), and explains diversity by dierent orderings in the strengths of these constraints (which is basically the only thing that is learnt).

5   Learnability Properties

The learnability of the produced systems by fresh agents confronted directly with the complete sound is an important question. Indeed, more generally, learnability of language has been the subject of many experiments, theories and debates. Experiments have shown for example that language acquisition is most successful when it is began early in life ([15]), which refers to the well-known concept of critical period ([13]). Also, learners of a second language typically have much more diculties than learners of a fist language ([9]). Until relatively recently, these facts were interpreted in favor of the idea that humans have an innate language acquisition device ([19]; [20]) which partly consists in pre-giving a number of linguistically specific constraint: for example, ([15]), argues that it is strong evidence for "maturationnaly scheduled   language specific   learning abilities". This view is also supported by a number of theoretical studies, like Gold's theorem ([10]), which basically states that in the absence of enough explicit negative evidence, one can not learn languages belonging to the superfinite class, which includes context free and context sensitive languages (but the applicability to human languages has been challenged, see ([8]).

Here we propose an alternative view, to which our model brings plausibility. It consists in explaining the fact that the learning skills of adults are lower than those of children by the fact that the brain resources needed to do so have already been recruited for other tasks or for a dierent language/sound system (see Rohde and Plaut, 1999 for a comparable view). Said another way, children are better to learn a completely new sound system than adults because their cognitive capabilities are less committed, whereas adults are already specialized. This is indeed what we observe in our model. To see that, a number of experiments were conducted in which on the one hand, some children agents had to learn a particular sound system, and on the other hand, adult agents had to learn a "second language" sound system. More precisely, in each experiment, first a society of agents was ran to produce a syllable system: after 15000 games, an agents was randomly chosen and called the teacher. This teacher was then used in the same game than described above, and with a second agent, the learner, except that here the teacher did not update its memory (he is supposed to know that he knows well the language as compared to the learner). The learner was

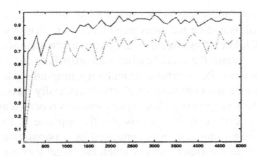

Fig. 3.   Evolution of success in interactions during the learning of an established sound system: top curve is when agent is a child (fresh agent) and bottom curve when it is an adult (it already knows another sound system)

each time in a first run a fresh agent (this models the child) and in a second run an agent taken from another society after 15000 games (which models an adult who knows already another sound system). This experiment was repeated 20 times. One example of success curve is on figure 1: the upper curve is the one for children learning success, and the lower curve for adults learning. Each point in the curve represents the mean success in last 100 games at a particular time t. The mean success after 5000 games of the 20 runs was of 97.3 percent for children against 80.8 percent for adults. This conform well to the idea of a critical period: adults never manage to learn perfectly another sound system. There is an explanation for that: whereas children start with a high plasticity in their inverse mapping memory (because they have no categories yet and so can freely delete and create many new items) and have no strong bias (in fact they are biased, as we will state in next paragraph, but not as much as adults) towards a particular zone of the syllable space, adults, on the contrary, are already committed to another sound system, and have more diculties to create new items in the appropriate zone of the syllable space because their skills resources (which are items in inverse mapping memory that are not prototypes of one of their previous language categories) are much lower. Of course, some of these category prototypes may be pruned, and thus freeing some resources, because they are unsuccessful for the new sound system. But in practice it seems that enough of them allow successful imitations of items in the new sound system, though imprecise, so that still not enough resources can be freed to resolve the remaining confusions. To conclude this paragraph, we see that our model fits very well with the idea that critical periods/second language learning eect need not a genetically programmed language specific mechanism to find an explanation, and that the more parsimonious idea of (un-)commitment of the cognitive system can account for it.

Now, we saw that children could actually learn nearly perfectly a sound system. This result is not obvious since they are faced directly to the complete sound system, in the contrary of the agents who co-built it: the building was incremental and the sound system complexified progressively, which does not mean that their job was easier since negotiation had also to take place, but it was dierent. An experiment was performed that shows on the one hand how non-obvious the task is and on the other hand has implications over a number of existing theories. Children/fresh agents were put in a situation of trying to learn a random syllable system: the adult/teacher was artificially built by putting in its categorical repertoire items whose articulatory programs where completely random (chosen among the complete set of combinatorially possible less-than-5-phonemes articulatory programs). This experiment was repeated again 20 times. Figure 2 shows the curves of 2 experiments: the top one is for child learning success when the target language was generated by a population of agents and the bottom one for child learning success when the target language was random. The mean success over the 20 experiments after 5000 games is 97.3 percent for "natural" sound systems and 78.2 percent for random sound systems. We see that children never learn reasonably well the random sound systems. This result is experimentally and functionally very similar to an experiment about syntax described in (Christiansen, 2000), in which human subjects were asked to learn small languages whose syntax was either the one of an existing natural language or a random/artificial one. They found that indeed subjects were much better at learning the language where the syntax was "natural" than the language where the syntax was "artificial". Deacon (1997) also made a point about this: "if language were a random set of associations, children would likely be significantly handicapped by their highly biased guessing".

This state of aair is in fact compatible with most of theories of language, which all basically suggest that human languages have many particular structures (that make them non-random) and that we are innately endowed with constraints that biases up us towards an easier learning of these languages, because they lead to the particular structure of languages. Now, where considerable disagreement comes in is again about the nature of these constraints and how they got there. On the one hand, the Chomskyan approach suggests that they are coded in a Universal Grammar genetically coded and linguistically specific, and consider language as a system mainly independent of its users (humans) who may have undergone biological evolution so as to be able to acquire and use it in an ecient way (this is suggested by [19]). This is not only true for syntax but also down to phonetics: this approach posits that we have an innate knowledge of what features (for example the labiality of a phoneme) and combination of features can be used in language ([5]). One of the problems with this approach is that the apparent "idiosyncrasies of language structure are hard to explain". On the other hand, a more recent approach considers that language itself evolved and its features were selected so as to fit to generic already existing learning and processing capabilities of humans (see for example ([7]), and that the coherent structures may have emerged through a process of self-organization at multiple levels ([24]). The fact that language evolved to fit to the primitive human brains ecological niche, and in particular to the brains of children, explains, as Deacon ([8]) puts it, why "children have an uncanny ability to make lucky guesses" though they do not possess innate linguistic knowledge. Again the present model tends to bring more plausibility to the second approach. Indeed, it is clear here that on the one hand innate generic motor, perceptual and cognitive constraints bias the way one explores and acquire parts of the syllable space, and on the

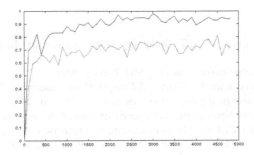

Fig. 4. Evolution of success in interaction during the learning of an established sound system by a child agent: top curve is when the sound system was generated with a population of agents with all constraints, bottom curve is when the sound system is completely random

other hand that the mechanism by which agents culturally negotiate which will be their particular sound system makes them select preferentially systems which allow easy imitation, hence easier learning. For instance, syllables that are very sensitive to noise will tend to be avoided/pruned since they lead to confusions. Also, syllable systems will tend to be coherent both with the process of exploration by dierentiation and the tendency to remember better easy items than dicult ones: given a part of a syllable system, the rest may be found quite easily by focusing the exploration on small variants of items of this part, and exploration is also made maximally ecient by focusing on easy parts.

6 Conclusion

We have presented an operational model of the origins of syllable systems whose particularity is the stress on embodiment and situatedness constraints or opportunities, which imply the avoidance of many shortcuts usually taken in the literature. It illustrates in details (and brings more plausibility) the theory which states that language originated in a cultural self-organized manner, taking as a starting point a set of generic non-linguistically specific learning, motor and perceptual capabilities. In addition to the demonstration of how an ecient communication system could be build with this parsimonious starting point and through cultural evolution, and to the fact that the produced sound systems had many structural similarities with human sound systems, we showed that the ability of children to learn sound systems so easily can be explained (contrarily to speculations of many Chomskyan researchers) by the evolutionary history of these sound systems, which were precisely shaped so as to fit the ecological niche formed by the brains and bodies of children, and not the other way around (as advocated by Chomskyan approaches to language). Yet, one has to note that we do not exclude that biological evolution driven by the need to adapt to a linguistic environment took a role; in fact it is very probable that genes (in particular those implicated in the development of the neural system) co-evolved with language, but, as Deacon puts it "languages have done most of the adapting".

References

1. Archangeli D., Langendoen T. Optimality theory, an overview, Blackwell Pulishers (1997).
2. Arkin, R. Behavior-based Robotics, MIT Press (1999).
3. Bizzi E., Mussa-Ivaldi F., Giszter S.Computations underlying the execution of movement: a biological perspective, Science, vol. 253, pp. 287-291 (1991).
4. de Boer, B. Investigating the Emergence of Speech Sounds. In: Dean, T. (ed.) Proceedings of IJCAI 99. Morgan Kauman, San Francisco. pp. 364-369 (1999).
5. Chomsky, N. and M. Halle (1968) The Sound Pattern of English. Harper Row, New york.
6. P. R. Cook, "Synthesis of the Singing Voice Using a Physically Parameterized Model of the Human Vocal Tract," Proc. of the International Computer Music Conference, pp. 69-72, Columbus, OH, 1989.
7. Christiansen, M., Using artificial language learning to study language evolution: Exploring the emergence of word order universals, in Language Evolution, Dessalles, Wray, Knight (eds.), Transitions to language, Oxford, Oxford University Press (2000).
8. Deacon T., The symbolic species, The Penguin Press (1997).
9. Flege J., Speech learning in a second language, In Ferguson, Menn, Stoel-Gammon (eds.) Phonological Development: Models, Research, Implications, York Press, Timonnium, MD, pp. 565-604 (1992).
10. Gold, E. Language identification in the limit. Information and Control 10, 447-474 (1967).
11. Hurford, J., Studdert-Kennedy M., Knight C., Approaches to the evolution of language, Cambridge, Cambridge University Press (1998).
12. Kirby, S., Syntax without natural selection: how compositionnality emerges from vocabulary in a population of learners, in Hurford, J., Studdert-Kennedy M., Knight C. (eds.), Approaches to the evolution of language, Cambridge, Cambridge University Press (1998).
13. Lenneberg, E. Biological foundations of language, New-York: Wiley (1967).
14. Lindblom, B., Phonological Units as Adaptive Emergents of Lexical Development, in Ferguson, Menn, Stoel-Gammon (eds.) Phonological Development: Models, Research, Implications, York Press, Timonnium, MD, pp. 565-604, (1992).
15. Long M. Maturational Constraints on Language Development, Studies in Second Language Acquisition 12, 251-285 (1990).
16. Lyon, R., All pole models of auditory filtering, in Lewis et al. (eds.) Diversity in auditory mechanics, World Scientific Publishing, Singapore (1997).
17. Massaro, D., Perceiving talking faces, MIT Press (1998).
18. MacNeilage, P.F., The Frame/Content theory of evolution of speech production. Behavioral and Brain Sciences , 21, 499-548 (1998).
19. Pinker, S., Bloom P., Natural Language and Natural Selection, The Brain and Behavioral Sciences, 13, pp. 707-784 (1990).
20. Piattelli-Palmarini, M., Evolution, selection and cognition: from "learning" to parameter setting in biology and in the study of language, Cognition, 31, 1-44 (1989).
21. Redford, M.A., C. Chen, and R. Miikkulainen Modeling the Emergence of Syllable Systems. In: Proceedings of the Twentieth Annual Conference of the Cognitive Science Society. Erlabum Ass. Hillsdale (1998).
22. Sakoe H., Dynamic programming optimization for spoken word recognition, IEEE Transaction Acoustic., Speech, Signal Processing, vol. 26, pp. 263-266 (1982).

23. Segui, J., Dupoux E., Mehler J. (1995) The role of the syllable in speech seg-mentation, phoneme identification, and lexical access, in Altman, (ed.), Cognitive Models of Speech Processing, Psycholinguistics and Computational Perspectives, MIT Press.
24. Steels, L., Synthesizing the origins of language and meaning using co-evolution, self-organization and level formation, in Hurford, Studdert-Kennedy, Knight (eds.), Cambridge University Press, pp. 384-404 (1998).
25. Oudeyer P-Y., The origins of syllable systems: an operational model, to ap-pear in the Proceedings of The International Conference on Cognitive Science, COGSCI'2001, Edinburgh, Scotland, (2001).
26. Oudeyer P-Y, Coupled Neural Maps for the Origins of Vowel Systems, to appear in the proceedings of the International Conference on Artificial Neural Networks, ICANN'2001, Vienna, Austria, Springer Verlag (2001).
27. Steels L., Oudeyer P-y., The cultural evolution of syntactic constraints in phonol-ogy, in Bedau, McCaskill, Packard and Rasmussen (eds.), Proceedings of the 7th International Conference on Artificial Life, pp. 382-391, MIT Press (2000).

Evolution Strategy in Portfolio Optimization

Jerzy J. Korczak [1], Piotr Lipiński[2], and Patrick Roger [3]

[1] Louis Pasteur University, LSIIT, CNRS, Strasbourg, France
jjk@dpt-info.u-strasbg.fr
[2] Louis Pasteur University, LSIIT, CNRS, Strasbourg, France
University of Wroc law, Institute of Computer Science, Wroc law, Poland
lipinski@lsiit.u-strasbg.fr
[3] Louis Pasteur University, LARGE, Strasbourg, France
roger@cournot.u-strasbg.fr

Abstract. In this paper an evolutionary algorithm to optimize a stock portfolio is presented. The method, based on Evolution Strategies, uses artificial trading experts discovered by a genetic algorithm. This approach is tested on a sample of stocks taken from the French market. Results obtained are compared with the Buy-and-Hold strategy and a stock index. Presented research extends evolutionary methods on financial economics worked out earlier for stock trading.

1   Introduction

Nowadays ever increasing attention is being paid to methods based on the principle of evolution. Evolutionary Computation has become a subject of general interest with regard to the capacity to solve complex optimization problems in science and technology [Back 1995], [S chwefel 1995], [Michalewicz 1996], [Eiben 1999a] and [Korczak 2001].

This paper presents an evolutionary approach to financial economics, more precisely to the optimization portfolio problem. It consists of minimizing, for a given level of the expected portfolio return, the value of the corresponding risk indicator. Since currently available analytical solutions were designed in restricted contexts, giving up restrictive assumptions would require completely new ecient algorithms which cannot be developed in the framework of classical methods.

Our approach combines the power of genetic algorithms ([Goldberg 1989], [Michalewicz 1994]) used to generate artificial trading experts, to the opportunities provided by Evolution Strategies which lead to the optimization of portfolio structures where individual trading experts' advice is integrated. The algorithm presented here is the result of extensive research in the application of artificial intelligence to stock trading, details of which being documented in [Korczak 1999] and [Korczak 2000a].

The paper is structured as follows. In section 2, a quick overview of the Markowitz portfolio theory is presented. Section 3 describes our approach to

P. Collet et al. (Eds.): EA 2001, LNCS 2310, pp. 156–167, 2002.

portfolio optimization, taking into account real market constraints. An evolutionary algorithm solving a given problem is presented in section 4. The approach is evaluated using real financial time series in section 5. The paper ends with some concluding comments.

2 Overview of Portfolio Theory

2.1 Introduction

The main goal of investors is to achieve optimal allocation of funds among various financial assets. Searching for an optimal stock portfolio, characterized by random future returns, seems to be a dicult task and is usually formalized as a risk-minimization problem under a constraint of expected portfolio return. Portfolio risk of is often measured in terms of the variance of returns but many other risk criteria have been proposed in the financial literature.

Portfolio theory can be traced back to the [Markowitz 1952] seminal paper and it is presented in an elegant way in [Huang 1988] or [Roger 1996].

2.2 Portfolio Optimization Problem

Consider a financial market in which N risky assets are traded; let $R = (R_1, R_2, \ldots, R_N)$ be the square-integrable random vector of their returns. Denote as $r = ER$ the vector of expected returns and V the corresponding covariance matrix which is assumed positive definite. A portfolio is a vector $x \in \mathbb{R}^N$ verifying $x \cdot 1 = 1$ where 1 is a N-component vector of ones. Hence x_i is the proportion of wealth invested in the i-th asset. Denote as X the set of all portfolios; for each $x \in X$, we define $R_x = x \cdot R$ as the portfolio return and then $x \cdot r = ER_x$ is the portfolio expected return.

For a fixed level e of expected return, $X_e = \{ x \in X : x \cdot r = e \}$ is the set of all portfolios leading to the desired expected return e. The optimization problem is then to find x such that:

$$\text{Risk}(x) = \min \{ \text{Risk}(x) : x \in X_e \}$$

where Risk(.) is the risk indicator (variance of returns in the Markowitz theory).

2.3 Ecient Frontier

A portfolio is called an ecient portfolio if it realizes the minimum variance among the portfolios having the same expected return. The set of ecient portfolios (when e varies) is called the ecient frontier.

In other words, a portfolio x is ecient if and only if it is the solution to:

$$\min \{ x \cdot V x : x \in X \}$$

under the constraints:

$$x \cdot r = e,$$

$$x1 = 1,$$

where e stands for the desired expected return.

Using the standard method of Lagranges coecients, the following solution is obtained:

$$x = \frac{1}{D}[BV^{-1}1 - AV^{-1}r] + e\frac{1}{D}[CV^{-1}r - AV^{-1}1]$$

where

$$A = rV^{-1}1 \qquad B = rV^{-1}r$$
$$C = 1V^{-1}1 \qquad D = BC - A^2$$

If a risk-free asset is traded, generating a known return r_0, the optimal solution, independent of e and called the market portfolio, becomes:

$$x_m = \frac{V^{-1}(r - r_0 1)}{1V^{-1}(r - r_0 1)}$$

Moreover, the market equilibrium is characterized by [Sharpe 1964]:

$$r_k - r_0 = {}_k(ER_m - r_0),$$

where $R_m = x_m R$ is the return on the market portfolio and ${}_k$ is defined as

$$_k = \frac{Cov(R_k, R_m)}{Var R_m}.$$

3 Evolutionary Approach to Portfolio Optimization

3.1 Basic Concepts

In spite of its wide diusion in the professional and academic worlds, the CAPM is often criticized for its artificial assumptions. Although it is an interesting theoretical model, its practical applications may often misfire.

In our previous work on the dierential evolution applied to the problem of portfolio optimization [Korczak 2000a], some artificial assumptions of the CAPM were rejected. More precisely, several operational constraints were introduced such as the imperfect divisibility of stocks, the existence of proportional transaction costs (at a rate c) and the restrictions on short selling.

In this paper, this approach is extended by considering an alternative measure of risk, emphasizing the downside risk, the semivariance ^{1} of returns, which was first suggested in the initial work of Markowitz.

---

^{1} A semivariance of a random variable X is defined as $SVar X = E(X - EX)^2$, where

$$(X - EX) = \begin{array}{ll} 0, & if\ 0\ X\ EX \\ X - EX, & if\ X\ EX\ 0 \end{array}$$

In the previous section, a portfolio was defined as proportions of wealth invested in various stocks. In our approach, the stock quantities are considered so as to take into account real market conditions such as transaction costs. For example, $x = (40, 30, 5, 25)$ means that the individual possesses 40 units of the first stock, 30 units of the second, and so on.

3.2 Artificial Trading Experts

For the purposes of this research, it is also assumed that artificial trading experts for each stock are based on technical analysis rules discovered by the genetic algorithm. In general, technical analysis assumes that future trends can be identified as a more or less complicated function of past prices. Using a trading rule is a practical way of identifying trends which, in turn and generate buying and selling signals.

Let S be the set of technical analysis trading rules used to take a trading decision on the market. Let M denotes the cardinality of S. On the basis of past prices, each rule generates a signal: to sell, to hold or to buy. For the sake of simplicity of computing these decisions will be replaced with real numbers 0 .0, 0.5 and 1 .0 respectively.

In the approach, an expert $e = (e_1, e_2, \ldots, e_M)$ is an M-dimensional binary vector. A i-th coordinate of the expert is equal to 1, if and only if the expert uses the i-th rule in the decision process to generate a buying or selling advice. Thus, there are 2^M possible experts, but only a few of them are usually ecient.

For example, $e = 001101$ means that the expert e generates an advice on the basis of rules numbered 3, 4 and 6.

In order to generate an expert advice, an arithmetic average d of active rules decisions is calculated as follows:

$$d = \frac{\sum_{i=1}^{M} e_i \, d_i}{\sum_{i=1}^{M} e_i},$$

where d_i denotes the decision of the i-th rule. Next, the obtained number d is transformed to a decision, i.e. a number 0 .0, 0.5 or 1.0. This can be done by means of a valuation function f and an earlier chosen threshold s $[0.0, 0.5]$ as follows:

$$f(d) = \begin{array}{l} 0.0, \text{ if } d \quad s \\ 0.5, \text{ if } s \quad d \quad 1 \quad s \\ 1.0, \text{ if } 1 \quad s \quad d \end{array}$$

Finally, an advice given by the expert is equal to f (d).

For example, $e = 001101$, two rules lead to buy and one leads to do nothing. Hence $d = 0.8333$. The final decision is to buy the stock as long as 1 s 0.8333, where s denotes the earlier chosen threshold.

The threshold can be referred to as the risk aversion coecient of the expert. For low levels of s the probability of doing nothing is high because the interval [s, 1 s] is large. Consequently, the strategy is conservative and the expert does not transact frequently. If the opposite is true, i.e., if   s is near 0.50, almost all decisions will be to buy or to sell.

Artificial trading experts are daily generated for each stock in the considered portfolio according to the process described in [Korczak 2000]. Decisions of these experts constitute the soul of a trading process presented in next section.

3.3 Trading Process

Let $a_t = (a_t^{(1)}, a_t^{(2)}, \ldots, a_t^{(N)})$ denotes the vector of expert advices at the end of day t. Let $S = 1, 2, \ldots, N$ and

$$S_t^{(b)} = i \quad S : a_t^{(i)} = 1.0 ,$$

$$S_t^{(h)} = i \quad S : a_t^{(i)} = 0.5 ,$$

$$S_t^{(s)} = i \quad S : a_t^{(i)} = 0.0 .$$

$S_t^{(b)}$ is the set of stocks which experts advice to buy, $S_t^{(h)}$ is the set of stocks which experts advice to hold and $S_t^{(s)}$ is the set of stocks which experts advice to sell. Superscripts (b), (h), (s) are abbreviations of 'buy', 'hold' and 'sell' respectively.

At the beginning of day   t + 1, a non negative number of each stock from $S_t^{(s)}$ will be sold and a non negative number of each stock from $S_t^{(b)}$ will be bought. Let $x_t = (x_t^{(1)}, x_t^{(2)}, \ldots, x_t^{(N)})$ denotes the vector made up of numbers of traded stocks.

For example, $x_t = (10, 20, 0, 12)$ means that 10 stocks of first stock are bought, 20 stocks of second stock are bought and 12 stocks of fourth stock are sold.

Certainly, the following constraints should be satisfied:

$$x_t^{(i)} \quad 0, \quad \text{for i} \quad S_t^{(b)},$$

$$x_t^{(i)} = 0, \quad \text{for i} \quad S_t^{(h)},$$

$$x_t^{(i)} \quad 0, \quad \text{for i} \quad S_t^{(s)}.$$

Moreover, a budget constraint as presented below should be fulfilled.

$$(1 + c) \quad p_{t+1}^{(i)} \quad x_t^{(i)} \qquad (1 \quad c) \quad p_{t+1}^{(i)} \quad x_t^{(i)},$$
$$i \quad S_t^{(b)} \qquad\qquad i \quad S_t^{(s)}$$

where $p_t = (p_t^{(1)}, p_t^{(2)}, \ldots, p_t^{(N)})$ denotes the vector of opening prices at day   t. This condition comes from the idea of self financing, which is discussed in the next section.

The process begins with a portfolio x_0 at time t_0. Let $X^{(1)}$ be a search space consisting of all portfolios, which can be obtained from x_0 at time t_1 according to the process presented above. The purpose is to find a portfolio x_1 $X^{(1)}$ minimizing the risk factor (i.e. semivariance) among the space $X^{(1)}$. By repeating this process, a sequence of trading decisions, which constitutes an investor strategy, can be obtained.

3.4   Idea of Self Financing

One of the main assumptions in this approach is the idea of self financing. All funds are invested at the beginning of the trading process and while the process is running, the funds can neither be added nor withdrawn. However, small amounts of money can be used to fulfill the equality as defined in the previous section.

The important question is what to do in the case where $S_t^{(b)} =$ or $S_t^{(s)} =$.
If $S_t^{(b)} =$, because the obtained funds cannot be invested elsewhere. Similarly, if $S_t^{(s)} =$, there will be no trading, because no funds are obtained.

However, a special risk-free asset is introduced which allows to store funds obtained in selling operations and makes it possible to carry out buying operations where there has been insucient funds obtained by selling transactions. In order to avoid a situation where all funds of the risk-free asset are invested on the first date, a threshold, which limits the percentage of funds available for investing on one day, is defined.

3.5   Financial Time Series

The approach has been validated using real data from the Paris Stock Exchange (Euronext). Every day, for every stock, the opening, maximum, minimum and closing prices are available, as are the transaction volume and the market index value (CAC40). In performance calculation the market index is used as a proxy for the market portfolio.

4   Evolution Strategy

This approach is based on Evolution Strategies, which are described in detail in [Schwefel 1995a] and [Korczak 2001]. In this section the modification introduced to the standard evolution strategy is presented.

In this approach, a portfolio is encoded as a real valued vector of dimension N, where N denotes the number of stocks included in the portfolio.

To evaluate the generated portfolios, various objective functions can be used. In the designed prototype, several functions, based on expected return and risk factors, are implemented. The available objective functions are the following:

$$F_1(x) = \frac{1}{1 + \ _1 \ \mathrm{SVar}(R_x)},$$

$$F_2(x) = \frac{1}{1 + \ _1 \ \mathrm{SVar}(R_x) + \ _2 \quad _x \quad _{x_0}},$$

$$F_3(x) = \frac{1}{1 + \ _1 \ \mathrm{Cov}(R_x, R_m) + \ _2 \quad _x \quad _{x_0}},$$

$$F_4(x) = \frac{1}{1 + \ _1 \ \mathrm{SVar}(R_x) + \ _2 \ \mathrm{Cov}(R_x, R_m) + \ _3 \quad _x \quad _{x_0}},$$

where x_0 denotes the initial portfolio, given by the user, R_m stands for the market return and x, x_0 stand for the   coecient of the considered portfolio x and the initial portfolio x_0 respectively. The factors 1, 2, 3 are used to tune the algorithm and to adjust the importance of each component of the objective function. The objective functions refer to some heuristics using parameters such as the   coecient. By introducing the dierence between the x of the generated portfolio and the x_0 of the portfolio of reference, we penalize the portfolio having x far away from x_0 of the reference to take into account the market risk. Nevertheless, the quality of a solution is defined in terms of expected return and risk of the portfolio on a test period as was mentioned in previous sections.

There are several methods of generating an initial population. The simplest method is random generating with uniform probability. It consists of   -times random choosing of an individual from the search space. The probability of choosing an individual should be the same for every individual in the search space.

The second method uses an initial portfolio given by a user as algorithm parameters. An initial population is chosen from the neighborhood of the given portfolio. It is done by generating a population of random modifications of the initial solution.

In the algorithm, common evolution operators such as reproduction and replacement are used.

In the process of reproduction, population of size   generates   descendants. Each descendant is created from   ancestors. Reproduction consists of three parts: parent selection, recombination and mutation, repeated   times.

Parent selection consists of choosing   parents from a population of size   . There are several commonly used methods of parent selection. The simplest method is random choosing with uniform probability. One of the most popular methods is random selection using the "roulette wheel", which means that the probability of choosing an individual is proportional to its value of the objective function.

Recombination consists of generating one descendant from   parents chosen earlier. The recombination operators described in [S   chwefel 1995a], such as global intermediary recombination, local intermediary recombination, uniform crossover and coping are incorporated into the system.

The approach uses a self-adaptive mutation which is presented in [Beyer 1995] and [S chwefel 1995a]. The parameters of the mutation are encoded in an individual together with a representation of the portfolio.

Each generated descendant has to undergo a process of verification in order to satisfy several constraints. An individual is accepted if the portfolio that it represents can be obtained from the initial portfolio in accordance with the trading process. In other cases, the individual is rejected, and the process of reproduction is repeated. As a result of this verification, ospring are obtained according to the trading process and the idea of self-financing is fulfilled.

In the replacement process, a new population of size   is chosen from an old population of size   and its   descendants.

The simplest method of replacement is a deterministic selection. According to this method, from (+   ) individuals   best survivals are chosen. But every individual can survive no more than   generations.

Apart from deterministic selection, the tournament selection can be used. To start with,   individuals are randomly chosen from the union of an old population and its ospring. From these   individuals, the best one is chosen for the new population. By repeating this process   times a new population is obtained.

Termination criteria include several conditions. The first condition is defined by the acceptable level of valuation function value. The second is based on the homogeneity of population, defied as a minimal dierence between the best and the worst portfolio. The third condition is defined as a maximal number of generations. The algorithm stops when one of them is satisfied. Readers interested in the programming aspects of the evolution-based strategy, can find more details in [Korczak 2001].

5   Case Study

The test concerns the Paris Stock Exchange, particularly the stocks belonging to its market index the CAC40. In this test, the 40 stocks are tracked over a period of about 4 years beginning January 2, 1997. Each stock time series contains the open, close, lowest and highest price, the trading volume and the value of the index at close of trading.

In our approach, an artificial trading expert is generated for each stock of the considered portfolio based on the genetic algorithm described in [Korczak 2000]. Expert trading decisions are respected in the portfolio evolution.

All experiments are carried out with the same financial parameters. Transaction costs are fixed at 0 .25 of the transaction value. The sell limit equals 50, i.e. during each transaction no more than 50 of the current number of stocks can be sold. Due to this limit, the trading risk is reduced; it is impossible to sell out all stocks at once. Moreover, no more than 50 of the capital of the risk-free asset can be invested at one time. The decision threshold, which is used to determine artificial trading expert decisions, was variable. These thresholds cover all strategies from conservative risk-aversive strategies (close to 0   .25) to highly active trading ones (close to 0   .45). The initial portfolio composition may be either randomly generated or arbitrarily defined by the user.

Table 1.   General Algorithm Parameters

Dimension of the search space	10 or 40
Size of a population	50 200
Number of ospring created by the population	100 400
Number of ancestors for each descendent	2 or 4
Maximal age which an individual can achieve	5, 10 or 20

Two general types of tests have been carried out. The first one refers to a portfolio constituted with 10 stocks randomly chosen among the stocks of the

CAC40 index. The purpose of the test was to evaluate the algorithm eciency for medium-size portfolios. The second type of tests refers to a portfolio consisting of all 40 stocks of the CAC40. The purpose of this was to compare the performance of each computed portfolio with that of the market portfolio approximated by the index return.

Each test was repeated several times during dierent time periods to avoid bias. In addition, dierent initial portfolios were used. In this paper, the detailed results are not presented because of the large amount of data, but the interested reader can find an extended report in [Lipiʹ nski 2001].

By selecting an initial portfolio and carrying out evaluations over the test period for each day of the test period, the optimal portfolio was discovered. The calculated portfolio for the next day was the optimal one, according to the constraints defined by expert decisions and the principle of self-financing. Moreover, according to the heuristics, the   coecient of these portfolios was relatively stable as compared to its initial value. In addition, the performance of the result was evaluated on the basis of expected return and risk, the latter being defined as the semivariance of the portfolio return.

To illustrate this process, two dierent initial portfolios are generated randomly (Table 2).

Table 2.   Initial Portfolios

Stock	Items	Value		Items	Value	
AGF	51	3111.00	4.54	19	1061.15	1.61
Alcatel	85	6162.50	9.00	93	5397.72	8.18
Alstom	7	183.05	0.27	85	2549.15	3.87
BNP	70	6993.00	10.22	71	6933.15	10.51
Carrefour	46	3850.20	5.62	74	5194.80	7.88
Danone	30	4668.00	6.82	51	6645.30	10.08
Eridania	75	7312.50	10.68	8	851.20	1.29
FranceTelecom	54	6555.60	9.58	76	11780.00	17.86
Legrand	97	18042.00	26.36	98	21550.20	32.67
Pinault	58	11571.00	16.90	19	3991.90	6.05
Total		68448.85			65954.57	

In order to assess the results, the final profit was compared with the profit achieved by the Buy-and-Hold (BH) strategy, which consists in keeping the initial portfolio unchanged during the whole test period. In most cases, the suggested strategy outperforms the simple BH strategy and market index. Although it is tightly linked to the test period and current trends of the market, repeating these experiments several times on dierent test periods confms the quality and the eciency of the proposed approach.

According to the described process of optimization following results have been obtained (Table 3).

Table 3.   Simulation Results

Date	BH Rate	Portfolio Rate	Date	BH Rate	Portfolio Rate
10-02-2000	0.07	1.03	05-18-2000	1.19	3.41
10-03-2000	-0.27	3.20	05-19-2000	-2.58	2.88
10-04-2000	-0.55	2.82	05-22-2000	-3.03	3.59
10-05-2000	-1.02	5.17	05-23-2000	-2.04	4.94
10-06-2000	-2.27	5.20	05-24-2000	-3.14	3.28
10-09-2000	-4.23	3.40	05-25-2000	-1.02	3.48
10-10-2000	-3.81	5.26	05-26-2000	-1.14	6.21
10-11-2000	-7.21	2.85	05-29-2000	-0.78	9.40
10-12-2000	-7.39	2.86	05-30-2000	0.28	13.48
10-13-2000	-5.84	4.77	05-31-2000	0.92	16.18

The presented two examples confm the eciency of the proposed algorithm. The first example shows that our approach can produce high profit even the BH rate is negative which means that prices of stocks are going down. Certainly, the obtained profit is greater during better market conditions, which can be observed on the second example.

A brief summary of all performed tests is presented in Table 4.

Table 4.   Summary of Results

Stocks	Length of test period	Number of tests	Number of results outperforming BH	Number of results outperforming index
10	20 days	10	7	2
10	20 days	10	8	3
10	20 days	10	8	3
10	20 days	10	7	3
10	20 days	10	9	4
10	60 days	10	6	2
40	20 days	10	7	1
40	60 days	10	4	0

Each case refers to a dicrent initial portfolio and test period. Every test has been repeated several times and the obtained results were compared with the BH strategy and the market index rate each time. Certainly, there is no connection between outperforming BH and the market index, but outperforming the index seems to be a more dicult task than beating BH.

Unfortunately, the simulation of the trading process of the portfolio consisting of all 40 stocks did not turn out to be better than the market portfolio. However, further research on configuring algorithm parameters should lead to an improvement of this score.

It is worth noting that the quality of the obtained results depends on the quality of the artificial trading experts generated earlier (see the detailed study of the expert performance in [Korczak 2000]). Moreover, the current market situation is also important, because when the prices of a large number of stocks are increasing, the obtained results are usually satisfactory, but they do not outperform the Buy-and-Hold strategy because of the assumed buy limit (50). In the inverse case, when prices of a large number of stocks are decreasing, the obtained profit will not be very high, but it is usually higher than the profit of the Buy-and-Hold strategy. In other words, the general appreciation of the performance of the approach cannot be easily assessed.

6   Conclusions

In this paper the evolutionary approach to the problem of portfolio optimization has been presented. The goals and constraints of the problem have been defined and an algorithm based on Evolution Strategies has been proposed. The approach rejects some artificial assumptions used in theoretical models such as perfect divisibility of stocks, and introduces transaction costs and other risk measures such as the semivariance. The approach has been evaluated and validated using real data from the Paris Stock Exchange.

In order to evaluate this approach, the obtained investment strategy has been compared with the Buy-and-Hold strategy. To reduce the time period bias on performance, several time series have been selected. The results have demonstrated that our evolutionary approach is capable of investing more eciently than the simple Buy-and-Hold strategy.

The evolutionary approach in stock trading is still in an experimentation phase. Further research is needed, not only to build a solid theoretical foundation in knowledge discovery applied to financial time series, but also to implement an ecient validation model for real data. The presented approach seems to constitute a practical alternative to classical theoretical models.

References

[Back 1995]   Back, T., Evolutionary Algorithms in Theory and Practice, Oxford University Press, New York, 1995.

[Beyer 1995]   Beyer H.G., Toward a Theory of Evolution Strategies: Self-Adaptation, Evolutionary Computation, 3 (3), 1995, pp. 311-347.

[Eiben 1999a]   Eiben, A.E., Rudolph, G., Theory of Evolutionary Algorithms: A Bird's Eye View, Theoretical Computer Science 229(1), 1999, pp. 3-9.

[Goldberg 1989]   Goldberg, D.E., Genetic Algorithms in Search, Optimization and Machine Learning, Addison-Wesley, 1989.

[Huang 1988]   Huang, C.F., Litzenberger, R., Foundations for Financial Economics, North-Holland, 1988.

[Korczak 1999]   Korczak, J., Approche g´ en´etique pour d´ ecouvrir un mod ele de march´e boursier, Actes de la Conf. d'Apprentissage, CAP, Palaiseau, 1999, pp. 9-16.

[Korczak 2000]   Korczak, J., Roger, P., Stock Timing with Genetic Algorithms, WP48, LARGE, Universit´ e Louis Pasteur, Strasbourg, 2000.

[Korczak 2000a]   Korczak, J., Roger, P., Portfolio Optimization using Dier-
                     ential Evolution, Zastosowania rozwiaza´   n informatycznych w
                     bankowo´sci, AE Wroc law, Poland, 2000, pp. 302-319.
[Korczak 2001]   Korczak, J., Lipi´ nski, P., Evolution Strategies: Principles and
                     Prototypes, Research Report 2005, LSIIT, CNRS, Universit´   e
                     Louis Pasteur, Illkirch, 2001.
[Lipiński 2001]   Lipi´nski, P., Portfolio Optimization Using Evolution Strate-
                     gies, Master of Computer Science Dissertation, University of
                     Wroc law, Wroc law, Poland, 2001.
[Markowitz 1952]   Markowitz, H., Portfolio Selection, Journal of Finance, 7, 1952,
                     pp. 77-91.
[Michalewicz 1994]   Michalewicz, Z., Genetic Algorithms + Data Structures = Evo-
                     lution Programs, Springer Verlag, New York, 1994.
[Michalewicz 1996]   Michalewicz, Z., Schoenauer, M., Evolutionary Computation,
                     Control and Cybernetics 26(3), 1996, pp. 307-338.
[Roger 1996]   Roger, P., L'evaluation des actifs financiers : mod   eles a temps
                     discret, De Boeck Universit´ e, 1996.
[Schwefel 1995]   Schwefel, H.-P., Evolution and Optimum Seeking, John Wiley
                     and Sons, Chichester, 1995.
[Schwefel 1995a]   Schwefel, H.-P., Rudolph, G., Contemporary Evolution Strate-
                     gies, Advances in Artificial Life, Springer, Berlin, 1995, pp. 893-
                     907.
[Sharpe 1964]   Sharpe, W., Capital Asset Prices : A Theory of Market Equi-
                     librium under Conditions of Risk, Journal of Finance, 19, 1964,
                     pp. 425-442.

Scatter Search for Graph Coloring

Jean-Philippe Hamiez [1] and Jin-Kao Hao [2]

[1] LGI2P, École des Mines d'Al es, EERIE
Parc Scientifique Georges Besse
F-30035 N'mes Cedex 01, France
hamiez@site-eerie.ema.fr
[2] LERIA, Universit´ e d'Angers
2, Bd. Lavoisier
F-49045 Angers Cedex 01, France
Jin-Kao.Hao@univ-angers.fr

Abstract. In this paper, we present a first scatter search approach for the Graph Coloring Problem (GCP). The evolutionary strategy scatter search operates on a set of configurations by combining two or more elements. New configurations are improved before replacing others according to their quality (fitness), and sometimes, to their diversity. Scatter search has been applied recently to some combinatorial optimization problems with promising results. Nevertheless, it seems that no attempt of scatter search has been published for the GCP. This paper presents such an investigation and reports experimental results on some well-studied DIMACS graphs.

1 Introduction

Scatter search [13,14] is an evolutionary approach related to the tabu search metaheuristic [12]. It is based on strategies proposed in the 1960s for combining decision rules and constraints. This approach has only been applied recently to a few optimization problems. Applications of this method include, e.g., vehicle routing [27] and unconstrained optimization [23]; see also [13] for more references.

Like other population-based methods, scatter search uses combination of configurations to generate new configurations which can replace others in the population. But the way combinations and replacements are made diers from the traditional strategies used in genetic algorithms. Combinations operate on multiple parents, and replacements rely on the improvement of a fitness function as well as the improvement of the population diversity. Furthermore, scatter search generally works with a small set of configurations and uses deterministic heuristics as much as possible in place of randomization to make a decision.

Graph k-coloring can be stated as follows: given an undirected graph G with a set V of vertices and a set E of edges connecting vertices, k-coloring

---

This work was partially supported by the Sino-French Joint Laboratory in Computer Science, Control and Applied Mathematics (LIAMA) and the Sino-French Advanced Research Programme (PRA).

P. Collet et al. (Eds.): EA 2001, LNCS 2310, pp. 168–179, 2002.

G means finding a partition of V into k classes $V_1, \ldots, V_k$, called color classes, such that no couple of vertices (u, v)   E belongs to the same color class. Formally, $V_1, \ldots, V_k$ is a valid k-coloring of the graph G = (V, E) if i [1..k] and u V_i, v $V_i / (u, v)$ E. The graph coloring problem (GCP for short) is the optimization problem   associated with k-coloring. It aims at searching for the minimal k such that a proper k-coloring exists. This minimum is the   chromatic number (G) of graph G.

k-coloring and the GCP are well-known   NP-hard problems [20] and only small problem instances can be solved exactly within a reasonable amount of time in the general case [5]. It is also hard even to approximate the chromatic number of a graph. In [25], it is proved that for some   0, approximating the chromatic number within a factor of   n is NP-hard. Indeed, one of the best known approximation algorithm [15] provides an extremely poor performance guarantee [1] of $O(n(\log \log n)^2 / (\log n)^3)$ for a graph with n vertices.

Graph coloring has many real applications, e.g., timetable construction [24] or frequency assignment [9]. There are many resolution methods for this problem: greedy constructive approaches, e.g., DSATUR [1] and the Recursive Largest First algorithm [24], hybrid strategies like HCA [4,8] and those proposed in [6, 26], local search meta-heuristics, e.g., simulated annealing [18] or tabu search [3, 16], neural network attempts [17], ... See [2] for a more extensive list of other references about the GCP.

Despite the fact that the literature on graph coloring is always growing, there exists, to our knowledge, no approach relying on scatter search for the GCP. The goal of our study is then to provide an experimental investigation of scatter search applied to the GCP. The paper is organized as follows: Sect. 2 recalls the general template of scatter search; Sect. 3 presents our first scatter search algorithm for graph coloring; next section gives the results we obtained on some of the well-known DIMACS benchmark graphs [19], before concluding.

2   General Design of Scatter Search

We briey recall here the components of scatter search; fundamental concepts and motivations are described in [13]. See also [14] for an exhaustive illustration of these components on a non-linear optimization problem.

1. Diversification Generation Method   . This step is used first to initialize the population and, eventually, to rebuild a subset of the population during the search. Configurations are built in order to respect a maximal diversity [2]. See [21], e.g., for an illustration of an appropriate generator for a 0-1 knapsack problem;

---

[1] The performance guarantee is the maximum ratio, taken over all inputs, of the color size over the chromatic number.

[2] One of the lessons provided in [22] suggests to "consider the use of   frequency memory to develop eective diversiÞation   ".

2. Improvement Method   . New configurations, obtained in step 1 or by combination, are improved in quality;

3. Reference Set Update Method   . Improved configurations are checked for replacing others in a reference set (RefSet for short) according to their quality (line 2 in Algorithm 1) or their diversity (line 3)   [3]. RefSet consists of b best evaluated configurations (BestSet) and d most diverse configurations (DivSet). See [23] for other update methods;

4. Subset Generation Method   . This step produces subsets of configurations (with two elements or more) from the reference set to be combined;

5. Configuration Combination Method   . Subsets of configurations built in step 4 are combined, generally using a problem-dependent combination operator. Laguna and Armentano [22] also suggest that "the use of   multiple combination methods can be eective."

Algorithm 1: Scatter search outline

begin

$\quad$ $P \leftarrow \emptyset$ /* Start with an empty population P */

$\quad$ **while** P is not full **do**

$\quad\quad$ Build a new configuration c with the *Diversification generation method*

$\quad\quad$ Apply the *Improvement method* to c to obtain c^*

$\quad\quad$ **if** $c^* \notin P$ **then** $P \leftarrow P \cup \{c^*\}$

$\quad$ Build *BestSet* by selecting / removing in P the b best evaluated configurations

$\quad$ Build *DivSet* by selecting in P the d most diverse configurations

$\quad$ $RefSet \leftarrow BestSet \cup DivSet$

[1] $\quad$ **while** *RefSet* contains new elements **do**

$\quad\quad$ **for** each subset i built by applying the *Subset generation method* on *RefSet* **do**

$\quad\quad\quad$ Use the *Configuration combination method* with i to obtain c_i

$\quad\quad\quad$ Apply the *Improvement method* on c_i to obtain c_i^*

[2] $\quad\quad\quad$ **if** $c_i^* \notin RefSet$ **and** evaluation of c_i^* is better than the worst evaluated element c_{worst} in *BestSet* **then**

$\quad\quad\quad\quad$ $BestSet \leftarrow BestSet \cup \{c_i^*\}; BestSet \leftarrow BestSet - \{c_{worst}\}$

[3] $\quad\quad\quad$ **else if** $c_i^* \notin DivSet$ **and** c_i^* add diversity to *DivSet* **then**

$\quad\quad\quad\quad$ $DivSet \leftarrow DivSet \cup \{c_i^*\}$

$\quad\quad\quad\quad$ Remove the element with minimal diversity in *DivSet*

end

Algorithm 1 gives an overall view of one single iteration of a generic scatter search procedure, and thus, the way the components are linked   [4]. The main loop 1 controls the termination of the procedure: the process stops when the population evolves no more in quality. Note that some scatter search procedures also use

---

[3] Note that some scatter search procedures only use the quality criterion to update RefSet . In this case, remove line 3.

[4] See [10] for helpful information on practical implementation.

the evolution of the RefSet diversity as a stop criterion. More iterations can be done by restarting loop 1 with a new set of configurations composed of the b best evaluated configurations found by the previous iteration and d new diverse configurations built using the diversification generation method.

3 Scatter Search for Graph Coloring

In this section we describe the five components of our scatter search procedure dedicated to the graph coloring problem. These components are organized in the same way as in Algorithm 1. Let us first describe some concepts useful for the understanding of the overall procedure and its composing elements.

Configuration: a configuration c is any partition $V_1, \ldots, V_k$ of V into k subsets. $V_i (i \in [1..k])$ is an independent set if $\forall u, v \in V_i, (u, v) \notin E$. c is a proper k-coloring if each $V_i \in c$ is an independent set. We will refer to partial coloring for configurations in which some vertices are not assigned a color.

Evaluation Function: two configurations can be compared in terms of quality using an evaluation (or a fitness) function $f: f(c) = |\{(u, v) \mid \exists V_i (V_i \in c, i \in [1..k]) / (u, v) \in E\}|$. In other words, f counts the edges having both endpoints in the same color class. Solving a k-coloring instance means finding a particular configuration c such as $f(c) = 0$.

General Resolution Strategy for Graph Coloring: k-coloring aims at finding a complete assignment of k colors to the vertices that satisfies all the constraints. Such an assignment is said consistent (proper k-coloring). The generalized GCP can then be stated as solving successive k-colorings with decreasing values of k until no proper k-coloring can be obtained.

3.1 Diversification Generation Method

We generate conicting configurations with k colors by means of random independent sets built using Algorithm 2.

Algorithm 2: A diversification generation method for the GCP

```
begin
    for i = 1 to k, by 1 do
        A ← ∅; Vᵢ ← ∅
        while V ≠ ∅ do
[1]         Choose (randomly) a vertex v ∈ V
            V ← V − {v}; Vᵢ ← Vᵢ ∪ {v}
            for each u ∈ V/(u, v) ∈ E do
                A ← A ∪ {u}; V ← V − {u}
        V ← A
    Put each v ∈ V in a color class such as it minimizes the conflicts over the graph
end
```

Randomization is used here (line   ??) only to insure diversity. Other mean-ingful choice rules can easily replace it. For instance, the vertices can be selected in decreasing order of their saturation degree, like in DSATUR [1].

3.2   Improvement Method

We use a tabu search algorithm to improve new configurations. This algorithm iteratively makes   best 1-moves, changing the current color of a conicting ver-tex[5] to another one, until achieving a proper coloring. "Best moves" are those, which minimize the dierence between the fness of the conguration before the move is made and the fitness of the configuration after the move is performed. In case of multiple best 1-moves, choose one randomly. A tabu move leading to a configuration better than the best configuration found so far, within the same execution of the improvement method or within the overall scatter search procedure, is always accepted (aspiration criterion   ).

The tabu tenure   l is dynamically computed by the formula proposed in [3]:

$$l =   E_c + random (g) \qquad (1)$$

where E_c is the set of conicting edges in configuration   c. random (g) is a function which returns an integer value uniformly chosen in [1..   g].   weights the number of conicting edges.

A move m can be characterized by a triplet (u, V_{old} , V_{new}), u   V, V_{old} and V_{new} being, respectively, the previous and the new colors of the conicting vertex u. So, when a move m is performed, assigning u to the color class V_{old} is forbidden for the next   l iterations by introducing the (u, V_{old}) couple in the tabu list.

The algorithm stops when a solution is obtained or when a maximum number of moves have been carried out without finding a solution. Algorithm 1 gives an outline of the procedure, which is extracted from an eective   generic tabu search [3] designed for various coloring problems (k-coloring, GCP, T-coloring and set T-coloring).

Algorithm 1:   A tabu search for graph coloring

begin
   | Let c be the configuration to improve
   | T L   / Initialize the tabu list   T L to empty /
   | c   c
   | while f (c)   0 and not Stop condition do
   | | Update c by performing a best 1-move   m (u, V_{old} , V_{new})
   | | T L   T L   (u, V_{old})
   | | if f (c) f   (c) then
   | | L c   c
end

---

[5] A vertex u V_i is said conicting if   v V_i/ (u, v)   E .

3.3 Reference Set Update Method

We use the same reference set update method as the one described in Algorithm 1 (Sect. 2):

1. An improved configuration c_{new} replaces another one in BestSet if its evaluation is better than the worst evaluated configuration c_{worst} in BestSet.
2. If c_{new} is not kept due to its quality (step 1), it is checked for diversity. c_{new} replaces the configuration with lowest diversity c_{worst} in DivSet if c_{new} has a higher diversity than c_{worst}.

To compute the diversity $D(c_1, c_2)$ between two configurations c_1 and c_2, we use the distance measure introduced in [8]. The distance between c_1 and c_2 is the minimum number of 1-moves necessary to transform c_1 into c_2. Let us call it the move distance (D_M).

Note that the Hamming distance $D_H(c_1, c_2)$ is not well adapted to compare, in terms of diversity, two configurations c_1 and c_2 of the GCP. This is due to the definition itself of this distance. For two strings (configurations) c_1 and c_2 of the same length, $D_H(c_1, c_2)$ is the number of dierent positions in the two strings. To illustrate the dierence between these two distances, let us consider the example of Fig. 1.

Left drawing gives a representation of the traditional assignment strategy in which colors are assigned to vertices. In Fig. 1(a), and according to the definition of the Hamming distance, $D_H(c_1, c_2) = 10$. So, c_1 is quite far from c_2. Right drawing shows a representation of the partition approach for the same configurations. In this case, the labeling of the colors is indierent. In Fig. 1(b), moving the vertex 7 from the class V_2 to V_3 in c_1 leads to c_2 (up to a permutation of the colors). Then, $D_M(c_1, c_2) = 1$.

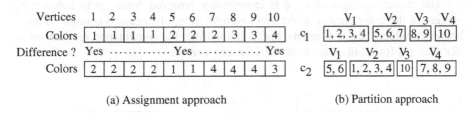

(a) Assignment approach   (b) Partition approach

Fig. 1. Comparing Hamming and move distances

Note that the diversity is a measure over all the configurations in RefSet. So, it is updated each time a new configuration is added to RefSet.

3.4 Subset Generation Method

The smallest subsets (denoted by 2-subsets) consist of all the couples of configurations in RefSet. Intermediate subsets (a-subsets, a $\in$ 3, 4) are built by augmenting each (a-1)-subset with the best evaluated configuration not included in the subset. Finally, c-subsets (c [5.. RefSet]) contain the best c elements. In the scatter search template, each subset is generated only once, while, in the context of genetic algorithms, combinations based on the same subset are allowed. See [10], from which this description is extracted, for motivations about this method.

3.5   Configuration Combination Method

Combination may be viewed as a generalization upon multiple parents of classical crossovers which usually operate with only two parents. We used a generalization of the powerful greedy partition crossover (GPX), proposed in [8] within an evolutionary algorithm. GPX has been especially developed for the graph coloring problem with results reaching, and sometimes improving, those of the best known algorithms for the GCP.

Given a subset p generated by the subset generation method, the generalized combination operator builds the k color classes of the ospring one by one. First, choose arbitrarily (and temporarily remove) a configuration c p. Temporarily remove from c a minimal set A_c of conicting vertices such as c becomes a partial proper k-coloring. Next, fill in a free color class $V_i (i [1..k])$ of the ospring with all conict-free vertices of the color class V^c c having maximum cardinality (break ties randomly) and remove these vertices from all configurations in p. Then, make A_c available for c and repeat these steps until the k color classes of the ospring contain at least one vertex. Finally, to complete the new configuration if necessary, assign to each free vertex in the ospring a color such that it minimizes the conicts over the graph.

The chosen configuration c is temporarily removed from p to balance the origins of the color classes of the new configuration; all removed configurations become enable when p is empty. Fig. 2 summarizes the combination mechanism. See also [8] for an illustration of GPX on two parents.

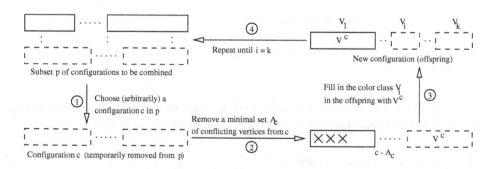

Fig. 2.   Generalization of GPX

We used randomization to give each configuration the same chance to be selected when building the color class V_i of the new configuration. Note that, if we choose configurations following the order of two dierent permutations   1 and   2 issued from the same subset   p, this may lead to two dierent osprings. Another possible way of choosing configurations could be to consider their costs.

4   Preliminary Results

We give in this section some preliminary results obtained with our scatter search algorithm (called SSGC hereafter). Results of SSGC are reported for some of the DIMACS benchmark graphs [19] [6] together with those of the best-known algorithms and those of a   generic tabu search   algorithm:

- Morgenstern [26], two local search algorithms based on particular neighbor-hoods (denoted by 3a and 3b in Table 1), and a distributed population-based algorithm (3c);
- Funabiki and Higashino [7], a descent algorithm with various heuristics mixed with a greedy construction stage and the search for a maximum clique (denoted by MIPS _CLR);
- Dorne and Hao [3], a generic tabu search (GTS for short) which solves successive k-colorings, k decreasing as long as a proper   k-coloring is found.

Columns 5-7 in Table 1 summarize the best-known results given in the above papers for some DIMACS graphs: the smallest coloring ever obtained, the method which produced such a coloring, and the best computing time required. Columns 1-4 recall the characteristics of each graph: its name, number of vertices and edges, and its chromatic number (or its best known lower bound when unknown).

Results of GTS are reported in Table 2 (columns 3-5), including mean number of moves. The last four columns give the results of our scatter search algorithm [7]. Time entries and number of moves and combinations are averaged over five to ten runs. A maximum of 100000 moves were allowed for the improvement method. We only make, for each execution, one iteration of scatter search, i.e., main loop 1 in Algorithm 1 was performed on a single reference set. The   and g parameters (Sect. 3.2) used in computing the dynamic tabu tenure   l (1) were empirically determined, respectively 2 and 10 at most. The size of the starting population was no more than 20 configurations and the   b and d parameters (Sect. 2) were both empirically fixed to 10 at most. For the at300   _20_0 and dsjc125.5 graphs, the improvement method was allowed to perform 1 of random walks, i.e.: moving a random vertex into a random color class. Information about computing time (in seconds) in Table 1 and 2 is only for indicative purpose because results were obtained on dierent machines. marks signal that no result is available.

---

[6] Available via FTP from: dimacs.rutgers.edu/pub/challenge/graph/benchmarks/.
[7] SSGC is implemented in C (CC compiler with -O5 ag) and runs on a Sun Ultra 1 (256 RAM, 143 MHz).

Table 1.   Best known results for some instances of the 2nd DIMACS challenge

Graph name	V	E		k	Method	Time (sec.)
school1	385	19095	14	14	MIPS_CLR	1
school1_nsh	352	14612	14	14	MIPS_CLR	1
dsjr500.1c	500	121275	84	85	3c	1240
r125.5	125	3838	36	36	3c	1
r250.1c	250	30227	64	64	3c	60
r250.5	250	14849	65	65	3c	181
r1000.1c	1000	485090	90	98	3c	1240
r1000.5	1000	238267	234	237	MIPS_CLR	3266
le450_15a	450	8168	15	15	3b	60
le450_15b	450	8169	15	15	3b	60
le450_15c	450	16680	15	15	3b	126
le450_15d	450	16750	15	15	3b	80
at300_20_0	300	21375	20	20	3a	1
dsjc125.5	125	3891	10	17	3c	14

Table 2.   Results of GTS and SSGC

Graph name	χ	GTS			SSGC			
		k	Moves	Time (sec.)	k	Moves	Combinations	Time (sec.)
school1	14	–	–	–	14	14	4	25
school1_nsh	14	–	–	–	14	18	9	31
dsjr500.1c	84	85	21000	70	85	103349	2	97
r125.5	36	36	147000	65	36	52	43	9
r250.1c	64	64	462	1	64	15497	146	349
r250.5	65	66	7800	6	65	180617	3604	3864
r1000.1c	$\geq$ 90	98	1623000	4500	98	50395691	484	–
r1000.5	$\geq$ 234	242	6027000	18758	240	20800000	189	–
le450_15a	15	15	103000	25	15	1064916	69	216
le450_15b	15	15	33000	8	15	677910	43	128
le450_15c	15	16	–	–	15	2135839	68	1091
le450_15d	15	16	–	–	15	8507576	280	3980
flat300_20_0	20	20	33000	17	20	732618	130	1849
dsjc125.5	$\geq$ 10	17	348000	136	17	6805080	1352	805

From Table 1 and 2 we can make several remarks. First, SSGC manages to reach the results of the best-known algorithms (column   k), except on the r1000.5 graph for which a 237-coloring has been published recently. The sophisticated algorithm used to reach this coloring includes, among other components, the search for a maximum clique. Nevertheless, SSGC obtains here a better coloring than GTS. Note that the previous best result for this graph was reported in [26] with 241 colors. Our scatter search approach also improves in quality on the results obtained with tabu search (GTS) on four graphs. This means that tabu search, the improvement method we used, surely benefits from the other general components of scatter search (Sect. 2).

Nevertheless, SSGC seems to be quite slow (regarding the number of moves) to reach the same results as those of GTS. This can partially be explained by the time spent in building the initial reference set, which include the improvement of all the configurations in the population. Furthermore, the generation method used in SSGC (random independent sets, see Sect. 3.1) diers from the one performed by GTS which is based on DSATUR [1]. It would certainly be interesting to follow the approach used in GTS to build the initial reference set. Another reason to the slowness of SSGC may be allotted with the diversity update step which occurs each time a new configuration c_{new} is inserted in $RefSet$. One possible way to speed up this step could be to regulate it according to a minimal diversity D_0 (fixed parameter or dynamically computed). In other words, c_{new} will be added to $DivSet$ only if its minimal diversity is greater than D_0.

Table 3.   Results of SSGC using a   descent   improvement method

Graph name		k	Moves	Combinations	Time (sec.)
school1	14	14	900	1	6
school1 _nsh	14	14	408	2	7
r125.5	36	36	7	66	7
r250.1c	64	64	83600	37499	–
r250.5	65	65	25	580	452

We also experimented scatter search with a less elaborated improvement method, the simplest   descent heuristic, which makes best 1-moves until achieving no improvement. This approach was only able to solve a few "easy" instances (see Table 3) implying that the global eciency of scatter search is greatly dependent on the particular eciency of its improvement algorithm.

5   Conclusions and Perspectives

We have developed a first adaptation of the scatter search method to the graph coloring problem. The preliminary results we obtained on some well-studied DIMACS graphs using this technique show that scatter search may be eective for some of them since SSGC reached most of the results of the best-known approaches. SSGC obtains also better results than those of a generic tabu search on some graphs.

At the same time, we observed that scatter search is quite slow due, essentially, to the numerous components it uses. Eorts must then be made to develop eective procedures to make the scatter search quicker. Another diculty is to identify meaningful rules when choices are needed.

We are now working on replacing some random choices made in SSGC by deterministic rules. A first modification have been carried out in the diversification generation method (Sect. 3.1) by selecting vertices to be included in the same independent set according to Br´ elaz heuristics [1]. Vertices are inserted in color classes in decreasing order of their saturation degrees. Ties are broken using the connection degree and, only if needed, randomness. Few results are available since tests are currently running. However, the modified procedure was able to solve the school1, school1 _nsh, and r125.5 graphs even while the improvement method was left out. In other words, solutions were found only by combining sets of configurations. This suggests that the generalization of the eective   greedy partition crossover   may be a suitable combination operator.

This research has highlighted the need for further investigation to make scatter search more eective for graph coloring. By using the   generic tabu search   in the initialization step for building the reference set, one may speed up the overall procedure. Another possible way could be to use maximal independent sets in place of selecting vertices at random to identify independent sets. Indeed, this later way shares the same objective as the   greedy partition crossover . Lessons provided in [22] could also be useful for further attempts. Finally, some choice rules proposed in [11] may be used as fast heuristics either to generate the initial population or to color free vertices after each combination.

References

1. Br´elaz., D.: New methods to color the vertices of a graph.   Commun. ACM   22(4) (1979) 251–256
2. Culberson, J.: Bibliography on graph coloring. Available on the wold wide web at http://liinwww.ira.uka.de/bibliography/Theory/graph.coloring.html (2000)
3. Dorne, R., Hao, J.K.: Tabu search for graph coloring, T-colorings and set T-colorings. In Voss, S., Martello, S., Osman, I.H., Roucairol, C. (editors)   Metaheuristics: advances and trends in local search paradigms for optimization   Kluwer Academic Publishers (1998) 77–92
4. Dorne, R., Hao, J.K.: A new genetic local search algorithm for graph coloring. Lecture Notes in Computer Science   1498   Springer-Verlag (1998) 745–754
5. Dubois, N., de Werra, D.: EPCOT: an ecient procedure for coloring optimally with tabu search.   Computers Math. Appl.   25(10/11)   (1993) 35–45
6. Ferland, J.A., Fleurent, C.: Object-oriented implementation of heuristic search methods for graph coloring, maximum clique, and satisfiability. In [19] (1996) 619–652
7. Funabiki, N., Higashino, T.: A minimal-state processing search algorithm for graph colorings problems.   IEICE Trans. Fundamentals   E83–A(7)   (2000) 1420–1430
8. Galinier, P., Hao, J.K.: Hybrid evolutionary algorithms for graph coloring.   J. Combin. Optim.   3(4)   (1999) 379–397
9. Gamst, A.: Some lower bounds for a class of frequency assignment problems.   IEEE Trans. Veh. Tech.   35(1)   (1986) 8–14
10. Glover, F.: A template for scatter search and path relinking. In Hao, J.K., Lutton, E., Ronald, E., Schoenauer, M., Snyers, D. (editors)   Artificial evolution . Lecture Notes in Computer Science   1363   Springer-Verlag (1998) 13–54

11. Glover, F.: Tutorial on surrogate constraint approaches for optimization in graphs. Tech. Report (February 2001)
12. Glover, F., Laguna, M.: Tabu Search. Kluwer Academic Publishers (1997)
13. Glover, F., Laguna, M., Mart´ , R.: Fundamentals of scatter search and path relinking. Control and Cybernetics   39(3) (2000) 653–684
14. Glover, F., Laguna, M., Mart´ , R.: Scatter search. In Ghosh, A., Tsutsui, S. (editors) Theory and applications of evolutionary computation: recent trends Springer-Verlag (to appear)
15. Halld´orsson, M.M.: A still better performance guarantee for approximate graph coloring. Inform. P rocess. Lett. 45 (1993) 19–23
16. Hertz, A., de Werra, D.: Using tabu search techniques for graph coloring.   Computing 39 (1987) 345–351
17. Jagota, A.: An adaptive, multiple restarts neural network algorithm for graph coloring. European J. Oper. Res.   93 (1996) 257–270
18. Johnson, D.S., Aragon, C.R., McGeoch, L.A., Schevon, C.: Optimization by simulated annealing: an experimental evaluation. II. Graph coloring and number partitioning. Oper. Res. 39(3) (1991) 378–406
19. Johnson, D.S., Trick., M.A. (editors): Cliques, coloring, and satisfiability: 2nd DI-MACS implementation challenge, 1993.   DIMACS Series in Discr. Math. and Theoretical Comput. Sci.   26 American Math. Soc. (1996)
20. Karp, R.M.: Reducibility among combinatorial problems. In Miller, R.E., Thatcher, J.W. (editors)   Complexity of computer computations   Plenum Press, New York (1972) 85–103
21. Laguna, M.: Scatter search. In Pardalos, P.M., Resende, M.G.C. (editors)   Handbook of applied optimization   Oxford Academic Press (to appear)
22. Laguna, M., Armento, V.A.: Lessons from applying and experimenting with scatter search. Tech. Report (March 2001)
23. Laguna, M., Mart´ , R.: Experimental testing of advanced scatter search designs for global optimization of multimodal functions. Tech. Report (August 2000)
24. Leighton, F.T.: A graph coloring algorithm for large scheduling problems.   J. Res. Nat. Bur. Stand.   84 (1979) 489 506
25. Lund, C., Yannakakis, M.: On the hardness of approximating minimization problems. Proc. 25th Annual ACM Symp. Theory of Comput.   (1993) 286–293
26. Morgenstern, C.A.: Distributed coloration neighborhood search. In [19] (1996) 335–357
27. Rego, C.: Integrating advanced principles of tabu search for the vehicle routing problem. Working paper, Faculty of Sciences, University of Lisbon (1999)

The Two Stage Continuous Parallel Flow Shop Problem with Limited Storage: Modeling and Algorithms

Thomas Bousonville

EuroBios, 191 Av Aristide Briand, F-94234 Cachan CEDEX
thomas.bousonville@eurobios.com

Abstract. Two stage continuous parallel ow shops with limited intermediate storage are common in process industry. Because of its computational complexity and continuous nature only special cases of the general problem have been solved to optimality so far. The focal point of this paper is the examination of appropriate indirect discrete representations that allow the application of evolutionary methods combined with local search. The results give insight into the most appropriate neighborhood structure and the usefulness of heuristic information for the guidance of the search process. In particular it is shown that the conceived Memetic Algorithm when submitted to a rigid time limit yields better results by using additional heuristic information.

1 Introduction

In this paper production systems as they are typical for the process industry will be studied. They can be found in chemical industry, but play also a major role in the production of consumer articles, such as the production of toothpaste or baby nutrition. Such systems consist principally of two stages: the mixing of the basic ingredients (making of a variant) and the filling or wrapping of the variants into dierent packages. The fial output (i.e. a variant associated with a specific package size) is called a SKU (stock keeping unit) and will in the sequel also be referred to as a product . A job then consists of the two tasks making and packing. Between the two stages there are buers in the form of tanks with limited capacities for work in progress. At each stage multiple machines are operating in parallel. There are usually also multiple storage facilities. A general outline of such a system is given in Figure 1.

Apart from mingling variants in the capacity bounded tanks, a second feature that separates the studied problem class from standard job shop or ow shop problems is their continuity. They are continuous in the sense, that a variant has not to be processed as a batch and will then be handed over to the next stage. Instead, any produced amount during the make stage is immediately (ideally we neglect ow times) available for storing and in sequence for packing. Of course there are also production systems of the sketched structure that operate in batch mode. For research on batch mode systems the reader is referred to e.g. [14,15].

P. Collet et al. (Eds.): EA 2001, LNCS 2310, pp. 180–191, 2002.

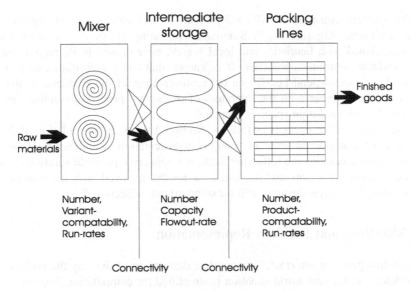

Fig. 1.   Outline of the production system

A typical optimization task for the studied application consists in constructing a minimal makespan schedule for a given demand of products. In practice the time horizon for such a schedule is about one week. Starting from the demanded amount of each single product the corresponding total volume of each base variant can be calculated.

In order to create a feasible schedule one has to determine processing times for each variant and product at the two stages. The continuous nature of the production system allows a   preemptive scheduling of the tasks. That means stopping a task at any point in time is allowed. A single, not interrupted production activity on one machine will in the following be referred to as an   operation .

Additionally to tank capacities the schedule is subject to a variety of other constraints: a machine can be non-available because of cleaning, changes of equipment, periodic maintenance. Not all machines of a stage might be able to process the complete range of products or variants. Especially the packing machines are often only able to process a special kind of package. Processing speed of a task may depend on the machine.

Because of the large number of constraints, only very special cases can be treated in a rigid linear programming approach. Even then, only small problems can be tackled successfully. Consider for instance the case where a storage tank can be connected to only one manufacturing and one packing unit at a time, each product has a dedicated machine and the size of an order does never exceed the storage capacity of a tank. This problem was studied by Jain and Grossmann [9]. To solve problems of this very special class, the authors are using a commercial MILP-solver on the basis of a MILP formulation, and conclude that only problems up to 15 jobs can be solved to optimality within a reasonable time.

For strongly constrained, NP-hard problems, stochastic search algorithms, such as Genetic Algorithms [7], Simulated Annealing [1] or Ant Systems [6], when combined with heuristics and local search, have proved to be among the best available solution approaches. It is known that these algorithms, known as metaheuristics, depend heavily on the representation of the problem, as the representation together with the operators of the search procedures define the way in which the solution space is searched.

The outline of the paper is as follows. Section 2 treats the formulation of the model and the important issue of problem representation is addressed. Based on these representations the following Section 3 conceives possible solution techniques. Section 4 presents and discusses the results achieved with the dierent approaches. The major findings will be summarized in Section 5.

2   Modeling and Problem Representation

This section gives an informal, but complete description of the way the problem is modeled, i.e. the real world situation is simplified for computational optimization. Then the possibilities of representation and search space configuration is discussed.

2.1   Problem Modeling

The production system introduced in the previous section consists of K_1 machines at the make stage and K_2 machines at the pack stage. There are dierent jobs i $1, .., J$ with respective demands d_i and base variants v_i. Each job i consists of a make task i^m for the first stage and a pack task i^p for the second stage. The processing speed of any job i on a machine k is denoted by r_{ki}. The changeover time needed for cleaning etc. between a job i and a job j on machine k is s_{kij}. A make task i^m can be split up into an arbitrary number of operations $o_{i^m 1}, 1$ $1, .., O^{i^m}$. The same holds for a pack task i^p (i.e. $o_{i^p 1}, 1$ $1, .., O^{i^p}$). Each operation $o_{i^m 1}$ (resp. $o_{i^p 1}$) is defined by start time $o^s_{i^m 1}(o^s_{i^p 1})$ and a duration $o^d_{i^m 1}(o^d_{i^p 1})$. A schedule can be described as a set of operations plus the machines they are scheduled on. For the storage of the intermediate variants there are T tanks. Each tank t has a limited capacity t^c.

The makespan is defined as the latest termination time over all operations of the pack stage:

$$\max_{i\ 1,..,J} \max_{1\ 1,..,O^{i^p}} o^s_{i^p 1} + o^d_{i^p 1}$$

The objective is now to find a set of operations that constitute a feasible schedule and minimize the makespan. A schedule is feasible, i the following holds:

1. no machine processes more than one operation at the same time
2. the sum of all operations of a task must equal the volume of that task
3. for two subsequent operations the changeover times are respected

4. at no time the capacity t^c of a tank t may be exceeded
5. tasks can only be executed on suitable machines
6. a pack operation can only start if there is an amount   0 of the required variant in the tank.

Note that this notation implicitly allows the parallel processing of one task on dierent machines.

2.2   Representation

In order to use metaheuristics and local search for scheduling, a representation of the schedule has to be found on which the neighborhood and other operators can be defined.

It is known that the choice of the representation is crucial to the performance of the metaheuristic. In strongly constraint problems the task of finding an intuitive and ecient representation is all but easy. Here, if we could fid a representation, guaranteeing that jobs are not processed in parallel on the same machine and that the sum of all job operations equals its total volume, it would still generally result in an infeasible schedule, because of the capacity restrictions on tanks. A schedule repair algorithm would be needed and turn down the performance of the search.

The second and more fundamental problem applying metaheuristics to the studied case is the possibly infinite number of solutions due to the possibility of dividing a task into an arbitrary number of operations. Generally, metaheuristics can be applied to combinatorial optimization problems with a finite number of possible solutions.

Thus instead of a direct representation of the schedule as a set of operations one could use a representation as a sequence of jobs in combination with a scheduler that maps this sequence onto a feasible schedule as it is illustrated in Figure 2. Such a representation allows the application of standard operators as they can be found in e.g. [2].

As the scheduler (Figure 3) is responsible for producing a feasible schedule it represents a crucial part of the algorithm. The constraints regarding machine compatibility, linkage between the resources and maintenance are taken into account when the earliest possible starting time for a job is calculated using EST (j) . Capacity constraints that may lead to a splitting of a job into more

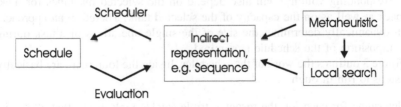

Fig. 2.   Program components

```
J               sequence of jobs
v[j]            volume of job  j
pos[j]          position of job  j  in J
EST (j)         returns earliest starting time of job      j
SP (j)          schedules the max amount of    j without interruption,
                returns the scheduled volume

procedure  Scheduler
begin
    while not empty   J
        j  :=  j  EST (j)   EST (i), pos [j]  pos  [i], i, j     J ;
            v[j  ] := v[j  ]   SP (j )
            if (v[j  ] = 0)
                J := J    j
    end;
end;
```

Fig. 3.   Creating a schedule from a job sequence

than one operation are then calculated by the function   SP (j) that determines for a given resource combination of a make and a pack machine as well as a tank and a job j its maximal schedulable amount without interruption.

2.3   Sequence of Jobs and Scheduler

By choosing a representation of the solution as a sequence of jobs, we leave the decision of when to split a task into operations up to the scheduler. Note that the splitting may not only be necessary to construct an   optimal or near optimal schedule, but to receive a   feasible schedule at all. Consider for example a job whose processing speed for the make task is by   a units per time higher than that of the pack task. Given a tank capacity of t^c the maximal concurrent production time would be t^c/a time units. Depending on the total volume of the job, this time may be too short to produce the job without interrupting the make process.

Consequently depending on the remaining space for intermediate storage a maximal time of concurrent make and pack can be calculated. This time (and the corresponding volume) will also depend on the selected machines for make and pack as well as on the capacity of the selected tank. Therefore an approach which dynamically determines the size of the single operations of a task during the composition of the schedule is required.

Figure 3 outlines the work of the scheduler. Inside the loop there are basically four steps to run through

1. determine for each job the resource triple (make, tank, pack) that allows it (or a part of it) to start earliest

2. select a job to be scheduled
3. determine the maximal volume of the job that can be scheduled without interrupting any of the tasks
4. schedule this volume of the job

During step 1 (determining the resource triple with the earliest possible starting time for each job) the machine restrictions and setup times are taken into account. It is important to note that the selection of a job is based on a mixture of a heuristic value (earliest availability) and the input sequence (in case of equal availability times). The final output of the scheduler depends on the initial order of jobs in the job sequence, but the heuristic directs the search process.

Let's also note that this way a non-delay schedule is constructed. A non-delay schedule is a schedule in which no machine is kept idle when there is an operation available for processing. For preemptive scheduling the optimal schedule is always a non-delay schedule [13]. Although the above scheduling heuristic is constructing a non delay schedule it does not guarantee that the optimal solution lies within the restricted solution space obtained by the indirect representation. For a simple counterexample imagine a problem with two jobs, two machines at each stage and two tanks. Let both jobs have equal processing speed on all machines. Then the jobs could be infinitely scheduled in parallel. If the total job processing time of job 1 is longer than that of job 2 less the setup time between the two jobs, then the optimal solution would divide job 1 for processing on dierent machines. The scheduler described above will not create this solution.

2.4   Sequence of Operations

As has been pointed out, the number of operations into which a task will be divided by the scheduler is generally not known in advance. But if

 - the processing speeds of a job are identical for all machines of the same stage and
 - all tanks have the same maximal capacity

the number of operations for a given problem instance produced with the above scheduler from Figure 3 will be invariant to the order in which the jobs appear in the sequence.

In that special case, the problem can alternatively be represented as a sequence of all job operations. Instead of choosing the next operation according to the heuristic value (earliest available) job, the selection now depends uniquely on the position in operation sequence. Steps 3 and 4 of the scheduler remain unchanged. This allows the exploration of a wider solution space, but the search may be less eective.

Just like the job sequence, this representation allows the application of standard operators, which will be discussed in the next section.

3   Solution Methods

In this section, possible search methods working on the two representations developed in Section 2 are presented.

3.1   Local Search

Local search methods for combinatorial problems are based on the definition of a neighborhood for a given solution. Desirable properties of a neighborhood are their coverage of the solution space, i.e. every possible solution can be reached from any other solution by a finite number of moves. Very important from a practical point of view is also the possibility of fast evaluation of a neighborhood. Often even large neighborhoods, quadratic or cubic in problem size, can be searched eectively by using local evaluation. Local evaluation is possible when the dierence in terms of solution quality only depends on the attributes of the solution changed by the local move.

Unfortunately this is not the case for the studied problem. Changing the order of jobs at any point in the sequence will also alter the schedule of any job that comes after that position in the job sequence (even if its position in the sequence remains unchanged). After every local move, the new neighbor has to be evaluated by rebuilding a feasible schedule applying the procedure given in Figure 3. As a practical consequence the size of a neighborhood is limited to be at most quadratic.

This criterion is fulfilled by the   transpose neighborhood, i.e. swap two adjacent jobs ((n   1) possibilities), and the   insert neighborhood, i.e. remove a job from one position and insert it at another position ((n   1)^2 possibilities), but not by the   block insert neighborhood, i.e. move a subsequence of jobs from one position to another (n(n + 1)(n   1)/ 6 possibilities).

Applying local search on a defined neighborhood means moving to a neighbor that improves the solution quality until all neighbors are inferior or equal. According to the way a neighbor is chosen the strategy may be called   first improvement (select the first improving neighbor) or   best improvement (select the neighbor with the highest improvement). Both strategies may lead to dierent local optima, and the one found by best improvement at the expense of higher computational eort is not necessarily better. As the evaluation of a solution is quite expensive the first improvement strategy is used to obtain the results presented in Section 4.

3.2   Genetic Algorithm

From the range of available metaheuristics we chose Genetic Algorithms (GA) for the series of experiments reported in this paper. Initial experiments with Simulated Annealing did not produce satisfactory results, which is probably due to the fact that high eort in solution evaluation makes it dicult to reach a quasi equilibrium at each stage of the cooling process within a reasonable time.

Instead of using a pure GA, the extended framework of a so called memetic algorithm [12] is applied. Memetic algorithms combine GA with local search. This approach has proven to be superior to the basic GA conception as it was initially proposed by Holland [8].

In the context of memetic algorithms the crossover and mutation operators are both considered as diversification strategies to escape from local optima found during the local optimization step [11]. Crossover, as a directed jump in the solution space tries to preserve valuable properties from both parents. As a scheduling problem is basically an ordering problem, the order crossover OX [5], which has been applied successfully on many problems of this nature, is chosen.

During the convergence of the search jumps initiated by the crossover operators will dynamically decrease as parents become more similar. This is not the case for mutation. In order to verify an impact of the jump distance, two dierent mutation operators were implemented: the exchange of two and four arbitrary chosen positions respectively. The operators are referred to as swap and 4exchange in the sequel.

Parents for reproduction are chosen using the stochastic universal sampling method [3] on the basis of the relative fitness of the individual in the population. As the local optimization evens out dramatic dierences in schedule quality a rank based ordering seems not necessary. All osprings are maintained for the next generation and the remaining places in the fixed size population are distributed among the best individuals of the parent population.

Charalambous and Hindi [4] also apply a Genetic Algorithm to the described problem, but their approach diers in two major points. First, the number and size of the operations is determined by a heuristic in advance, and second they do not include local search into their GA. Unfortunately it was not possible to exchange the problem instances and hence do benchmarks for reasons of confidentiality.

4   Computational Results

The results we report in this section are based on an industrial problem instance with the following characteristics: demand for 57 products composed of 20 variants, 3 machines at the make and 7 at the pack stage, as well as 5 tanks. The processing on the pack stage is heavily constrained enabling only one or two pack machines to process a given job. The make machines work with dierent speeds according to the variant, but the processing speed of one variant is equal on all machines. The same is true for the pack stage. As the tanks all have the same maximum capacity the conditions for using both representations (job and operation sequence) for the problem are fulfled. For dierent products and variants there are setup times on the respective stage as well as for the tanks.

Besides the construction and verification of an overall solution method, one objective of the presented work is the assessment of the contributions that are made by the dierent parts of the hybrid solution technique. We therefore compare:

Table 1.   Comparison of computational results

		MA		RLS	
		insert	transpose	insert	transpose
	best	382559.0	393131.0	390014.0	396231.0
JB	mean	387718.4	399788.8	393231.0	399367.9
	stddev	2437.8	3730.4	1880.9	1872.1
	best	392447.0	422527.0	395550.0	459171.0
OP	mean	398004.3	442405.7	400389.3	468302.8
	stddev	2852.5	12596.4	2571.0	5417.9

Table 2.   Ranking of the solution approaches

Strategy	Excess over best solution
JB+MA evolutionary search + heuristic + local opt	0
JB+RLS heuristic + local opt	1.9
OP+MA evolutionary search + local opt	2.6
OP+RLS repeated local opt	3.4

- job (JB) vs. operation (OP) representation (highlighting the inuence of a heuristically directed search),
- the memetic algorithm (MA) vs. repeated local search (RLS) and
- insert vs. transpose neighborhood.

In order to have a fair comparison each configuration was given a time limit of 1000 seconds to find a solution. This value was derived from a typical practical situation. Taking into account the stochastic nature of the techniques each configuration was run ten times.

For the available time budget a GA configuration with population size 10, crossoverrate 0.4 and mutationrate 0.3 was adjusted empirically. Regarding the two tested mutation operators, swap and 4exchange, no signifiant dierence in solution quality was registered.

Table 1 shows the best and average results as well as their standard deviation for the eight compared configurations. It is clearly visible that the insert neighborhood despite the much higher eort for local optimization is in all cases superior in comparison to the transpose neighborhood.

The meaning of the other results might be easier to interpret, if we associate the abbreviations with the underlying concept. This is done in Table 2 where the best results of each strategy are ordered according to their excess over the best solution found over all runs. Additionally the development of the best solutions for each strategy is shown in Figure 4.

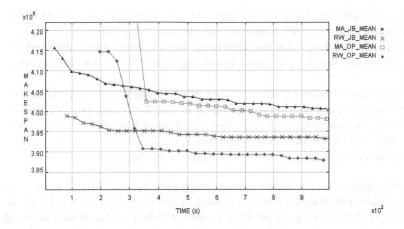

Fig. 4.   Development of solution quality in time

Clearly the combination of heuristic information with the genetic search engine and the local optimization achieves the best and most robust results. Even the average value of this combination is better that the best result of any run of the other procedures. Answering the question which component apart from local search is most important, we note that the heuristic together with the local search gives better results than the GA plus the local optimization without exploiting any heuristic information. What reasons are there for this at first sight quite deceptive behavior of the GA? Certainly the vast and probably rugged search space in conjunction with the rigid time limit and the long solution evaluation times, allowing only a rather small population size and few generations (about seven), are not ideal conditions for an evolutionary search method. The evolutionary process has not enough time and material to exploit what Goldberg [7] called its implicit parallelism. Figure 4 shows that the runs with the Genetic Algorithm framework yield still improvements during the final phase of the search. Thus further improvements might be expected for an extended time limit or a more ecient representation.

5   Conclusion

The paper presents a real world scheduling problem as it can be encountered in the process industry. The problem is computationally complex and the numerous constraints request a exible modelling and solution approach.

The problem does not imply a natural representation that favors the application of combinatorial search procedures. Therefore an indirect representation that is transformed into a feasible schedule by a scheduler is proposed. Based on the conceived representations a heuristic and a Genetic Algorithm in conjunction with local search are compared. The performance results of the compared local search alternatives are in favor of the insert neighborhood. The results also show that given a rigid time limit and the high solution evaluation costs the Memetic Algorithm without additional heuristic information has diculties to converge to a high quality solution. On the other hand, when combined with the heuristic the evolutionary search is the most promising strategy even for short running times.

Further improvements concern a more ecient implementation of the local search structure using data structures similar to the 'don't look bit' strategy for the traveling salesman problem [10]. This would allow bigger population sizes for the embracing GA. On the other hand a metaheuristic strategy like ant systems that explicitly use heuristic information might be promising, too.

Acknowledgements.   This work was supported by the "Metaheuristics Network", a Research Training Network funded by the Improving Human Potential programme of the CEC, grant HPRN-CT-1999-00106. The information provided is the sole responsibility of the authors and does not reect the Community's opinion. The Community is not responsible for any use that might be made of data appearing in this publication.

References

1. Aarts, E. and Korst, J.:   Simulated Annealing and Boltzmann Machines.   Chichester et al.: John Wiley Sons (1989)
2. Anderson, E.J., Glass, C.A. and Potts, C.N.: Machine Scheduling. In Aarts, E., Lentra, K. (Eds.),   Local Search in Combinatorial Optimization   . John Wiley Sons. 361-414 (1997)
3. Baker, J.E: Reducing Bias and Ineciency in the Selection Algorithm. In Greffenstette (Ed.),   Procee dings of the Second International Conference on Genetic Algorithms , (pp. 14-21), Hillsdale, NJ: Lawrence Erlbaum Associates. (1987)
4. Charalambous, C. and Hindi, K.S.: Applying GAs to Complex Problems: The Case of Scheduling Multi-Stage Intermittent Manufacturing Systems.   Procee dings of the Second Int. GALESIA Conference   (1997)
5. Davis L.: Applying Adaptive Algorithms to Epistatic Domains. In Joshi Aravind (Ed.),   Procee dings of the International Conference on Artificial Intelligence   , (pp. 162-164). Los Altos, California:Morgan Kaufmann Publishers (1985)
6. Dorigo, M., Di Caro, G. and Gambardella, L.: Ant Algorithms for Discrete Optimization.   Artificial Life , 5, 137-172 (1999)
7. Goldberg D. E.:   Genetic algorithms in search, optimization and machine learning   . Reading, Mass.: Addison-Wesley (1989)
8. Holland, J.H.:   Adaptation in Natural and Artificial Systems   . Ann Arbor: University of Michigan Press (1975)
9. Jain, V. and Grossmann, I.E.: A Disjunctive Model for Scheduling in a Manufacturing and Packing Facility with Intermediate Storage.   Optimization and Engineering , 1, 215-231 (2000)
10. Johnson, D.S. and McGeoch, L.A.: The traveling salesman problem: a case study in local optimization. In E. H. L. Aarts and J. K. Lenstra (Eds.),   Local Search in Optimization , (pp. 215-310). New York:John Wiley and Sons (1997)
11. Merz, P. and Freisleben, B.: Memetic Algorithms. In D. Corne, M. Dorigo and F. Glover (Eds.),   New Ideas in Optimization , (pp. 217-294). London: McGraw-Hill (1999)

12. Moscato, P.: On Evolution, Search, Optimization, Genetic Algorithms and Martial Arts: Towards Memetic Algorithms , C3P Report 826, Pasadena, CA: California Institute of Technology (1989)
13. Pinedo, M.: Scheduling: Theory, Algorithms, and Systems , Englewood Clis, NJ: Prentice Hall, 1995
14. Pinto, J.M. and Grossmann, I.E.: A continuous time mixed integer linear programming model for short term scheduling of multistage batch plants.   Ind. Eng. Chem. Res., 34 (9), 3037-3051 (1995)
15. Pinto, J.M. and Grossmann, I.E.: Assignment and sequencing models for the scheduling of process systems.   Annals of Operations Research , 81, 433-466 (1998)

SAT, Local Search Dynamics and Density of States

Mériéma Bélaidouni and Jin-Kao Hao

LERIA, Universit´ e d'Angers
2 bd Lavoisier, 49045 Angers Cedex 01, France

Abstract. This paper presents an analysis of the search space of the well known NP-complete SAT problem. The analysis is based on a measure called "density of states" (d.o.s). We show experimentally that the distribution of assignments can be approximated by a normal law. This distribution allows us to get some insights about the behavior of local search algorithms.

1 Introduction

In the last decade, many studies intended to understand the dynamics of local search methods in order to contribute to the development of new eective methods for hard problems like the problem of satisfiability (SAT) [7]. Several authors look at the problem structure for an explanation of the behavior of local search methods. Thus studies on the number of solutions [5], the backbone fragility [20], the number and the arrangement of local optima [9,23], provide a more complete picture of the structure of SAT instances.

However, these studies concern only the computing time for a (local) search algorithm to find an optimal solution, which is only one aspect of local search dynamics. Now, there are many other interesting aspects concerning the dynamics of local search. One of them is the question of the quality of solutions found by a local search algorithm. This important feature was the main object of the autocorrelation measure [11,22]. Many authors verifid the eect of autocorrelation on heuristics performance. Kaufmann [11] and Weinberger [22] show that a downhill algorithm produces solutions of a better quality on NK landscapes if the autocorrelation is increased. Similar conclusions was obtained by Manderick [13] with genetic algorithms applied to NK and TSP landscapes, also by Angel and Zissimopoulos [2] with Simulated Annealing applied to the Graph Bipartioning Problem (GBP).

The work presented in this paper concerns also the quality of solutions and its relation with properties of problems. We are interested in the study of the long tail phenomenon observed for several heuristic search algorithms [18]. More particularly, we will investigate an interesting and intriguing eect about this phenomenon. Indeed, our experiments show that the costs generated by a local search algorithm for a given problem instance stagnate invariably within a particular cost interval, independently of the starting point. This behavior cannot

P. Collet et al. (Eds.): EA 2001, LNCS 2310, pp. 192–204, 2002.

be explained using autocorrelation but it can be explained using a new measure of problem structure called density of states (d.o.s).

The density of states counts the number of configurations for each cost value of an optimization problem instance. The definition can be extended, to decision problems. Applied to SAT problem for example, d.o.s gives the number of truth assignments for each number of unsatisfied clauses. This measure contains not only the number of solutions [5], but also the number of assignments that satisfy all clauses except one, then the number of assignments that satisfy all clauses except two and so on,...until the number of assignments that satisfy no clauses. This measure is studied in biology [8,16], physics [12] and optimization [4,16]. Its approximation can be carried out analytically on some problems [11], by enumeration on small instances [6,12] or by approximation on large instances [4, 12,16].

This paper undertakes the study of the 3-SAT problem using d.o.s. It shows through benchmark instances that d.o.s follows, in good approximation, a normal law. This information (given by d.o.s) turns out to be very informative to explain and predict the dynamics of local search methods. In particular, it allows us to give some explanations to the intriguing behavior with the long tail, that we observe in this paper on a Metropolis algorithm.

The article is organized as follows: Section 2 presents the experiment concerning the long tail with a Metropolis algorithm. Section 3 defines the density of states (d.o.s). Section 4 measures d.o.s for various SAT instances and shows its relation with Metropolis dynamics. Section 5 concludes and gives some further directions.

2   Metropolis and the Long Tail

This section presents an intriguing experiment with Metropolis. The aim of this experiment is to answer the following general questions: Where does a local search begin? where does it end? and does it follow a particular direction (do costs go up or down)? First, let us present the Metropolis algorithm used in the experiment.

2.1   Metropolis

Metropolis [15] method is at the heart of Simulated Annealing [1] which is a method widely used in combinatorial optimisation. For a minimization problem, Metropolis at temperature T (noted MTR(T)) is defined as follows: the process starts with a random configuration s of the search space S; then the process moves from the configuration s to a neighboring configuration s , with a probability p defined by:

$$p = \begin{cases} 1 & \text{if } \quad f \quad 0 \\ e^{\quad f/T} & \text{if } \quad f \quad 0 \end{cases} \qquad (1)$$

where   f = f(s)   f(s), and T is a given temperature.

2.2   Experiments with the Long Tail

The following experiments shed light on a behavior never presented before about dynamics of local search. Let   F   be a SAT formula,   S   the set of truth assignments, f   the cost function that associates to each truth assignment   s   S   the number of unsatisfied clauses in the formula   F   and v   the neighborhood relation used in this paper. Two configurations are neighbors according to   v   if they dier by the value of one variable of an unsatisfied clause. The experiments consist in running once   Metropolis during   N   moves   from two very dierent initial assignments:

– MTR(T)-I: from a random assignment,
– MTR(T)-II: from a satisfying (optimal) assignment.

Let us consider the instance uf250-01 from SATLIB library [10]. This instance is satisfiable. A solution can be obtained for a reasonable amount of time using Metropolis at dierent temperatures. Temperature   T = 25 is one of these successful temperatures.

Our first experiment consists in running Metropolis at temperature   T = 30 with a random initial assignment during   N = 20 .000 moves. Figure 1 (left) gives the evolution, through time, of the costs generated by MTR(30)-I. We notice that MTR(30)-I starts its evolution with a cost value around   f   120. After a while, all generated cost values oscillate in the cost interval [2   , 12]. This last phase generates the so-called (long) tail.

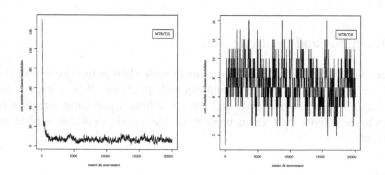

Fig. 1.   Evolution of cost values through time for MTR(30)-I (left, from a random assignment) and MTR(30)-II (right, from a satisfying assignment).

Our second experiment consists in running Metropolis at temperature   T = 30 during N = 20 .000, but this time, we start the search with a satisfying assignment (for example the solution found with MTR(25)). Metropolis is designed such that it does not stop when it encounters a satisfying solution. It stops only when the running time (in terms of number of moves) is over. Figure 1 (right) gives the evolution, through time, of costs generated by MTR(30)-II. We observe that MTR(30)-II which is initialized at an optimal cost   f = 0 looses this attribute through time and degrades continually its good starting assignment to finally oscillate in the same cost interval as the first experiment: [2 , 12].

These experiments are intriguing because they show that the costs generated by Metropolis are strongly attracted by the assignments of some particular cost interval (the cost interval [2 , 12] for the instance used in the above experiment) independently of its starting assignment.

Now, let us try to understand why MTR(T) stagnates around a particular cost interval independently of the starting assignment. We believe that this particular behavior find its source (origin) in the density of states of the instance. Therefore, we propose in this paper to examine the relation between the d.o.s and the above long tail phenomenon.

2.3 Process Cost Density

To establish the link between the d.o.s and Metropolis behavior, we need an intermediate notion which is the process cost density (p.c.d). The notion of p.c.d is not new, it is known as the equilibrium density for Metropolis [1]. The process cost density is simply the distribution of costs generated by a search strategy 'after' an infinite running time. Of course, this density is not tractable but it can be approximated in some cases [17]. In what follows, we present an algorithm to approximate the p.c.d of a search strategy   (T) for SAT:

```
process cost density approximation

Data  : F :SAT formula, p: noise, N: Maxip,
  (T = t): search strategy with vector parameters T at t
begin
     run  once    with vector of parameters T at t on        F  during N ips ;
     collect the set of generated costs ;
     count the number      N (f ) of assignments having the cost value      f ;
     approximate the process cost density by the frequency distribution        N(f)/N
end ;
Result  : p.c.d
```

Applying this algorithm to the costs generated in the above experiments, we obtain the p.c.d of MTR(30)-I and the p.c.d MTR(30)-II.

Figure 2 (left) displays the p.c.d of MTR(30)-I, whereas Figure 2 (right) displays the p.c.d of MTR(30)-II. The curves obtained have a bell shape. They are centered at f   7 and cover the cost interval [2,12]. Our tests show that both curves belong to the same bell shaped distribution.

Consequently, it appears that MTR(30)-I and MTR(30)-II leave their initial cost (high and low) to coincide with a bell shaped distribution. This behavior has been observed for MTR(T) at all temperatures.

2.4 Some Facts about p.c.d

In the last section, we have established the link between Metropolis behavior and p.c.d. Now, let us make several remarks concerning p.c.d. First, as p.c.d concerns the probability of occurrence of each cost, it does not depend on their

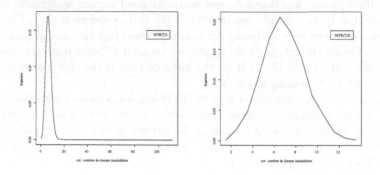

Fig. 2.   Process cost density for MTR(30)-I left and MTR(30)-II right.

order of occurrence in the run. Therefore, p.c.d is not dependent on a particular run but is common to all runs. Our experiments on SAT instances confirm this remark, in very good approximations. Second, the approximation of p.c.d needs a suciently long run. If running time is too short the p.c.d, approximation is not suciently accurate. The p.c.d approximation is improved by increasing the sample size N.

Up to now, we have linked the long tail behavior of Metropolis with the process cost density and analyzed some aspects of this density. The next step is to understand the relation between p.c.d and d.o.s. Next sections are dedicated to this issue.

3   Approximating Density of States

Process cost density of a search strategy depends on three factors 1) the strategy and its parameters, 2) the given problem instance and 3) the neighborhood relation used by the local search strategy.

In this section, we are concerned by the approximation of d.o.s for dierent classes of SAT instances. The aim is to 1) answer the question 'Where does the local search begin?' and 2) show how p.c.d is correlated to d.o.s for the Satisfiability problem?

Let F be a Satisfiability formula,   S the set of assignments and   f the cost function which associates to each formula   F and truth assignment   s S   the number of unsatisfied clauses. Assuming the independence of clauses satisfaction. The random variable   nuc "number of unsatisfied clauses" follows a binomial law

$$nuc \quad B(nc, p)$$
$$P[nuc = k] = C_{nc}^{k} \cdot q^{k} \cdot (1 \quad q)^{nc \quad k} \tag{2}$$

where nc is the number of clauses,   q the probability for the unsatisfiability of one clause. Thus,

$$E(nuc) = nc.q \qquad (3)$$

and,

$$(nuc) = \overline{nc.q.(1 \quad q)} \qquad (4)$$

where E(.) and   (.) are the mean and standard deviation. For k-SAT instances, we have $q = \frac{1}{2^k}$. For instances with variable lengths of clauses,   q is averaged over all clause lengths $k_1,...,k_l$.

$$q = \frac{\sum_{i=1}^{i=l} n_i.q_i}{\sum_{i=1}^{i=l} n_i} \qquad (5)$$

and

$$q_i = \frac{1}{2^{k_i}} \qquad (6)$$

where n_i is the number of clauses of length k_i.

The binomial deduction is inspired by a study on the random MAX-CSP [21]. In the case of MAX-CSP, the study showed a perfect concordance between theory and practice [3] .

In general, the assumption of satisfaction independence for the clauses is not reliable. This question is discussed in details in the next section.

4   Experiments

In this section, we fst aim to approximate d.o.s on dierent kinds of SAT instances, to determine the law of distribution. Secondly, we show how d.o.s. inuences 1) the cost of the initial truth assignment used by the local search method and 2) its process cost density.

4.1   Approximating Density of States

Experiments are realized on one instance of each family of SATLIB library [10] (see Table 1 and Table 2). Density of states is approximated by two dierent methods:

1. Theoretical (analytical) method: under the assumption of clause satisfaction independence, we apply the formulas (1) to (6).
2. Experimental method: we use random selection which consists of taking assignments from the assignments set   S in a random and independent manner. The sample size is between 1 .000 and 50 .000 depending on the studied formula.

Table 1.   D.o.s of random SAT instances according to analytical approximation and random selection.

Instances		n	nc	$B(nc, p)$		RandomSelection	
				μ	σ	μ95%Conf.	σ95%Conf.
(a)	uf100-01	100	430	53.75	6.85	[53.66, 53.75]	[6.77, 6.83]
(b)	uf 200-01	200	860	107.5	9.70	[107.5, 107.61]	[9.72, 9.80]
(c)	uf 250-01	250	1065	133.12	10.80	[133.06, 133.20]	[11.16, 11.26]
(d)	f600	600	2550	318.75	16.70	[318.62, 318.83]	[16.75, 16.90]
(e)	f1000	1000	4250	531.25	21.56	[531.18, 531.44]	[21.48, 21.67]
(f)	RTI_k3_n100_m429_1	100	429	53.62	6.85	[53.60, 53.67]	[7.01, 7.07]
(g)	BMS_k3_n100_m429_1	100	272	34	5.45	[33.97, 34.03]	[4.56, 4.60]
(h)	CBS_k3_n100_m429_b50_1	100	429	53.62	6.85	[53.59, 53.68]	[6.78, 6.84]
(i)	jnh201	100	800	45.63	6.55	[45.62, 45.88]	[6.43, 6.61]
(j)	aim-200-3_-yes1-1	200	680	85.25024	8.63	[85.23, 85.31]	[6.07, 6.13]

Random SAT.   All instances used in Table 1 are random and satisfiable formulas generated by models of Satisfiability problem.   n is the number of variables and nc the number of clauses.

Instances from (a) to (h) are 3-SAT formulas. So, d.o.s estimation can be carried out by formulas (2) and (3) with   q = $\frac{1}{8}$. Instances (i) and (j) are variable length formula. So, we apply formulas (2) and (3) with q calculated by formulas (4) and (5). We have obtained for example for uf100-01 a mean of   = 53 .75 and a standard deviation of   = 6 .85.

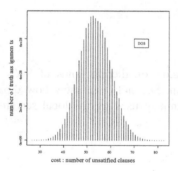 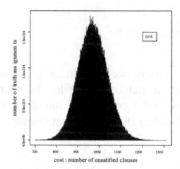

Fig. 3.   (left) d.o.s as approximated by random selection for uf100-01. (right) d.o.s as approximated by random selection for   hanoi 4.

Now, it remains to confirm the binomial distribution and its parameters by a dierent method. Concerning the shape, the experimental method described above leads for uf100-01 to a normal distribution as shown in Figure 3 (left). For all other instances (with   nc   1000) the agreement with normality has been tested successfully (using a standard normality test). This normality result agrees perfectly with analytical formula since it is possible to approximate a binomial by a normal under certain conditions (conditions satisfied in this case).

Concerning the distribution parameters, the experimental results for all families of random SAT instances of SATLIB are presented in Table 1. We observe that the mean and standard deviation approximated by the experimental method agree with the mean and standard deviation approximated by the analytical method. Thus, according to the experimental method, uf100-01 founds a mean in the confidence interval [53.66,53.75] and a standard deviation in the confidence interval [6.77, 6.83], and the analytical method founds the estimated mean at 53.75 and the estimated standard deviation at 6.85. We remark that the mean belongs to the confidence interval, whereas the standard deviation is sometimes outside the confidence interval without being very far. The dierence in the estimation of the standard deviation varies with the instance.

Extension to other instances.   The following experiments concern satisfiable formula of SATLIB that are issued from other optimization problems (instances (k) to (q) in Table 2). Again, we apply the analytical formula (even if the independence condition is likely unsatisfied) to calculate the mean and standard deviation. Table 2 gives the mean and standard deviation estimated by the analytical method.

Moreover, we approximate mean and standard deviation by random selection. Concerning the shape, experiments show that for hanoi4 the normality remains true as shown in Figure 3 (right). The normality agreement has been tested successfully for all other instances (using a standard normality test). Concerning the distribution parameters, we observe, surprisingly, that the estimated mean found by the analytical method agrees with the confidence interval found by the experimental method, and this for all instances. However it is not the case for the standard deviation. Indeed the analytical and experimental methods found very dierent values.

Table 2.   D.o.s of random SAT instances according to analytical approximation and random selection.

	Instances	n	nc	$B(nc,p)$		RandomSelection	
				μ	σ	μ95%Conf.	σ95%
(k)	flat150-1	450	1680	401.2495	17.47614	[400.54, 401.98]	[36.20, 37.21]
(l)	ais10	181	3151	681.7787	23.114	[681.55, 682.76]	[97.91, 98.77]
(m)	bwlarge.c	3016	50457	10988.78	92.71	[10987, 10991]	[127.61, 128.73]
(n)	logistics.c	1141	10719	2012.814	40.43	[2011, 2013]	[127.61, 128.73]
(o)	ssa7552-038	1501	3575	767.60	24.551	[767.43, 767.74]	[24.34, 24.55]
(p)	par16-1-c	317	1264	173.25	12.22	[173.20, 173.32]	[9.79, 9.88]
(q)	hanoi4	718	4934	970.42	27.92	[969.80, 970.65]	[68.00, 68.60]

To conclude this section we recall its main results: 1) d.o.s for Satisfiability can be approximated by a normal law; 2) the approximated mean of the analytical approximation and the experimental approximation coincide; 3) the approximated standard deviation of the analytical method and the experimental approximation do not coincide for SAT instances issued from other optimization problems.

However, as the experimental method (random selection) cannot cover all costs in a reasonable amount of time, the accuracy of the above results may be discussed. Improving the experimental estimation of d.o.s needs more elaborate sampling techniques based on Metropolis algorithm [16], for example. Notice that such a sampling technique is usually very time consuming.

Now, let us see the implications of normality of d.o.s for local search. One first point is that finding an assignment that satisfies the maximum number of clauses may be as dicult as fiding an assignment that satisfis the minimum number of clauses since d.o.s is symmetric. Other implications are presented in what follows.

4.2   Does d.o.s Explain the Initial Costs?

At this point, one can answer the question 'Where does local search begin?'. Indeed, as d.o.s is normal, random initial assignments have very probably a cost around the mean. For the experiment of Section 2.2 with uf250-01, the initial random cost was 120 which coincides with the mean and the standard deviation of d.o.s (= 133 .12 and   = 10 .80). This random initial assignment cost is not that bad, since it is between very good and very bad costs. Another point is that it is very unlikely to select by chance an optimal (satisfying) assignment, since the associated probability is the smallest over all costs.

4.3   Does d.o.s Explain the Process Cost Density?

After explaining the eect of d.o.s on the random initial assignment, we look at the relation between the process cost density of the local search strategy and the density of states of the given instance. We propose to link these densities by comparing their distributions and their parameters (and   in our case). In this paper, we will establish the link between p.c.d of Metropolis and d.o.s.

Metropolis.   To establish the link between p.c.d of Metropolis and d.o.s., we will show that the mean of the p.c.d. goes from the mean of d.o.s to the best costs with the temperature decrease, and standard deviation of the p.c.d starts at the standard deviation of d.o.s and diminishes with the temperature decrease. Our experiments concern Metropolis at three dierent categories of temperatures. 1) t   a very high temperature 2) t_{opt} a temperature that finds an optimal assignment in a reasonable amount of time and, 3) an intermediate temperature t_{opt} T t   .

As shown by Figure 4 the process cost density of MTR(T) is a bell shaped distribution (or slightly asymmetric when p.c.d mean approaches the near op-timal area). The mean becomes lower and lower as the temperature decreases (T=1000, T=60 then T=40). Thus, p.c.d is closer and closer to optimal cost ar-eas and so it has more and more chances to reach a satisfying solution. Similarly, the standard deviation becomes smaller and smaller as temperature decreases. This implies that neighboring assignments are more and more similar, which is

Table 3.   Approximation of p.c.d for MTR(T) at dierent temperatures

Ins--tances	MTR(T) $T > t_\infty$		MTR(T) $T = t(t_{opt} < t < t_\infty)$		MTR(T) $T \approx t_{opt}$	
	μ95%Conf.	σ95%Conf.	μ95%Conf.	σ95%Conf.	μ95%Conf.	σ95%Conf.
(a)	[49.36, 49.44]	[6.61, 6.67]	[13.22, 13.24]	[3.34, 3.36]	[3.47]	[1.55]
(b)	[98.28, 98.40]	[9.38, 9.47]	[24.66, 24.69]	[4.65, 4.67]	[3.97]	[1.70]
(c)	[121.05, 121.18]	[10.69, 10.79]	[28.25, 28.28]	[5.10, 5.12]	[4.42, 4.43]	[1.84]
(d)	[292.12, 292.33]	[16.99, 17.13]	[72.18, 72.20]	[7.90, 7.94]	[53.83, 53.98]	[11.22, 11.32]
(e)	[486.41, 486.67]	[20.74, 20.92]	[123.02, 123.08]	[10.24, 10.29]	[20.60, 20.63]	[4.55, 4.56]
(f)	[48.83, 48.92]	[6.60, 6.66]	[11.74, 11.76]	[3.22, 3.24]	[3.24]	[1.59]
(g)	[31.96, 32.02]	[4.47, 4.50]	[11.11, 11.12]	[2.81, 2.83]	[3.46, 4.47]	[1.52, 1.53]
(h)	[49.22, 49.31]	[6.62, 6.67]	[11.88, 11.90]	[3.10, 3.12]	[3.31, 3.34]	[1.93, 1.94]
(i)	[41.73, 41.81]	[6.18, 6.23]	[10.56, 10.58]	[2.94, 2.96]	[2.76, 2.78]	[1.65, 1.67]
(j)	[81.61, 81.64]	[5.96, 5.99]	[36.01, 36.04]	[4.60, 4.62]	(12.25)*	(3.96)*
(k)	[301.38, 301.74]	[28.87, 29.13]	[55.42, 55.46]	[6.14, 6.17]	[11.05, 11.06]	[2.88, 2.89]
(l)	[228.15, 228.42]	[42.46, 42.65]	[10.47, 10.49]	[2.73, 2.74]	[3.60, 3.61]	[1.08, 1.09]
(m)	[4667.90, 4673.71]	[466.63, 470.74]	[137.38, 137.46]	[12.97, 13.02]	[10.40, 10.42]	[3.77, 3.78]
(n)	[1075.96, 1076.24]	[71.44, 71.63]	[96.64, 96.69]	[8.80, 8.84]	[24.88, 21.90]	[4.46, 4.48]
(o)	[711.94, 712.03]	[23.28, 23.35]	[206.35, 206.43]	[12.97, 13.02]	[28.15, 28.17]	[5.01, 5.03]
(p)	[163.44, 163.56]	[9.71, 9.80]	[49.00, 49.04]	[6.63, 6.66]	[11.61, 11.62]	[1.89, 1.90]
(q)	[660.37, 660.94]	[45.98, 46.38]	[54.86, 54.91]	[7.82, 7.85]	[11.10, 11.11]	[2.67, 2.68]

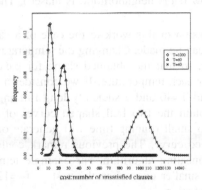

Fig. 4.   MTR(T) p.c.d for dierent (decreasing) temperatures on formula uf200-01.

not in favor of the search process. The bell shaped distribution and the trends for both mean and standard deviation are confirmed for all instances as shown by Table 3.

Now, let us examine the relation between p.c.d of MTR(t) and d.o.s. Firstly, the p.c.d of MTR(t) coincides exactly with the p.c.d of the biased random walk[1]. This is because the temperature is so high that it has no eect: the condition of selection is always satisfied.

Secondly, the p.c.d of the biased random walk is generally dierent from the p.c.d of the unbiased random walk[2]. This dierence is discussed in [3]. Also, the

---

[1] Biased random walk starts at a random initial assignment and moves from an assignment to another by ipping a variable within an unsatisfied clause.

[2] Unbiased random walk starts at a random initial assignment and moves from an assignment to another by ipping a random variable.

p.c.d of the unbiased random walk coincides exactly with d.o.s. This point is also discussed in [3].

Consequently, the p.c.d of MTR(t) is slightly dierent from d.o.s. The dierence between p.c.d of MTR(t) and d.o.s corresponds to the dierence between biased and unbiased random walk. This bias is confirmed by experiments: for example the uf-100-01 has an estimated mean in the confidence interval [53.66,53.75] and an estimated standard deviation in the confidence interval [6.77,6.83] (see Table 1) but MTR(t) has an estimated mean in the confidence interval [49.36,49.44] and an estimated standard deviation in the confidence interval [6.61,6.67] (see Table 3).

In conclusion, there exists an infinity of MRT(T) p.c.d that cover the cost area from d.o.s mean to optimal costs. All these densities are bell shaped curves (or slightly asymmetric when p.c.d mean approaches the near optimal area). The mean of MTR(T) p.c.d starts at the mean of d.o.s (or slightly below if the neighborhood is biased). Then it diminishes as temperature decreases. Similarly, the estimated standard deviation of MTR(T) p.c.d starts at the standard deviation of d.o.s (or slightly below if the neighborhood is biased). Then it diminishes as temperature decreases.

Concerning the extension of this work to the case of a varying temperature interesting observations can be made. Changing the temperature allows to jump from a curve to another, so that the obtained shape for p.c.d depends on the running time allowed for each temperature. If we consider a schedule with two temperatures T=1000 and T=40 and a suciently long running time for each temperature, we will obtain the two bell shaped curves of Fig. 4. Otherwise the temperature with too small running time will generate only a piece of the corresponding bell shaped curve. The previous remark could be used to tune objectively simulated annealing algorithm. For example temperature T=1000 should be changed to a smaller one, when the cost   f =120 is reached because the process starts its stagnation.

To extent the previous conclusions to other local search methods, we have repeated all the experiments of this study with WSAT [19]. WSAT is one of the best methods for finding solutions to satisfiable formulas. Our experiments concern WSAT(p) with noise parameter from p=1 to p=0.5 on SATLIB instances. We have observed that WSAT(p) presents the same behavior as Metropolis. Indeed WSAT(p) reach the same particular cost interval whatever is the starting solution. Even a near-optimal one. This cost interval can be related to the p.c.d of WSAT(p). We have also observed that the p.c.d of WSAT(p) is a bell-shaped curve whose mean and standard deviation become smaller and smaller as noise decreases. The mean of p.c.d is between the mean of d.o.s and the optimal cost, whereas the standard deviation of p.c.d is between the standard deviation of d.o.s and zero. Therefore, the conclusions concerning the dynamics of Metropolis remain true for WSAT. We conjecture these conclusions would be true for evolutionary algorithms applied to SAT.

5   Conclusions and Perspectives

In this paper, we have analyzed the search space of SAT problem. We have shown on random and structured SAT formulas that the density of states approaches a normal law. This distribution sheds light on some interesting questions related to local search behavior. First, we understand why local search methods like Metropolis and WSAT are attracted by some cost intervals independently of the cost of the initial assignment. In fact, these costs correspond to an equilibrium cost interval which is determined by the density of states, the neighborhood relation and the local search method itself and its parameters. Second, we learn that the random initial assignment will have a cost around the mean of the d.o.s.

In our future work, we want to reinforce the proof of the relation between p.c.d and d.o.s. We also want to see what happens to p.c.d of WSAT beyond the optimal noise value 0 .5 (i.e. between 0 .5 and 0). Indeed, in [14] the authors show that the number of resolved formulas decreases. Also, we want to understand the relations between local optima and the dynamics of local search.

Acknowledgments.   This work was partially supported by the Sino-French Joint Laboratory in Computer Science and Applied Mathematics (LIAMA) and Sino-French Advanced Research Program (PRA).

References

1. E. Aarts and J. Korst. Simulated Annealing and Boltzmann Machines.   John Wiley and Sons , 1989.
2. E. Angel and V. Zissimopoulos. Autocorrelation Coecient for Graph Bipartitioning Problem.   Theo. Computer Sience , 91: 229–243, 1998.
3. M. Belaidouni and J. K. Hao. An Analysis of the Configuration Space of the Maximal Constraint Satisfaction Problem.   Proc. of PPSN VI , LNCS 1917: 49–58, 2000.
4. J. Bresina, M. Drummond and K. Swanson. Search Space Characterization for Telescope Scheduling Application.   Proc. of AAAI , 1994.
5. D. A. Clark, J. Frank, I.P Gent, E. MacInttyre, N. Tomov and T. Walsh. Local Search and the Number of Solutions,   Proc. of CP , 119–133 1996.
6. J. Cupal, I. L. Hofacker and P. F. Stadler. Dynamic Programming Algorithm for the Density of States of RNA Secondar Structures.   Computer Sciences and Biology, 184–186, 1996.
7. S. A. Cook. The Complexity of Theorem proving procedures.   Proc. of ACM Symp. Theo. Comput. , 151–158, 1971.
8. J. Cupal, C. Flamm, A. Renner and P.F. Stadler. Density of States, Metastable States and Saddle Points Exploring the Energy Landscape of an RNA molecule. Proc. Int. Conf. Int. Sys. Mol. Biol. , 88–91, 1997.
9. J. Frank, P. Cheeseman, and J. Stutz. When Gravity Fails: Local Search Topology. (Electronic) Journal of Artificial Intelligence Research.   , 7: 249–281, 1997.
10. H. H. Hoos and T. Stutzle   Proc. of SAT, IOS Press. , 283–292 , 2000.
11. S.A. Kauman. Adaptation on Rugged Fitness Landscapes,   Lectures in the Sciences of Complexity SFI Studies in the Sciences of Complexity, ed. D. Stein, Addison-Wesley Longman , 527–618, 1989.

12. S. Kirkpatrick and G. Toulouse. Configuration Space Analysis of Travelling Salesman Problem. J. Physics , 46: 1277–1292, 1985.
13. B. Manderick, B. de Weger, and P. Spiessens. The Genetic Algorithm and the Structure of the Fitness Landscape.   Proc. of ICGA , 143–150, 1991.
14. D. McAllester, B. Selman and H. Kautz. Evidence for Invariants in Local Search. Proc. of AAAI , 321–326, 1997.
15. N. Metropolis, A.W. Rosenbluth, M.N Rosenbluth and A.H. Teller. Equation of State Calculations by Fast Computing Machines.   Journal of Chemical Physics , 21: 1087–1092, 1953.
16. H. Rose, W. Ebeling and T. Asselmeyer. Density of State   s - a Measure of the Diculty of Optimization Problems.   Proc. of PPSN IV , LNCS 1141: 208–226, 1996.
17. G.R. Schreiber and O.C. Martin. Procedure for Ranking Heuristics Applied to Graph Partitionning.   Proc. of MIC , July 1997.
18. B. Selman, H.A. Kautz. An Empirical Study of Greedy Local Search for Satisfiability Testing. Proc. of AAAI , 1993.
19. B. Selman, H.A. Kautz and B. Cohen. Noise Strategies for Improving Local Search. Proc. of AAAI , 1994.
20. J. Singer, I. P. Gent and A. Smaill. Backbone Fragility and Local Search Cost Peak. Journal of Artificial Intelligence Research   , 235–270, May 2000.
21. G. Verfaille. CSP Aleatoires: Estimation de l'Optimum,   Research Notes , 1998.
22. E. Weinberger. Correlated and Uncorrelated Fitness Landscape and How to Tell the Dierence.   J. Biol. Cybern. , 63: 325–336, 1990.
23. M. Yokoo. Why Adding More Constraints Makes a Problem Easier for Hill-Climbing Algorithms: Analyzing Landscapes of CSPs.   Proc. of CP , LNCS 1330: 357–370, 1997.

A Multiobjective Evolutionary Algorithm for Car Front End Design

Olga Rudenko [1], Marc Schoenauer [1], Tiziana Bosio [2], and Roberto Fontana [2]

[1] Centre de Math´ ematiques Appliqu´ ees, Ecole Polytechnique,
91128 Palaiseau Cedex, France
roudenko,marc @cmapx.polytechnique.fr
[2] Department of Statistical Methodologies, Fiat Research Center (CRF)
Strada Torino 50, 10043 Orbassano (To), Italy
t.bosio,r.fontana @crf.it

Abstract. The aim of this study is to find the optimal structural geometry of the front crash member of a car of minimal mass that optimally satisfies all operational conditions. The mechanical domains that have been considered are crash , acoustic (dynamic) and static . They are summed up by 9 objective functions, resulting in a 10-objective optimization problem. However, this problem is further turned into minimizing the mass while maximizing the internal energy (crash objective), subject to constraints on the 8 objectives that arise from the acoustic and static domains. The dimension of the objective space of this constrained problem is much lower than that of the original 10-objective problem. This significantly reduces convergence time, while decreasing decision making eorts among solutions obtained though pareto-based multiobjective optimization.

Nevertheless, since the computation of the structural responses is based on a very time-consuming FEM crash analysis, direct computation of the fitness within an evolutionary algorithm is impossible: The response of car front members is computed using an approximative evaluation that had been identified during the BE96-3046 European project (CE $)^2$: Computer Experiments for Concurrent Engineering.

Thanks to this approximation, very good results are obtained in a reasonable time using a Pareto elitist evolutionary algorithm based on NSGA-II ideas, combined with an infeasibility objective approach for constraint handling.

Introduction

The paper deals with a design optimization problem which consists in optimizing 10 competing objectives for a car front end part. Taking in account the prevalence of crash and mass objectives over the other criteria, and in order to simplify results interpretation and to reduce the computation time, the initial problem is reformulated as a 2-objective constrained optimization problem.

Multi-objective optimization is characterized by the non-uniqueness of optimal compromises between all objectives called Pareto optimal solutions. The

P. Collet et al. (Eds.): EA 2001, LNCS 2310, pp. 205–216, 2002.
c Springer-Verlag Berlin Heidelberg 2002

notion of Pareto optimality is only first step towards the practical solution of a problem, which involves some decision process to choose a single compromise solution from the Pareto set according to some preference information. The main goal of the multiobjective optimization is to find a good and complete approximation of the Pareto set that will later ensure the possibility of making an informed and justified decision about definitive design.

Because they process a set of potential solutions in parallel, Evolutionary Algorithms seem to be particularly suited for identifying the Pareto set of a multiobjective problem. But the number of evaluations these algorithms have to perform before converging (usually several thousands), make them intractable when the CPU-time required by a single evaluation is of the order of several hours. Hence, statistical methods for design and analysis of computer experiments has to be used to identify emulators of costly objectives. Emulators are simple but highly adaptive mathematical functions that approximate simulators accurately, but run thousands of times faster.

This paper presents the optimization results obtained by a Pareto based genetic algorithm coupled to emulators of mechanical simulations for a problem of Car Front End Design.

In section 1, the mechanical problem of the car front end optimization is introduced: the parametrisation of the car front end and the objectives, corresponding to the responses in dierent domains, are given. Section 2 introduces the trends in approximation of fitness functions in Evolutionary Algorithms, and presents the DACE approximation technique used for the fast calculation of the car front end responses. In section 3.1 the choice of the optimization method as well as the choice of constraint handling approach are justified, and their most important features are detailed. Sections 4.2 and 4.3 present the Pareto solutions of the 2-objective unconstrained and constrained problems respectively. Those solutions are favorably compared with the results obtained during the (CE)^2 project [1].

1   The Car Front End Model

1.1   Design Variables

The full model of the car front end considered in this paper is defined by 13 design variables, and is described in figure 1.

The 13 design variables are related to the shape and the size of the beam and they can be categorized into thickness variables and shape variables . The allowed range for each thickness variable is suggested by the technological details of metal sheets stamping. Ranges on shape variables are chosen in order to achieve a significant variation in the main section.

1.2   Responses

The problem at hand involves 10 objectives which are computed using results from three dierent mechanical simulations.

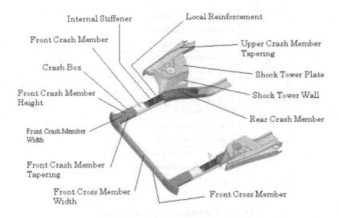

Fig. 1.   Beam model

More precisely, the objectives are:

- mass : a straightforward computation from the thickness and shape variables using the mechanical data on the materials involved (kept fixed in this paper).
- crash : the criterion involved is internal energy absorbed by the front end after the impact – to be maximized; Note that this simulation is by far the most computationally intensive;
- acoustic : the mean and minimum dynamic stiness at shock tower and engine mounts attachments are two more objectives to be maximized;
- static : a rather standard simulation, from which 2 dierent static strength responses and 4 dierent static stiness responses

However, all objective do not have the same priority in the eyes of the designer. In particular, the mass and the internal energy are by far the most important criteria.

Note that some criteria are only inuenced by part of the 13 design variables. Figure 2 gives the list of these causal relationships.

Simulation codes based on a FEM analysis with a shell description of the geometry have been prepared to explore the mechanical behavior in the 3 dierent domains. But the requested CPU-time was 6 hours for crash, 1.5 hours for acoustic and 0.5 hours for static. Hence using actual simulation is not possible in the framework of Evolutionary Computation, as standard runs involve a few thousand fitness calculations. This is why approximations of all simulations have been developed and used during the whole optimization process.

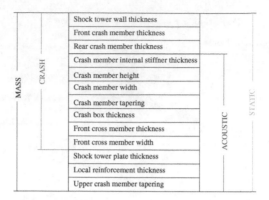

Fig. 2.   Parameters and missions

2   Optimization and Approximation

2.1   State of the Art

In real-world engineering problems, computing the objective function(s) is generally the main source of computational cost. Hence many researchers have suggested to replace the objective function(s) with some approximation the cost of which is very low.

Some examples of the objectives are gathered in an initial phase (either purposely randomly generated, or from past experiments). These examples are used to build an approximate function for each of the objectives. These approximations are then used in place of the actual objectives during the evolutionary optimization.

Two important issues have to be considered: what mathematical model to choose for the approximation; and what strategy to use during the evolutionary optimization.

Early work embedding approximation into evolutionary optimization had chosen some simple linear approximation [10,7]. But of course linear models are inappropriate for most real-world (non-linear) problems. Two very popular models have been widely used, namely Neural Networks [12] and, particularly in Mechanical Engineering, krieging [13].

Grefenstete's pioneering work [6,7] used objective function approximation to tune the parameters of the genetic algorithm, and use the the resulting optimal parameter set to run a GA with the actual objective. But the most popular approach used in Engineering is to mix both the approximation and the actual computation: once a new population has been generated using the approximate fitness, part of the population is evaluated using the actual function (generally some of the best individuals according to the approximation). While avoiding a too large bias due to the error coming from the approximation, such strategy allows to gradually refine the approximation as the population is moving toward areas of the search space that have probably not been sampled by the initial training examples.

However, when the cost of a single evaluation is really high (e.g. several hours, as for the crash experiment used in the present work), even a few computations of the actual objective during the evolutionary optimization is beyond reach, and the same approximate fitness function has to be used throughout the optimization – and is computed using all archived examples of the actual functions. A minimal sanity check consists in computing the actual values of the objectives for the few solutions that are retained as solutions in the end of the optimization.

2.2 The DACE Model

For present problem, the emulators have been built using the   DACE approximation method for all three simulations (crash, dynamic and static). The method is based on the krieging of spatial statistics [14]. The assumption is that any response y can be written as

$$y = m(,x) + V(,,),$$

where $x = (x_1,...,x_n)$ is the set of design variables   y depends on, m is a linear regression model with (vector) parameters   , and V is a stochastic process with unknown covariance dependent on certain (vector) parameters   and . These parameters are estimated using the Maximum Likelihood principle. In the case of computer experiments, where there is no a priori measurement error, the error parameter   is set to 0, and the model is an exact interpolator of experimental data. Technical, theoretical and practical issues of emulator-building are described in more detail in [1].

3   Multiobjective Constrained Optimization

3.1   Multiobjective Evolutionary Algorithms

A number of MEAs have been suggested in the last decade[18,3]. The recent EMO'01 was the first conference entircly dedicated to the Multiobjective Evolutionary Algorithms (MEAs). The primary reason for this increased interest is the ability of these algorithms to find multiple optimal compromise solutions in one single run.

Multiobjective problems aim at simultaneously optimizing dierent objective functions. When those objectives are contradictory, there does not exist a single solution, and compromises have to be made.

A solution x is said to Pareto-dominate   another solution   y if x is not worse than y with respect to all criteria, and is strictly better than   y with respect to at least one criterion. The set of all points of the search space that are not Pareto-dominated is called the   Pareto-set of the multiobjective problem at hand: it represents the best possible compromises with respect to the contradictory objectives.

Solving the multiobjective problem amounts to choose one solution among those non-dominated solutions, and some decision arguments have to be given. Whereas any optimization method can be used to get one single solution from the Pareto front by aggregating the dierent objectives with some predefied multiplicative factors, MEAs are to-date the only algorithms that actually give a sample of the Pareto front, allowing decision makers to choose one of them with more complete information.

MEAs are very similar to standard Evolution Algorithms, and the basic generation loop is the same [3]: select some parents for reproduction; apply some variation operators   (e.g. crossover, mutation) to those selected parents to generate ospring; replace some parents with some ospring. The dierence lies in the Darwinian-like phases (selection and replacement) where, in most MEAs, the notion of Pareto-dominance replaces the usual order relationship "has better fitness than".

However, in order to obtain the best possible sampling of the Pareto front, MEAs also use two techniques borrowed to standard evolutionary computation: Pareto elitism and diversity preserving techniques.

Elitism consists in preserving in the populations some of the the best individuals encountered so far along evolution. Of course, within the frameworks of MEAs, "best individuals" is replaced by "non-dominated individuals". Recent results [17] clearly show that elitism can speed up the performance of the MOEAs significantly, also it helps to prevent the loss of good solutions ones they have been found.

A crucial problem for multiobjective evolutionary approaches is to preserve diversity of solutions. Evolutionary Computation has worked out many niching techniques for that purpose but the majority of them require a user-defined parameter: this is true for classical niching techniques like sharing, with a sharing radius [8] or for more recent multi-objective specific techniques like squeezing, with the "squeeze factor" [2]. The NSGA-II [4] approach, detailed in section 3.3, has been chosen mainly because it is free of such parameter.

3.2   Constraint Handling

The car front end optimization involves 10 objectives (see section1.2). In the present approach, however, only the two most critical objectives will be considered, and the other objectives will only be tested against some limit values: from the point of view of optimization, this amounts to treat them as constraints, thus greatly reducing the dimension of the objectives space and, consequently, probably accelerate convergence, and greatly facilitate the interpretation of the results.

Various constraint handling techniques have been proposed for Evolutionary Computation [9]. Some of them actually borrowed ideas from   . . . multiobjective optimization [5]. However, having already decided to decrease the number of objectives, this was not the way to go. Another very popular constraint handling technique is the penalization method, but again it requires the a priori definition of the penalty parameters. Hence it was finally decided to use the infeasibility objective method, that combines the constraint violations to give a single measure of an individuals infeasibility which is then treated as an objective in the pareto ranking. This approach is detailed in section 3.4.

The resulting problem is thus a 3-objective optimization problem: minimize the Mass, maximize the Internal Energy and minimize the Infeasibility Objective. In order to direct the optimization towards the feasible solutions, it is necessary to use the goal attainment method , which gives the priority to every feasible individual over the best infeasible one.

3.3 NSGA-II

Selection and replacement in NSGA-II [4] are based on two hierarchically ordered criteria: in order to compare two individuals, their domination ranks are first compared, smallest one being the best. If their domination ranks are equal, the best is the one situated in the sparsest objective space region, as measured by the crowding distance .

Standard tournament selection is used. The replacement is a deterministic selection from the merged parent and ospring populations.

Domination rank.   The domination rank is based on the Pareto dominance map of the population: individuals that are not dominated by any other individuals are given rank 1. They are removed from the population and the non-dominated among remaining individuals are given rank 2. This process continues until all the individuals are ranked.

Crowding distance.   The crowding distance is a measure of the density of solutions in objective space: for each objective, sort every group of individuals of the same rank according to that objective only; The partial crowding distance for that objective for a given individual is the distance between the two points surrounding the individual in these sorted lists. The total crowding distance is the sum of these partial crowding distances over all objectives. It can also be viewed as the largest centroid enclosing the individual at hand without including any other point of the same rank.

This measure is not based on any user-defined parameter. However, since it requires sorting the population according to each of the objectives, it becomes computationally heavy for problems with many objectives.

3.4   The Infeasibility Objective Approach

The infeasibility approach [16] amounts to turn the whole set of constraints into a single objective. More precisely, let $c_j(x)$ be the violation of constraint j (i.e. $\max(0, g_j(x))$ for inequality constraint g_j, and $h_j(x)$ for equality constraints h_j. Then the solutions infeasibility is taken as the normalized sum of all violations

$$c(x) = \sum_{j=1}^{m} \frac{c_j(x)}{c_{max,j}}.$$

The solutions infeasibility $c(x)$, is to be minimized, a null value ensuring that all constraints are satisfied.

Normalizing the constraint violations (by dividing by the scaling factor $c_{max,j}$), is mandatory since dierent constraint might have very dierent scales. In the current implementation, that scaling factor is taken as the maximum value of the corresponding constraint violation found in the initial population. If no infeasible solutions were found, the scaling factor is set 1; however the focus of this approach is on the solution of highly constrained problems that are unlikely to yield any feasible solution from a randomly generated population.

4   Results

4.1   Experimental Conditions

A real uncoded evolutionary algorithm has been used in this study. Every individual is a vector from $\mathbb{IR}^{13}$ representing the design variables .

The following variation operators have been applied: intermediate crossover with probability 0.7 and Gaussian auto-adaptive mutation with one standard deviation per variable, with probability 0.05. Replacement and selection described in section 3.3 have been used with tournament size 2.

In the unconstrained case, population size was set to 150 and maximum number of generations to 100; In the constrained case, population size was 700 and maximum number of generations 400.

4.2   Two-Objective Problem without Constraints

First, the NSGA-II method has been applied to solving the simple two-objective optimization problem of minimizing mass and maximizing internal energy. The region of interest is limited by   Mass   36 and IE   30000. On Figure 3, the Pareto solutions, obtained using NSGA-II, are compared to the results of the aggregating approach, ECCOMAS values [1].

This Pareto front contains 23 solutions that dominate the ECCOMAS-point for both objectives and 32 solutions which belong to the region   Mass   36 and IE   30000 and which are non-dominated by ECCOMAS result.

However this unconstrained problem has no any practical relevance, the results are useful as a reference when turning to the constrained problem, pointing out how much the dierent constraints limit the optimization of the main responses.

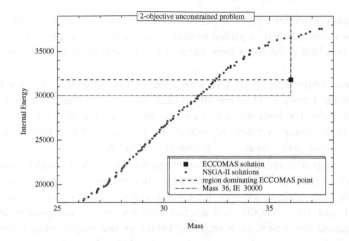

Fig. 3.   Objective space

4.3   Two-Objective Constrained Problem

A much more realistic case is the constrained two-objective problem presented in section 3.2: the objectives are still to minimize the mass and maximize the internal energy, but limiting the search to the region described by Table 1: every constraint value is a little bit stronger than the ECCOMAS value of the corresponding response and actually that was the principle of their choice. Doing so ensures that every feasible solution of the constrained problem (if any) will be better than the ECCOMAS one for these 8 objectives.

Table 1.   Constraints on the 8 secondary objectives

Objective	Response name	Constraint value	ECCOMAS optimal solution	
max	Mean dynamic stiness	35000	33735.3	N/mm
max	Min dynamic stiness	4100	4001.2	N/mm
min	Max equiv. stress for bump	360	367.4	Mpa
min	Shock towers displacements for bump	2.3	2.353	mm
min	Torsion for turn	0.02	= 0 .02	degrees
min	Max equiv. stress for rebound-max brake turn	203	206.5	Mpa
max	Shock towers stiness	255	249.6	Kg/mm
max	Torsional stiness	250000	243204.4	Kgm/rad

The method used during (CE)² project [1] was an additive aggregation of the 10-objectives: the result of a global optimization method (a genetic algorithm) was taken as initial point for a local method (Powel) used to refine the genetic solution.

Solutions obtained by NSGA-II with Infeasibility Objective Approach are presented in the Figure 4. They are all feasible, and most of them dominate the ECCOMAS point for both mass and energy and, consequently, for all the 10 objectives of the initial problem. Moreover, a quite large set of Pareto solutions satisfies the constraints Mass   36 and IE   30000.

The next step after emulator-based optimization should be the calculation of true (given by simulators) objective values for "interesting" candidate solutions. Clearly, it is unrealistic to ask that for all the 328 points of Figure 4 verifying Mass   36 and IE   30000. And it's not only a matter of computational cost, but also because the dierences in response values of two neighboring solutions is very small: for isntance, there is no point in making choice between   Mass = 34.4188,   IE = 30583   and   Mass = 34.4214,   IE = 30593.7. On the other hand, thanks to the ecient preserving diversity technique of NSGA-II, the set of solutions is a very good sampling of the Pareto front. So, by picking up one Pareto solution at, say, every 100g interval on the mass axes, one obtains a nice uniform representation of the region   Mass   36, IE   30000 by 18 quite dierent good compromises for the problem of Car Front End Design.

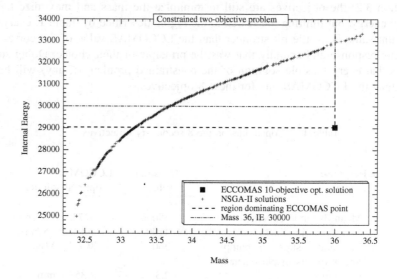

Fig. 4.   Objective space

4.4   Discussion

Such a high number of Pareto solutions comes from the size of NSGA-II population (700 individuals) that proved mandatory to reach as quickly as possible feasible region and to explore it eciently. This work deals with a design problem, that deserves large eorts before making defiitive decision (as millions copies of such part will be later actually assembled). Hence algorithms giving high quality solutions are prefered to algorithms giving solultions more rapidly, but with less performances.

Applying Pareto based method allowed one to obtain suciently rich variety of Pareto solutions of high quality, dominating the previous best one obtained during $(\text{CE})^2$ project for all 10 objectives.

5   Future Work

In theory, the Pareto-optimal solutions of the 10-objective problem would give us all the best compromises between disciplines under consideration. In practice, a good sampling of a 10-dimensional Pareto front is almost impossible to obtain unless we use enormous populations – and hence encounter enormous computing times.

Moreover, the chances are high that every point on that Pareto front actually has one criterium for which it is better than our constrained solutions while at the same time it will have   at least   one criterium value that will make it totally uninteresting as a possible solution.

In another words, by chance, we are not searching for a good approximation of the whole Pareto front. We are interested only in the part of it such that Mass   36, IE   30000, all other criteria values being at least not much worse than the constraints values given by Table 1.

Hence, the next step we foresee is to introduce some target value to attire the population to the region of interest. Such an approach has been successfully used in some other applications [15]. But two-objective constrained optimization results will be very useful while choosing target value and while analyzing 10-objective optimization's candidate solutions.

However, the interpretation of results of an 10-objcetive optimization is still an open problem, whatever the actual method used to get those results. Moreover, such interpretation is probably problem-dependent. This is one of the main reasons that explain why researchers and engineers using Pareto-based methods generally avoid dealing with such a high number of objectives by transforming some of them into constraints – just like what has been done in present work.

References

1. Bates, R., Bosio, T., Fontana, R. and al. (2000) Computer Experiments for Concurrent Engineering.   European Congress on Computational Methods in Applied Sciences and Engineering

2. Corne, D. W., Knowles, J. D. and Oates, M. J. (2000). The Pareto Envelope-Based Selection Algorithm for Multiobjective Optimization. In Schoenauer et al. (Eds.) Parallel Problem Solving from Nature (PPSN VI)   , pages 839-848. Springer Verlag LNCS Vol. 1917.
3. Deb, K. (2001). Multi-Objective Optimization Using Evolutionary Algorithms   . John Wiley.
4. Deb, K., Agrawal, S., Pratap, A. and Meyarivan, T. (2000). A Fast Elitist Non-dominated Sorting Genetic Algorithm for Multiobjective Optimization : NSGA-II. In Schoenauer et al. (Eds.)   Parallel Problem Solving from Nature (PPSN VI)   , pages 849-858. Springer Verlag LNCS Vol. 1917.
5. Fonseca, C. M. and Fleming, P. J. (1998) Multiobjective Optimization and Multiple Constraint Handling with Evolutionary Algorithms - Part 1: A Unified Formulation.   IEEE Transactions on Systems and Cybernetics - Part A: Systems and Humans . 28 No.1, pages 26-37.
6. Grefenstette, J. J. and Fitzpatrick J.M. (1985). Genetic search and approximate function evaluation. In Grefenstette, J. J. (Eds),   Procee dings of the 1st International Conference on Genetic Algorithms   , pages 160–168. Morgan Kaufman
7. Grefenstette, J. J. (1995). Predictive Models Using Fitness Distributions of Genetic Operators, In L. Darrell Whitley and Michael D. Vose (Eds),   Foundations of Genetic Algorithms 3 , pp. 139-162. Morgan Kaufmann.
8. Goldberg, D.E. and Richardson, J. (1987). Genetic Algorithms with sharing for multi-modal function optimization. In Grefenstette, J.J. (Eds),   Procee dings of 2nd ICGA , pages 41–49, Lawrence Erlbaum Associates.
9. Michalewicz, Z, and Schoenauer, M. (1996). Evolutionary Algorithms for Constrained Parameter Optimization Problems.   Evolutionary Computation , 4(1):1–32.
10. Mosetti, G. and Poloni C. (1995). Aerodynamic Shape Optimization by means of Hybrid Genetic Algorithms, In 3rd International Congress on Industrial and APplied Mathematics, Hamburg.
11. Obayashi, S. (1997). Pareto Genetic Algorithm for Aerodynamic Design Using the Navier-Stokes Equations. In uadraglia, D., P´   eriaux, J., Poloni C., and Winter G. (Eds) Genetic Algorithms and Evolution Strategies in Engineering and Computer Sciences , pages 245–266. John Wiley.
12. Poloni, C. and Pediroda V. (1997). GA coupled with computationaly expensive simulations: tools to iprove eciency, In uadraglia, D., P´   eriaux, J., Poloni C., and Winter G. (Eds)   Genetic Algorithms and Evolution Strategies in Engineering and Computer Sciences , pages 267–288. John Wiley.
13. Ratle, A. and Berry, A. (1998). La r´   eduction du rayonnement acoustique des structures par un design optimal utilisant un algorithme g´   enétique. Actes du 66i eme congr es de l'Acfas, Montreal.
14. Sacks, J. W., Welsh, T. Mitchell and Wynn, H. (1989). Design and Analysis of Computer Experiments.   Statistical Science 4   (4).
15. Sbalzarini, I. F., Muller, S., Koumoutsakos, P. (2001). Michrochannel Optimization Using Multiobjective Evolution Stratagies. In [18].
16. Wright, J. and Loosemore, H. (2001). An Infeasibility Objective for Use in Constrained Pareto Optimization. In [18].
17. Zitzler, E., Deb, K. and Thiele, L. (2000). Comparison of Multiobjective Evolutionary Algorithms: Empirical Results.   Evolutionary Computation , 8(2): 173-195.
18. Zitzler, E., Deb, K., Thiele, L., Coello Coello, C. A., and Corne, D. editors (2000). Evolutionary Multi-Criterion Optimization (EMO 2001)   , Springer Verlag, LNCS series Vol. 1993.

EASEA Comparisons on Test Functions: GALib versus EO

Evelyne Lutton [1], Pierre Collet [2], and Jean Louchet [3]

[1] Projet Fractales INRIA, B.P. 105, 78153 Le Chesnay cedex, France,
Evelyne.Lutton@inria.fr , http://www-rocq.inria.fr/fractales/
[2] EEAAX – CMAPX École Polytechnique, 91128 Palaiseau cedex, France,
Pierre.Collet@Polytechnique.fr , http://www.eeaax.polytechnique.fr
[3] ENSTA, 35 Boulevard Victor, 75011 PARIS, France,
Louchet@ensta.fr , http://www.ensta.fr/louchet

Abstract. The EASEA [1] language (EAsy Specification of Evolutionary Algorithms) was created in order to allow scientists to concentrate on evolutionary algorithm design rather than implementation. EASEA currently supports two C++ libraries (GALib and EO) and a JAVA library for the DREAM. The aim of this paper is to assess the quality of EASEA-generated code through an extensive test procedure comparing the implementation for EO and GALib of the same test functions.

1 Introduction

Evolutionary algorithms are dicult to implement because of their inherent complexity: the programmer needs to create a data structure and evolve a population, using a problem-specific evaluation function and genetic operators involving choices which may be decisive to the outcome of the algorithm. Moreover, Evolutionary Algorithms are mainly used to solve or optimise complex applications in technical fields sometimes remotely connected to computer science, and scientists needing Evolutionary Algorithms do not always have the skills to implement them.

EASEA (EAsy Specification of Evolutionary Algorithms) is a language specially designed to hide away implementation complexity: the user is only asked to provide problem-specific procedural code (namely the fitness function and the crossover, mutation and initialisation operators).

While most research in the Evolutionary community is devoted to enriching the evolutionary paradigm with new features and concepts [7,11,15,16], we have chosen a pragmatic, application-oriented approach with the development of the EASEA language. It comes with an EASEA compiler which converts .ez specification files into C++ files or JAVA files, relieving the user from the burden of programming the evolutionary algorithm.

---

[1] Research funded by the European Commission IST Programme 1999-12679 (Future and Emerging Technologies).

P. Collet et al. (Eds.): EA 2001, LNCS 2310, pp. 219–230, 2002.

2   Presentation of EASEA

In theory, a good enough specification language would allow to implement an algorithm using any library capable of implementing the described evolutionary algorithm. Therefore, the EASEA compiler had to be able to generate code for several libraries, not only to prove its generality, but for other reasons as well:

- All libraries have dierent features. Let us consider GALib, EO and DREAM:

   GALib is extensively used although its limited exibility betrays its old design (only one mutator and Xover, no tournament replacement,   ...) EO oers a full object-oriented template approach allowing it to be much more versatile, although it is still a young library (v0.9xxx) and it does not naturally support distribution over islands, for instance, DREAM is written over a fully distributed architecture, but it is still a rough prototype, so its evolutionary library is still very minimal.

   By supporting many dierent underlying libraries, EASEA users have access to a superset of all available features. If a user needs a feature absent from a library, he is directed towards the library hosting the feature.
- Supporting dierent libraries promotes communication between research projects: A team using GALib may recompile an EASEA file which was created for an EO environment. Results of dierent teams can then be compared on identical machines and environments.
- Similarly, appending an EASEA description of an algorithm to a research paper or to a web page will allow any user to recompile the program in his own environment provided one of the supported libraries is installed.

When the EASEA project started within the EVO-Lab research action, in January 1999, the GALib [5] C++ library was chosen as a first target, as it was already stable and used by many programmers around the world. The first EASEAv0.1 compiler for GALib was released in september 1999.

Then, the European EO C++ library [3] began to converge towards more stable versions, and EO was chosen as a second target. The EASEA Millennium Edition (v0.6) was the fist release (in January 2001) which could indierently generate code for either GALib or EO.

Finally, the EASEA v0.7 prototype is now able to produce JAVA code for the DREAM (Distributed Resource Evolutionary Algorithm Machine [6]).

3   Time for Tests

As a consequence, the same algorithm described in a   .ez file can be automatically converted into a source file using the GALib C++ library, the EO C++ library or the DREAM JAVA library. This unique tool raises many natural questions, among which:

1. Is the quality of the EASEA code generation equivalent for the dierent libraries ?

2. What about comparing the respective performances of dierent libraries ?
3. Is there any dierence between an EASEA-coded algorithm and the same algorithm coded by a human programmer using the same library ?

Extensive tests must be elaborated to answer these basic questions.

3.1   Choosing Tests

EASEA can create code for two C++ libraries and one JAVA library. As a starting point, it seems more sensible to compare the two C++ implementations, rather than to introduce other unknown factors by adding a JAVA/C++ competition, allowing us to use exactly the same compiler with the same options.

Full competition: EO and GALib are totally dierent libraries, with dierent features. A way to compare both libraries would be to use freely all available features of each library to show that solving a particular problem takes X seconds using GALib with GALib-specific algorithms and parameters, and Y seconds using EO with EO-specific algorithms and parameters. Making such a comparison would be very dicult, as it would be necessary to fid the optimal way of solving the problem with GALib and demonstrate it is optimal a very dicult task as there are clearly a great number of dierent ways to implement a same problem and do the same with EO, before it is possible to compare execution times between the two libraries. Moreover, this would still rely on the quality of the code generation of the EASEA compiler: Matthew Wall may find a much more ecient way to code the problem as he knows his GALib library inside out, and the same could be said for Marc Schoenauer with the EO library, for instance. Therefore, such a test would have mixed two issues: the quality of the EASEA code generation and the capacities of the two dierent libraries to solve the problem. EO winning over GALib would have meant that an EASEA-EO implementation would be faster than an EASEA-GALib implementation, even though it could be possible that a Schoenauer-EO implementation would be slower than a Wall-GALib implementation.

Competition on common features: Such a competition tries to compare comparable things as we have already decided to do when we chose not to compare a C++-based library and a JAVA-based library. The idea is to pick up common features between the libraries and compare them over a set of common test functions, using the same   .ez source file. The possibility that Wall-GALib or Schoenauer-EO implementations give much better results than EASEA-GALib or EASEA-EO implementations is then greatly reduced, as all parameters, all operators (selection, replacement) and all algorithms are imposed beforehand. In such conditions, the EASEA code generation process is very close to human-code generation as EASEA uses man-made templates. In fact, the EO template file was programmed by Marc Schoenauer, which means that apart from the fitness function and mutation/crossover operators (specific to the test function),   an EASEA-EO user actually has his evolutionary algorithm coded by Marc Schoenauer . As a conclusion, although the result of this testbench will indeed compare the EASEA-EO and EASEA-GALib implementation of the same evolutionary algorithm, the fact that they are so close to human implementation enables to use EASEA to actually compare the performances of the libraries.

4   Weierstrass-Mandelbrot Test-Functions

Irregularity has been experimentally and theoretically identified as an EA ﬁiculty"factor. This is why fractal"functions, such as the Weierstrass-Mandelbrot ones, have been used in [9,8] to experimentally confirm theoretical analysis of irregularity inuence on deception, taking advantage of their uniform regularity over the search space. Hölder exponents have been established as a relevant measure of irregularity and deception, and as a basis to many fractal analysis methods, especially in the domain of signal analysis.

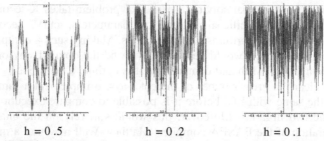

h = 0 .5   h = 0 .2   h = 0 .1

Fig. 1.   1D Weierstrass Test functions with increasing irregularities, the horizontal line represents the maximal value (attained in 0.)

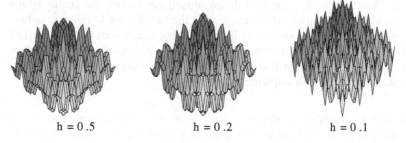

h = 0 .5   h = 0 .2   h = 0 .1

Fig. 2.   2D Weierstrass Test functions with increasing irregularities, the horizontal line represents the maximal value (attained in 0.)

Weierstrass-Mandelbrot functions, which are defined as:

$$W_{b,h}(x) = \sum_{i=1} b^{-ih} \sin(b^i x) \text{ with } b > 1 \text{ and } 0 < h < 1$$

depend on a parameter h, which can be viewed as being the global Hölder exponent of the function (it is also equal to 2 − d, where d is the "fractal dimension" of the graph of the function). Weierstrass-Mandelbrot functions are very irregular for small values of h, and become smoother as h tends to 1, see figures 1 and 2.

Therefore, we used these functions as controlled regularity test functions in the experiments presented below. In the case of maximisation, we compute an upper bound of $W_{b,h}$, which is $MaxVal = \sum_{i=1} b^{ih}$, and thus maximise:

$$WM_{b,h}(x) = MaxVal \quad W_{b,h}(x) \quad \text{with } b \quad 1 \text{ and } 0 \quad h \quad 1$$

$W_{b,h}(x) = 0$, thus $WM_{b,h}(x)$ is always positive and is maximal at 0. with value $MaxVal$. In the minimisation case, we directly use $W_{b,h}(x)$.
Similarly, for a 2D search space we maximise:

$$WM2D_{b,h}(x) = 2 \quad MaxVal \quad W_{b,h}(x) \quad W_{b,h}(y) \quad \text{with } b \quad 1 \text{ and } 0 \quad h \quad 1$$

or minimise $W_{b,h}(x) + W_{b,h}(y)$.

4.1 Experimental Results

The following experimental settings (see   weiermax.ez   in appendix) were used:

- real-encoded genome x   [1., 1.],
- plus stragegy (population sizes and number of generation specified for each table),
- tournament selection, with tournament size 2 (labeled "T"), or Roulette Wheel, with no scaling (labeled "RW"),
- barycentric crossover,
- uniform mutation of radius   (no label), or log normal self-adaptive gaussian mutation (labeled "ad"), [13,14]

Results in tables 1 to 8 show the mean value over 20 runs.

5   Other Test-Functions

We used classical functions from   http://www.geatbx.com/docu/fcnindex. html conveniently scaled so that their global optimum be naught (see tables 5 to 7),

6   Conclusion and Future Work

The numerous tests conducted in this paper allow to answer some of the questions raised in section 3:

1. All in all, both libraries give very comparable results (although EO's appear to be slightly more accurate see AckleyPath maybe due to dierent implementations of the random number generator). We were not able to explain the results of Rosenbrock 500 and Griewangk 500 where EO results are much worse than GALib (see tables 5 and 6).
   This leads us to one of the main conclusion of this paper, which has shown on a significant number of tests that   EASEA does indeed create comparable evolutionary algorithms using GALib and EO out of the same source file   .

Table 1.   Comparison of GALib and EO performances on 1D Weierstrass test functions
to be maximised . Population size (50+40) for 50 generations (2050 evaluations). For
comparison purposes, all fitness values are normalised so that the maximum is 1. for
each test function. This table shows that EO is slightly slower than GALib, although
the CPU time is much more variable, for comparable results.

		GALib		EO	
h	Algorithm	best value Mean( )	CPU time Mean( )	best value Mean( )	CPU time Mean( )
0.5	T	0.9996(0.0017)	0.2435(0.0073)	1.0000(0.0000)	0.2540(0.0092)
0.5	RW	1.0000(0.0000)	0.2480(0.0112)	1.0000(0.0000)	0.2675(0.0109)
0.5	T+ad	1.0000(0.0000)	0.2605(0.0097)	1.0000(0.0000)	0.2700(0.0077)
0.5	RW + ad	1.0000(0.0000)	0.2625(0.0083)	1.0000(0.0000)	0.2830(0.0105)
0.4	T	1.0000(0.0001)	0.2610(0.0054)	1.0000(0.0000)	0.2755(0.0175)
0.4	RW	1.0000(0.0000)	0.2615(0.0096)	1.0000(0.0000)	0.2825(0.0141)
0.4	T+ad	1.0000(0.0000)	0.2750(0.0059)	1.0000(0.0000)	0.2835(0.0142)
0.4	RW + ad	1.0000(0.0000)	0.2790(0.0070)	1.0000(0.0000)	0.2965(0.0085)
0.3	T	0.9902(0.0141)	0.2680(0.0081)	0.9912(0.0123)	0.2760(0.0139)
0.3	RW	0.9959(0.0096)	0.2675(0.0070)	0.9956(0.0103)	0.2860(0.0086)
0.3	T+ad	0.9917(0.0126)	0.2850(0.0087)	0.9943(0.0110)	0.2890(0.0118)
0.3	RW + ad	0.9955(0.0085)	0.2805(0.0074)	0.9963(0.0079)	0.3000(0.0089)
0.2	T	0.9998(0.0002)	0.2615(0.0079)	0.9996(0.0009)	0.2780(0.0150)
0.2	RW	0.9998(0.0002)	0.2635(0.0073)	0.9998(0.0002)	0.2845(0.0120)
0.2	T+ad	0.9999(0.0001)	0.2775(0.0077)	0.9998(0.0002)	0.2850(0.0112)
0.2	RW + ad	0.9997(0.0002)	0.2720(0.0060)	0.9998(0.0003)	0.2950(0.0092)
0.1	T	0.9998(0.0002)	0.2595(0.0080)	0.9998(0.0001)	0.2800(0.0110)
0.1	RW	0.9999(0.0001)	0.2600(0.0063)	0.9998(0.0002)	0.2880(0.0133)
0.1	T+ad	0.9998(0.0002)	0.2840(0.0066)	0.9999(0.0001)	0.2895(0.0112)
0.1	RW + ad	0.9999(0.0002)	0.2800(0.0077)	0.9999(0.0001)	0.3015(0.0135)

2. This result allows to compare EO and GALib performance:
   - The EO engine appears to be faster on tournaments than GALib but
     slower on RouletteWheels.
   - Genome manipulation is much slower with EO than with GALib (due
     to the extensive use of templates by the EO library, according to Marc
     Schoenauer) confirmed by a constant overhead for a given genome size.
     Therefore, genome length impacts EO's performance much more than
     GALib's.

As a side eect, these tests show that the usual statement that the fness
function accounts for 90 of the calculation time of an evolutionary algorithm needs to be qualified . This can easily be seen on figur e 3 : on dimension
500 where 16200 sphere   evaluations use 3.31 seconds of the 75.6 seconds of
the EO adaptive mutation algorithm (4.37 of the total time). Griewangk
(also shown on figure 3) only reaches 80 in the best case (GALib non-
adaptive) and the decisive 90 value is only attained for Weierstrass 500
(904s for the EA vs 839s for evaluation only) Therefore, the overhead induced by the library is far from being negligible on problems using millions
of very fast evaluations (scheduling, ...).

Table 2.   Comparison of GALib and EO performances on 2D Weierstrass test functions to be maximised   . Population size (150+120) for 100 generations (12150 evaluations). For comparison purposes, all fitness values were normalised so that the maximum is 1. for each test function. This table shows that EO is faster than GALib on tournaments, but slower on RouletteWheels, for comparable results.

h	Algorithm	GALib		EO	
		best value Mean( )	CPU time Mean( )	best value Mean( )	CPU time Mean( )
0.5	T	0.9998(0.0005)	2.6240(0.0213)	1.0000(0.0000)	2.5835(0.0467)
0.5	RW	1.0000(0.0000)	2.6305(0.0150)	1.0000(0.0000)	2.7830(0.0517)
0.5	T+ad	0.9999(0.0002)	2.8035(0.0096)	1.0000(0.0001)	2.6960(0.0443)
0.5	RW + ad	1.0000(0.0001)	2.8040(0.0086)	1.0000(0.0000)	2.8845(0.0565)
0.4	T	0.9996(0.0012)	2.7900(0.0148)	0.9995(0.0018)	2.7480(0.0662)
0.4	RW	1.0000(0.0002)	2.8055(0.0150)	1.0000(0.0000)	2.9395(0.0329)
0.4	T+ad	1.0000(0.0001)	2.9775(0.0126)	0.9999(0.0002)	2.8555(0.0439)
0.4	RW + ad	0.9998(0.0002)	2.9645(0.0092)	0.9998(0.0001)	3.0620(0.0275)
0.3	T	0.9943(0.0072)	2.9015(0.0250)	0.9988(0.0023)	2.7915(0.0524)
0.3	RW	0.9993(0.0031)	2.8950(0.0112)	0.9983(0.0047)	3.0050(0.0285)
0.3	T+ad	0.9975(0.0045)	3.0345(0.0107)	0.9966(0.0061)	2.9020(0.0312)
0.3	RW + ad	0.9974(0.0009)	3.0180(0.0093)	0.9958(0.0051)	3.1380(0.0331)
0.2	T	0.9992(0.0011)	2.8435(0.0276)	0.9994(0.0005)	2.7880(0.0417)
0.2	RW	0.9985(0.0013)	2.8785(0.0115)	0.9989(0.0012)	2.9880(0.0273)
0.2	T+ad	0.9985(0.0014)	2.9965(0.0146)	0.9983(0.0014)	2.8780(0.0339)
0.2	RW + ad	0.9965(0.0020)	2.9875(0.0109)	0.9970(0.0020)	3.0815(0.0255)
0.1	T	0.9995(0.0003)	2.8460(0.0227)	0.9992(0.0006)	2.8395(0.0474)
0.1	RW	0.9988(0.0006)	2.9040(0.0227)	0.9988(0.0005)	3.0115(0.0300)
0.1	T+ad	0.9990(0.0005)	3.0445(0.0206)	0.9988(0.0007)	2.9135(0.0508)
0.1	RW + ad	0.9986(0.0008)	3.0410(0.0089)	0.9988(0.0007)	3.1175(0.0417)

Table 3.   Normalised results on a home-made Random Search on 1D Weierstrass functions for 2050 evalua-tions (as for table 1). GALib and EO find better results than the RS. A rel-atively constant overhead of nearly 0.3 seconds is added by both libraries.

h	best value Mean( )	CPU time Mean( )
0.5	0.9820(0.0096)	0.0593(0.0016)
0.4	0.9712(0.0100)	0.0525(0.0013)
0.3	0.9643(0.0072)	0.0416(0.0007)
0.2	0.9996(0.0004)	0.0283(0.0007)
0.1	0.9999(0.0001)	0.0153(0.0003)

Table 4.   Normalised results on a home-made Random Search on 2D Weierstrass functions for 12150 evalu-ations (similar to table 2). CPU times are still very small but on 12150 eval-uations, both libraries show a roughly constant overhead of nearly 3 seconds

h	best value Mean( )	CPU time Mean( )
0.5	0.5447(0.0072)	0.2021(0.0010)
0.4	0.4469(0.0061)	0.1472(0.0016)
0.3	0.3526(0.0034)	0.0895(0.0009)
0.2	0.2581(0.0005)	0.0422(0.0005)
0.1	0.1381(0.0001)	0.0126(0.0008)

The general conclusion of this paper is that EASEA allowed to create compa-rable EAs using dierent evolutionary libraries out of the same   .ez files, showing that the concept is working apparently correctly. This sound basis allows to infer

Table 5.   Comparison on a set of test functions to be   minimised , with a dimension
(genome size)   3. Population size (200+160) for 100 generations (16200 evaluations).
The selfadaptive gaussian mutation is much slower although it gives much better results
in some cases. On standard tournaments, EO is constantly 5 to 7 seconds slower than
GALib, while on selfadaptive mutations, where much more work is done on the genome,
EO is in average 6 seconds slower on small genomes (dim 3) and 18 seconds slower on
large genomes (dim 500), showing that the overhead of EO over GALib due to genome
manipulation (selfadaptation) varies from 0 to 12 seconds.

			GALib		EO	
n	Function	Alg.	best value Mean( )	CPU time Mean( )	best value Mean( )	CPU time Mean( )
3	Sphere	T	0.000(0.000)	1.268(0.059)	0.000(0.000)	6.138(0.124)
3	Sphere	T+ad	0.000(0.000)	2.912(0.070)	0.000(0.000)	13.46(0.110)
10	Sphere	T	0.000(0.000)	1.398(0.044)	0.000(0.000)	6.300(0.077)
10	Sphere	T+ad	0.000(0.000)	3.631(0.030)	0.000(0.000)	13.66(0.166)
50	Sphere	T	0.079(0.019)	1.804(0.009)	0.077(0.016)	6.502(0.053)
50	Sphere	T+ad	0.000(0.000)	8.007(0.043)	0.003(0.001)	18.60(0.096)
100	Sphere	T	0.328(0.051)	2.239(0.026)	0.339(0.052)	6.911(0.069)
100	Sphere	T+ad	0.046(0.058)	13.55(0.089)	0.082(0.025)	24.91(0.078)
500	Sphere	T	3.429(0.382)	5.627(0.076)	3.261(0.293)	10.91(0.076)
500	Sphere	T+ad	2.832(0.728)	57.48(0.125)	2.979(0.521)	75.62(0.158)
3	AckleyPath	T	0.000(0.000)	1.361(0.032)	0.000(0.000)	6.606(0.098)
3	AckleyPath	T+ad	0.000(0.000)	2.884(0.068)	0.000(0.000)	13.47(0.129)
10	AckleyPath	T	1.038(0.826)	1.557(0.015)	0.915(0.672)	6.454(0.047)
10	AckleyPath	T+ad	0.057(0.251)	3.753(0.068)	0.000(0.000)	13.91(0.072)
50	AckleyPath	T	6.139(0.617)	2.239(0.046)	6.043(0.569)	7.114(0.074)
50	AckleyPath	T+ad	4.249(1.289)	8.435(0.080)	4.048(0.759)	19.18(0.137)
100	AckleyPath	T	7.796(0.386)	2.996(0.039)	7.633(0.381)	7.910(0.080)
100	AckleyPath	T+ad	6.632(1.062)	14.26(0.069)	6.403(0.900)	25.88(0.167)
500	AckleyPath	T	10.02(0.364)	9.030(0.082)	9.908(0.390)	14.34(0.069)
500	AckleyPath	T+ad	10.04(0.611)	60.86(0.160)	9.885(0.596)	79.47(0.279)
3	Griewangk	T	0.003(0.004)	1.403(0.037)	0.002(0.003)	7.021(0.308)
3	Griewangk	T+ad	0.000(0.001)	2.892(0.073)	0.002(0.003)	15.04(0.651)
10	Griewangk	T	0.283(0.157)	1.598(0.069)	0.126(0.088)	6.566(0.085)
10	Griewangk	T+ad	0.053(0.048)	3.728(0.101)	0.012(0.011)	14.23(0.169)
50	Griewangk	T	8.242(2.461)	2.547(0.041)	8.129(1.240)	7.378(0.055)
50	Griewangk	T+ad	1.054(0.149)	8.776(0.084)	1.190(0.123)	19.36(0.092)
100	Griewangk	T	29.19(4.416)	3.744(0.041)	28.94(5.486)	8.561(0.048)
100	Griewangk	T+ad	5.175(4.182)	15.02(0.031)	8.097(2.884)	26.32(0.103)
500	Griewangk	T	299.8(25.30)	12.95(0.061)	720.0(31.58)	18.15(0.066)
500	Griewangk	T+ad	239.2(74.12)	64.79(0.146)	702.7(55.35)	83.03(0.157)

that more specific features of a library are equally well implemented by EASEA,
which should therefore be considered as a useable EA specification language.

Future work on EASEA testbenches will try to evaluate the quality of JAVA
implementations on the DREAM[6] and try to elaborate significant tests on specific features. A consequence of the present work is that potential users can find

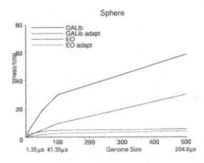 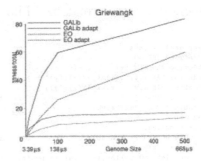

Fig. 3.   Percentage of CPU time spent on fitness computation with respect to total CPU time, below the horizontal axis, CPU times for one fitness computation are specified.

many implementation examples on the EASEA web page[1], where the EASEA Millennium Edition (v0.6c) compiler and its manual are available.

References

1. EASEA Millennium Edition   (v0.6c) page: http://www-rocq.inria.fr/EASEA/   .
2. EVONET home page: http://www.evonet.polytechnique.fr   .
3. EO home page: http:/eodev.sourceforge.net   .
4. P. Stearns,   ALex   AYacc home page: http://www.bumblebeesoftware.com   , Bumblebee Software Ltd.
5. M. Wall, GAlib home page: http://www.mit.edu/people/moriken/doc/galib   .
6. B. Paechter, T. Baeck, M. Schoenauer, A.E. Eiben, J.J. Merelo, and T. C. Fogarty, "A Distributed Resource Evolutionary Algorithm Machine," Proc. of CEC 2000.
7. I. Landrieu, B. Naudts, "An Object Model for Search Spaces and their Transformations,"EA'99 conference, Springer Verlag LNCS 1829, France, 1999.
8. B. Leblanc and E. Lutton, "Bitwise regularity and GA-hardness", ICEC 98, May 5-9, Anchorage, Alaska.
9. E. Lutton and J. L´ evy V´ehel, "H older functions and Deception of Genetic Algorithms", IEEE trans. on Evolutionary computation, Vol 2, No 2, pp. 56-72, 1998.
10. Z. Michalewicz, "Genetic Algorithms + Data Structures = Evolution Programs", Springer Verlag, 1992.
11. N. J. Radclie, Forma Analysis and Random Respectful Recombination," ICGA'91, pp. 222-229, 1991.
12. N. J. Radclie and P. D. Surry, Fitness variance of formae and performance prediction," FOGA'95, pp. 51-72, Morgan Kaufmann publ., 1995.
13. H.-P. S chwefel, "Collective phenomena in evolutionary systems",   31st annual meeting int. society for general system research   , Vol 2, pp. 1025-1033, Budapest, 1987.
14. H.-P. S chwetel, "Numerical Optimisation of Computer Models". John Wiley Sons, New-York, 1981. 1995 - 2nd edition.
15. P. D. Surry and N. J. Radclie, Formal Algorithms + Formal Representation = Search Strategies," PPSN'96, Springer Verlag LNCS 1141, pp. 366-375, 1996.
16. P. D. Surry, "A Prescriptive Formalism for Constructing Domain-Specific Evolutionary Algorithms," PhD thesis, Univ. of Edinburgh, 1998.

Appendix :   weiermax.ez   File

```
\User declarations :
  #define ITER 50
  #define Abs(x) ((x) < 0 ? -(x) : (x))
  #define MAX(x,y) ((x)>(y)?(x):(y))
  #define MIN(x,y) ((x)<(y)?(x):(y))
  #define SIGMA  0.1                        /* mutation parameter */
  double h=0.5, MaxTheo=0.;
\end
\User functions:
  double WM(double h,double y){        /* Weierstrass-mandelbrot function */
    double val=0., b=2.;
    for (int i=0;i<ITER;i++) val += pow(b,-(double)i*h) * sin(pow(b,(double)i)*y);
    return (MaxTheo - Abs(val));
  }
\end
\Initialisation function:
  if (argc>1) h = (double)atof(argv[1]);
  else {fprintf(stderr,"Holder exponent h = ? \n");  (void)scanf("%lf",&h);}
  for (int i=0;i<ITER;i++) MaxTheo+= pow(2.,-(double)i*h); // a majoration for the WM function
  fprintf(stderr,"Holder exponent h = %f    Global maximum = %f at 0.\n",h,MaxTheo);
\end
\User classes :
  GenomeClass { double x; }
\end
\GenomeClass::initialiser : // "initializer" is also accepted
  Genome.x=random(-1.,1.);
\end
\GenomeClass::crossover :
  double alpha = (double)random(0.,1.); // barycentric crossover
  if (&child1) child1.x = alpha*parent1.x + (1.-alpha)*parent2.x;
  if (&child2) child2.x = alpha*parent2.x + (1.-alpha)*parent1.x;
\end
\GenomeClass::mutator : // Must return the number of mutations
  Genome.x +=SIGMA*(double)random(-1.,1.);
  Genome.x = MAX(-1.,MIN(1.,Genome.x));                    // to stay inside [-1,1]
  return 1;
\end
\GenomeClass::evaluator : // Returns the score
  return (WM(h,Genome.x));
\end
\GenomeClass::display :
  fprintf(stderr,"Best value = %f at x = %f\n",WM(h,Genome.x),x);
\end
\Default run parameters :          // Please let the parameters appear in this order
  Number of generations : 50       // NB_GEN
  Mutation probability : 1         // MUT_PROB
  Crossover probability : 1        // XOVER_PROB
  Population size : 50             // POP_SIZE
  Selection operator : Tournament // RouletteWheel, Deterministic, Ranking, Random
  Offspring population size : 80% // 40%
  Replacement strategy : Plus      // Comma, SteadyState, Generational
  Discarding operator : Worst      // Best, Tournament, Parent, Random
  Evaluator goal : Maximise        // Minimise
  Elitism : On                     // Off
\end
```

Table 6.   Comparison of GALib and EO performances on a set of test functions to be minimised , dimension   3. Population size (200+160) for 100 generations (16200 evaluations). Results are fairly comparable, (although often slightly better for EO), showing that both libraries work the same way on the same settings. However, results are strangely worse and more erratic (huge sigma) for EO than for GALib on Griewangk 500 and Rosenbrock 500.

n	Function	Alg.	GALib		EO	
			best value Mean( )	CPU time Mean( )	best value Mean( )	CPU time Mean( )
3	Rastrigin	T	0.049(0.216)	1.313(0.046)	0.001(0.004)	7.317(0.212)
3	Rastrigin	T+ad	0.000(0.000)	2.735(0.074)	0.049(0.216)	14.81(0.1855)
10	Rastrigin	T	4.374(2.145)	1.517(0.039)	3.229(1.333)	6.667(0.087)
10	Rastrigin	T+ad	3.134(2.801)	3.577(0.078)	2.039(1.315)	14.39(0.146)
50	Rastrigin	T	179.5(18.89)	2.197(0.042)	172.5(20.70)	7.310(0.091)
50	Rastrigin	T+ad	23.35(6.926)	8.382(0.076)	32.86(8.383)	19.61(0.182)
100	Rastrigin	T	548.7(32.55)	2.945(0.017)	528.2(37.45)	8.081(0.070)
100	Rastrigin	T+ad	143.3(29.55)	14.20(0.071)	199.5(37.33)	26.12(0.135)
500	Rastrigin	T	4035(84.44)	8.971(0.086)	4026(83.80)	14.43(0.072)
500	Rastrigin	T+ad	2677(656.5)	60.78(0.135)	2629(364.2)	79.33(0.211)
3	Rosenbrock	T	1.221(0.019)	1.314(0.024)	1.217(0.013)	6.287(0.061)
3	Rosenbrock	T+ad	1.207(0.000)	2.754(0.089)	1.207(0.001)	13.38(0.146)
10	Rosenbrock	T	9.563(0.882)	1.452(0.034)	10.02(0.454)	6.381(0.069)
10	Rosenbrock	T+ad	9.051(0.528)	3.661(0.074)	9.115(0.331)	13.82(0.098)
50	Rosenbrock	T	94.01(10.05)	1.930(0.033)	99.54(11.49)	6.775(0.038)
50	Rosenbrock	T+ad	50.10(0.870)	8.153(0.063)	52.95(1.987)	18.76(0.115)
100	Rosenbrock	T	252.0(24.50)	2.475(0.042)	248.7(24.03)	7.344(0.053)
100	Rosenbrock	T+ad	126.7(10.53)	13.77(0.091)	144.6(18.93)	25.12(0.107)
500	Rosenbrock	T	1942(196.9)	6.676(0.083)	3809(170.8)	11.96(0.047)
500	Rosenbrock	T+ad	1846(305.4)	58.53(0.128)	3036(538.8)	76.72(0.195)
3	Schwefel	T	52.77(67.20)	1.345(0.029)	35.72(54.22)	6.702(0.279)
3	Schwefel	T+ad	41.45(56.49)	2.706(0.080)	22.56(45.34)	14.61(0.203)
10	Schwefel	T	1518(242.1)	1.600(0.032)	1626(236.3)	6.800(0.144)
10	Schwefel	T+ad	1466(192.5)	3.640(0.060)	1505(314.7)	14.97(0.276)
50	Schwefel	T	12885(1010)	2.510(0.043)	13412(859.6)	7.747(0.090)
50	Schwefel	T+ad	10372(1255)	8.670(0.081)	13163(2078)	20.91(0.411)
100	Schwefel	T	29442(1580)	3.580(0.042)	30894(1609)	8.860(0.129)
100	Schwefel	T+ad	24860(3573)	14.79(0.018)	27651(4078)	27.85(0.402)
500	Schwefel	T	182339(4546)	11.99(0.077)	183457(3638)	17.74(0.128)
500	Schwefel	T+ad	157329(14000)	63.82(0.149)	176812(13422)	83.69(0.492)
3	Weierstrass	T	0.331(0.303)	6.628(0.412)	0.312(0.345)	12.36(0.205)
3	Weierstrass	T+ad	0.242(0.262)	8.000(0.149)	0.601(0.338)	20.04(0.150)
10	Weierstrass	T	5.493(2.093)	18.32(0.167)	5.299(1.509)	24.31(0.397)
10	Weierstrass	T+ad	4.999(1.938)	20.42(0.061)	6.525(1.288)	32.69(0.245)
50	Weierstrass	T	48.87(7.055)	85.58(0.132)	37.85(4.086)	90.48(0.114)
50	Weierstrass	T+ad	41.26(5.506)	91.81(0.089)	42.21(3.516)	103.7(0.117)
100	Weierstrass	T	93.46(11.15)	168.7(0.115)	87.36(4.366)	173.5(0.099)
100	Weierstrass	T+ad	89.67(5.229)	181.0(0.089)	94.11(5.757)	192.7(0.183)
500	Weierstrass	T	539.3(17.25)	838.6(3.744)	524.9(7.661)	838.8(0.133)
500	Weierstrass	T+ad	547.4(17.44)	894.5(0.147)	553.0(14.10)	904.6(0.940)

Table 7.   Simple random search results for an equivalent number of function evaluations (16200) as in experiments of tables 5 and 6 (minimisation   ). This table shows that the libraries' overhead are constant for a given genome size: approximately 1.3 second for GALib for dimension 3, against 6 to 7 seconds for EO, and 60 seconds for GALib for dimension 500 and adaptive mutation against 73 seconds for GALIb. Here again, the libraries did their job by giving much better results than a simple random search.

n	Function	best value Mean( )	CPU time Mean( )
3	Sphere	0.0018(0.0013)	0.0220(0.0040)
10	Sphere	0.4588(0.0859)	0.0720(0.0040)
50	Sphere	8.9766(0.4934)	0.3390(0.0083)
100	Sphere	21.8519(0.7584)	0.6740(0.0080)
500	Sphere	140.8237(1.9817)	3.3190(0.0405)
3	AckleyPath	4.7354(0.9722)	0.0885(0.0036)
10	AckleyPath	16.6226(0.5800)	0.1775(0.0043)
50	AckleyPath	20.3691(0.0972)	0.6960(0.0049)
100	AckleyPath	20.7216(0.0458)	1.3435(0.0243)
500	AckleyPath	21.0645(0.0145)	6.5055(0.0619)
3	Griewangk	0.1525(0.0638)	0.0555(0.0050)
10	Griewangk	390.2570(5.1864)	0.2140(0.0111)
50	Griewangk	781.3926(42.4660)	1.0760(0.0196)
100	Griewangk	1969.6368(65.2537)	2.2330(0.0635)
500	Griewangk	15265.3389(139.6805)	10.8215(0.0576)
3	Rastrigin	3.2252(1.0738)	0.0415(0.0036)
10	Rastrigin	65.8087(6.6322)	0.1330(0.0046)
50	Rastrigin	651.1677(17.0618)	0.6550(0.0050)
100	Rastrigin	1449.1632(26.2853)	1.3025(0.0043)
500	Rastrigin	8349.9958(71.7607)	6.5110(0.0030)
3	Rosenbrock	1.8106(0.3473)	0.0285(0.0036)
10	Rosenbrock	1934.5497(69.1475)	0.0910(0.0030)
50	Rosenbrock	8781.0553(627.0606)	0.4390(0.0030)
100	Rosenbrock	24720.0889(1134.4771)	0.8830(0.0046)
500	Rosenbrock	194602.7597(4972.7807)	4.3490(0.0195)
3	Schwefel	60.1297(45.7104)	0.0600(0.0000)
10	Schwefel	1869.5170(159.9681)	0.1955(0.0050)
50	Schwefel	15452.0258(516.0859)	0.9660(0.0049)
100	Schwefel	34211.7308(579.2089)	1.9310(0.0030)
500	Schwefel	192109.9503(1375.4008)	9.6355(0.0050)
3	Weierstrass	0.8275(0.1597)	5.0175(0.0043)
10	Weierstrass	7.5288(0.4296)	16.7660(0.0482)
50	Weierstrass	57.4280(1.1064)	83.7745(0.2206)
100	Weierstrass	125.1542(1.6515)	168.5590(2.6685)
500	Weierstrass	693.9525(5.7427)	839.6965(0.1667)

Evolving Objects: A General Purpose Evolutionary Computation Library

M. Keijzer [1], J.J. Merelo [2], G. Romero [2], and Marc Schoenauer [3]

[1] Danish Hydraulic Institute
mak@dhi.dk
[2] GeNeura Team, Depto. Arquitectura y Tecnolog´   a de Computadores
Universidad de Granada (Spain)
todos@geneura.ugr.es   , http://geneura.ugr.es
[3] CNRS and Ecole Polytechnique, France
marc@cmapx.polytechnique.fr

Abstract.   This paper presents the   evolving objects library (EOlib), an
object-oriented framework for evolutionary computation (EC) that aims
to provide a exible set of classes to build EC applications.   EOlib design
objective is to be able to evolve any object in which fitness makes sense.
In order to do so, EO concentrates on interfaces; any object can evolve if
it is endowed with an interface to do so. In this paper, we describe what
features an object must have in order to evolve, and some examples of
how EO has been put to practice evolving neural networks, solutions to
the Mastermind game, and other novel applications.

1   Introduction

Evolutionary Algorithms (EAs) are stochastic optimization algorithms based on
a crude imitation of natural Darwinian evolution. They have recently become
more and more popular across many dierent domains of research, and people
coming from those external domains face a dicult dilemma: either they use an
existing EA library, and then have to comply to its limitation, or write their own,
which represent a huge work, and generally leads to   ... some other limitations
that their authors are not even aware of, mainly because these scientists are not
closely related to recent EA research.

For instance, evolving any kind of objects, (e.g. Neural Networks), has been
a dicult matter, mainly due to the lack of exibility of current evolutionary
computation libraries with respect to the representation used and the variation
operators that can be used on that representation. Most libraries (such as [41,
42]; see [19] for a comprehensive list) allow only a few predefined representations.

Evolving other types of data structures hence often has to start by   attening
them to one of the usual representations, such as a binary string, oating point
array or LISP tree. In the case of NNs, for instance, this representation has
to be decoded to evaluate the network (e.g. on a training set in the case of a
regression problem), but it sometimes lacks precision (e.g. in the case of binary
string representation), or expressive power: a string, whatever its shape, is a

P. Collet et al. (Eds.): EA 2001, LNCS 2310, pp. 231–242, 2002.
c Springer-Verlag Berlin Heidelberg 2002

serialization of a complex data structure, and evolution of a string using standard string-based variation operators makes keeping actual building blocks together more dicult than the evolution of a structure more closely representing neural nets, such as two arrays of weights together with biases for 3-layer perceptrons, or, more generally, an array of objects representing   ... neurons.

Similarly, most existing libraries propose only a limited range of ways to apply Darwinian operators to a population (e.g. limited to some proportional selection and generational replacement), or/and generally a single method for applying dierent kinds of variation operators to members of those population (e.g. limited to sequentially applying to all members of the population one crossover operator and one mutation operator, each with a given probability). However, there are numerous other ways to go, and the strong interaction among all parameters of an Evolutionary Algorithm makes it impossible to a priori decide which way is best.

This paper presents   EOlib, a paradigm-free evolutionary computation library, which allows to easily evolve any data structures (objects) that fulfills a small set of conditions. Moreover, algorithms programmed within   EOlib are not limited to basic existing EC paradigm like Genetic Algorithms, Evolution strategies, Evolutionary Programming or Genetic Programming, be it at the level of population evolution or variation operator application. Indeed, while all of the above do exist in EO, original experiments can easily be performed using   EOlib building blocks.

The rest of the paper is organized as follows: section 2.1 briey introduces EAs and the basic terminology, and also presents the state of the art in EA libraries. Section 3 presents Evolving Objects, a representation-independent, paradigm-independent, object-oriented approach to Evolutionary Computation. The rest of the paper discusses the   EO class library structure in section 4 and surveys some of the existing applications in section 5. Finally, section 6 concludes the paper and presents future lines of work.

2   Background

2.1   Evolutionary Algorithms

This section will briey recall the basic steps of an EA, emphasizing the interdependencies of the dierent components. The problem at hand is to optimize a given objective function over a given search space. A population of individuals (i.e. a P-uple of points of the search space) will undergo some artificial Darwinian evolution, in which the fitness of an individual is directly related to the values the objective function takes at this point.

After a (generally random) intialisation of the population, the generation loop of the algorithm is described in Figure 1

- Stopping criterion   (and statistics gathering): The simplest stopping criterion is based on the generation counter   t (or on the number of function evaluations). However, it is possible to use more complex stopping criteria,

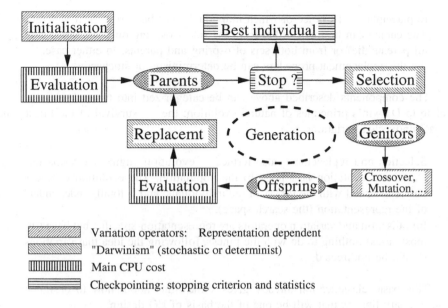

Variation operators: Representation dependent

"Darwinism" (stochastic or determinist)

Main CPU cost

Checkpointing: stopping criterion and statistics

Fig. 1.   Sketch of an Evolutionary Algorithm

which depends either on the evolution of the best fitness in the population along generations (i.e., measurements of the gradient of the gains over some number of generations), or on some measure of the diversity of the population.

- Selection   : Choice of some individuals that will generate ospring. Numerous selection processes can be used, either deterministic or stochastic. All are based on the   fitness of the individuals, directly related to the objective function. Depending on the selection scheme used, some individuals can be selected more than once. At that point, selected individuals give birth to copies of themselves, the genitors.
- Application of variation operators   : To each one of these copies some operator(s) are applied, giving birth to one or more ospring. These operators are generally stochastic operators, and one usually distinguish between crossover (or recombination) and mutation operators:
    crossover operators are operators from E^k (in most cases, k = 2) into E, i.e., some parents exchange genetic material to build up one ospring   [1].
    mutation operators are (generally) stochastic operators from   E into E.
- Evaluation   : Computation of the fitnesses of all newborn ospring. As mentioned earlier, the fitness measure of an individual is directly related to its objective function value. Note that in any real-world application, 99 of the total CPU cost of an EA comes from the evaluation part.

---

[1] Many authors define crossover operators from   E E into E E (two parents generate two ospring), but no signifiant dierence was ever reported between both variants.

– Replacement   : Choice of which individuals will be part of next generation.
The choice can be made either from the set of ospring only (in which case
all parents die) or from both sets of ospring and parents. In either case,
the this replacement procedure can be deterministic or stochastic.

The components described above can be categorized into two subsets, that
relate to Darwin's principles of natural evolution: the   survival of the fittest and
small undirected variations .

– Selection and replacement, also termed   evolution engine , describe the way
Darwinian evolution is applied to the population, The evolution engine is
only concerned with the fitness of the individuals, and is totally independent
of the representation (the search space).
– Initialisation and variation operators are representation-specific, but have (in
most cases) nothing to do with the fitness, following the idea that variations
should be undirected.

This basic classification already gives some hints about how to design an
evolutionary library, that will be one of the basis of EO design.

2.2   EA Libraries

A look at the Genetic Algorithms newsgroup FA [19] shows scores of
freeware EA libraries; but another look at the GA newsgroups (such as
news:comp.ai.genetic also show that very few people actually use them. The rule
is home-brew libraries. Most libraries are too hard to use, too restrictive (for
instance, restricted to only one EC paradigm), or just plain bad products.

A product stands out among the rest: Matthew's GALib [42], a widely used
evolutionary computation library, which includes several paradigms, several representations, and a good deal of variation operators. However, it lacks exibility
in a number of areas.

First, the choice of existing representations is also limited to arrays (of bits,
integers or oating point, or any combination), although it can be expanded
by sub-classing. However, evolving a neural network, for instance, would mean
squeezing it into an array.

Second, it only allows for two variation operators for each genome: mutation
and crossover (besides the initialization operator). Moreover, those operators
are always applied sequentially, and the only degree of freedom in that respect
are the probabilities of application. Hence, for instance, the popular experiment
involving an equidistributed random choice of several dierent mutations is not
straightforward. Similarly, there is no simple way to implement Evolution Strategy operators (self-adaptive mutation, or global recombination [2].

Last but not least, only scalar fitness is implemented, which makes it dicult
to add constraint handling techniques in a generic way, and almost impossible
to do multi-objective optimisation.

3   Evolvable Objects

The library introduced in this paper,   EOlib, has exibility designed from the ground up. This exibility owns everything to the object-oriented design: every data structure, every operator, every statistic computing routine is an object.

3.1   Data Structures

Any data structure can be evolved, if at least one variation operator is provided for such structures. A few pre-existing representations already exist, from the humble bitstring, up to and including GP parse trees and multilayer perceptrons.

What features does a data structure need to be evolvable within EO? It should be initialisable ; selectable   and replicable ; and either mutable   or combinable . These properties will be used as computational analogs for the three criteria for evolution outlined by Maynard-Smith [26], namely heredity, variability and fecundity and will be examined in turn:

- Initialisability   : This property, while essential in an EA, does not really have a natural counter-part in any of the biological models of evolution (of course, we don't consider here creativism as a model for evolution). It is generally also given little attention in existing libraries, as standard procedures exist for standard representations. However, even such standard procedures can be questionable in some situations [21]. Whatever, in EO, initialisers are themselves objects, which allows one to use more than one initialisation procedure, a common feature in GP for instance [3].
- Selectability   : One of the main components of Darwinian evolution is   natural selection , sometimes also seen as   survival of the fittest . In EO, like in all EA libraries, all objects are attached a fitness, and that fitness is used to perform such a selection. However, fitnesses in EO are not limited to scalar fitness (see section 3.2 below).
- Replicability   : It should be possible to obtain (possibly imperfect) copies of an object, be it by itself or through the use of other objects (replicators). This has a close analogy with the   Criterion of Heredity . It should also be possible to create objects from scratch, using   object factories .
- Mutability   : It is the first possible implementation of Maynard-Smith's   Criterion of Variability , that states that the genotype copying process has imperfections, thus ospring are not equal to the parent(s). Mutation increases the diversity of a population. Mutation operators, or   mutator , can change an Evolving Object in one or several ways, but the inner workings of the mutation need not be known from outside, neither a particular representation will be needed in order to mutate. The client can only be guaranteed that the object will change in some (generally stochastic) way.
- Combinability   : Another possible variation operator combine two or more objects to create a new one (in a similar way to GA's   crossover). This is not always possible, but when it is, the operation generally decreases diversity, in the sense that it makes the objects in the population more similar to

each other (although in some cases, such as binary crossover non-respectful of gene boundaries or the Distance Preserving Crossover of Merz and Freisleben [16], it could increase diversity). As it happens with mutation, the exact inner workings of recombination does not need to be known by the client. These objects will usually be called   combinators   or maters . One way to ensure a minimal meaningfulness of maters is to follow some of the rules of forma recombination [31]. Since in practice it's generally impossible for combiners to follow all of them [11], each combiner should follow at least one. Combinability can thus serve both as a heredity component and a variability component, this depending on the exact nature of the combination. Balancing heredity and variability is known in the field of Evolutionary Computation as the exploration/exploitation dilemma.

The good news is that most problems solved by computer can be implemented in data structures having these characteristics, including evolutionary algorithms themselves, which have been evolved already by Fogel and coworkers [15], Baeck [1] and Grefenstette [17]; indeed, in the EO framework, algorithms can be just another object, and multilevel evolutionary algorithm can be naturally fitted within the EO framework.

3.2   Fitness Function

The fitness in EAs is the only way to specify what represent the natural environment in natural evolution. In most EA libraries, unfortunately, fitness is limited to one single scalar value, and natural selection hence ends up being based on comparisons of those scalar values. However, such choice is very restrictive, and does not make provision for other selection mechanisms, such as selection based on constraints, based on several objectives, or more complex co-evolution processes involving either one population of partial solutions [9] or several competing or cooperating populations [30].

In EOlib, fitness can be of any type (more technically: all Evolutionary Objects are templatized over the fitness type), which opens the door to many other types of EAs. Of course, scalar (real-valued) fitness is still the most widely used, and most popular selections and replacements for real-valued fitness are available. But it is also possible to use fitnesses that are vectors of real numbers, and to design multi-objective [12] or generic constraint-handling selectors [29]. For instance, NSGAII selection [13] and adaptive segregated constraint handling [4] are already implemented in EO.

3.3   Variation Operators

Variation operators in EO are objects that exist outside the genotypes they act on: hence any number of variation operators can be designed for the same evolving data structure. Besides, variation operators can take any number of inputs and generate any number of outputs, allowing for instance easy implementation of orgy operators [14] or ES global recombination operators [2]. Furthermore, being separate objects, variation operators can own some private data: for instance, a special selector for choosing the mate of a first partner can be given to a crossover operator, allowing sexual preferences to be taken into account, as in [36,20]; all these private parameters can then be modified at run-time, allowing easy implementation of e.g. the standard deviation of Gaussian mutations in Evolution Strategies, either following the well-known one-fifth rule [32] or using self-adaptation mechanisms [2].

Variation operators can be combined in dierent ways. Two basic constructs exist: the sequential combination, in which variation operators are applied one after the other to the whole population with specific rates (as in Simple Genetic Algorithms for instance); the proportional combination, that chooses only one operator among the ones it knows, based on relative pre-defined weights. Furthermore, those two ways of combining variation operators can be recursively embedded. For instance, a very popular combination of operators is to mix dierent crossovers and dierent mutations within the Simple GA framework –which amounts to a sequential recombination of a proportional choice among the available crossovers and a proportional choice among the available mutations. Note that these constructs, being themselves objects, can be evolved at run-time, e.g. modifying the dierent rates based on past evolution.

3.4   Evolution

Evolution engines can be given in dierent ways: Of course, most popular engines (e.g. Generational GA, Steady-State GA, EP, both ES+ and ES, strategies) are available. But also, all parameters of an evolution engine can be specified in great details: the selection operator and its parameters, the number of ospring to generate, the proportion of strong elitism (best individuals are copied onto the next generation regardless of ospring), the replacement procedure (whether it involves the parents or not), the weak elitism (replace the worst individual in the new population by the best parent if the best fitness is decreasing),   ...Hence new evolution engines can be defined simply by assembling existing EO building blocks.

4   Technical Description

All the EO ideas have been put in practice in the   EOlib class library, an Open Source C++ library which is available from   http://eodev.sourceforge.net , together with all facilities of open project in SourceForge: several mailing lists, CVS access to the source tree, bug reporting,   ...The current version is 0.9.1, the leading zero in the version indicates that it is not yet complete.   EOlib needs an ANSI-C++ compliant compiler, such as the Free Software Foundation   gcc (in Linux, other Unix avors or the CygWin environment for Win95/98/NT); most classes also work with commercial compilers such as Microsoft's Visual C++ 6.0.

Besides the "evolutionary classes" mentioned in the previous section, general facilities for EC applications, such as check-pointing for stopping and restarting applications, multiple statistics gathering, graphic on-line representation that uses gnuplot in Linux are also provided. Moreover,   EOlib is open: using existing tutorial template files, implementing one's own new statistics and displaying it on-line, for instance, is straightforward.

There are two ways to use EOlib. The most frequent case is when your representation is already defined in EO (be it bitstrings, real-valued vector or parse-trees), and you simply want to optimize a given fitness function. The only thing that has to be programmed is that fitness computation function, and all other components of the algorithm can simply be input as program parameters.

On the other hand, using an ad hoc representation requires coding the basic representation-dependent parts of the algorithm: the definition of the representation, the initialisation and the variation operators (see section2.1)   . . . and the fitness function, of course. Similarly, testing a new selection for instance can be done by simply plugging it into an existing EO program, everything else being unchanged. Template files are provided in the tutorial section to help the programmer write his/her own components.

One further plan is to provide an   object repository , so that if something is programmed using EOlib, the object classes can be immediately posted for everyone to use them. One major outcome would be to improve the reproducibility of EC results: whereas a paper is written using   EOlib, the source code of all experiments would be available, and further research could actually use it as a starting point. A link could be added with a paper repository, such as the one the European Evolutionary Computation Network of Excellence, EvoNet, is designing.

5   Applications

EO, so far, has been applied to a number of dierent areas. The great exibility of the library has been used to implement complex representations (e.g. multi-layer perceptron, Voronoi diagrams,   . . .), together with their specific variation operators, multi-objective optimization, specific constraint handling mechanisms, hybrid mutation operators, . . .

- Evolving multilayer perceptrons   [8] no binary or oating point vector representation was used; the objects that were evolved were the multilayer perceptrons themselves. The EO class was used for the population-level operators, but new diversity-generation operators had to be designed: add or eliminate a hidden layer neuron, hidden layer crossover, and mutate initial weights. The back-propagation training algorithm was also used as mutation operator [7]. This application is available also from   http://geneura.ugr.es/pedro/G-Prop.htm .
- Genetic Mastermind.   In the case of the game of Mastermind [27], a GA was programmed to find the hidden combination, improving results ob-

tained in previous implementations.The subject of evolution were the mastermind solutions themselves. The variation operators were also adapted to these objects: a permutation and a   creep operator, which substituted a number (color) by the next, and the last by the first. A huge improvement was obtained; the algorithm explored only 25 of the space that was explored before [5], that is, around 2 of the total search space, and thus obtained solutions much faster. The game can be played online at http://geneura.ugr.es/jmerelo/GenMM   ; the code can be downloaded from the same site.

- Evolution of fuzzy-logic controllers [33]: bidimensional fuzzy-logic controllers were evolved to approximate two-variable functions; variation operators added and subtracted row and columns, and changed values of precedents and consequents. The evolved object approximated the function, and besides, found a proper number of rows and columns for the controller.

- Evolution of RBF neural nets [34]: data structures representing RBFs with diverse radii in each dimension are evolved; variation operators add and subtract RBFs, and change the position of the centers and the value of the radii. Evolved RBFs are usually smaller and more accurate than other found by trial-and-error or incremental procedures.

- Evolutionary voice segmentation [28]: the problem consists in finding the right division of a speech stream, so that dierent words, phrases, or phonemes can be separated; EO evolves segmentation markers, with very good results. In this case, the evolved data structure are deltas with respect to a linear segmentation.

- As a plug-in to EOlib, a visualization tools that uses Kohonen's Self-Organizing Map [25] has been presented in [35]. This tool presents, after training, a two-dimensional map of fitness to the attened, one-dimensional vector representation of a chromosome, allowing to assess the evolutionary process by checking that it has explored eciently the search space.

- A parallel version of EOlib using MPI and PVM is in development; the MPI version has been tested on several benchmark problems [6].

- EOlib has been applied to image segmentation in [39,40], which applies genetic algorithms to a stripe straightening algorithm used to process and then compress fruit y embryo images.

- A dicult problem of car engineering, in which the very costly objective function has been replaced by a surrogate cheap model, has been recently tackled using the a combination of multi-objective and constraint-handling techniques (see [37], submitted to the same conference).

- A hybrid surrogate mutation operator has been implemented and tested for parametric optimization. The first results, also submitted to the same conference, are very promising [38].

- Topological optimum design of structures has been a long-time research of on of the authors [24]. However, it was recently ported into EO framework [18] as it is basically a multi-objective problem (minimizing both the weight of the structure and the maximal displacement under a given loading). Within EO, it has been possible to really compare both approaches, as they use

exactly the same representation and variation operators (including the way they are applied).
- Adaptive Logic Programming   [22] A variable length chromosome was used to steer a path through a logic program in order to generate (constrained) mathematical expressions. Using EO, it was possible to compare the results with a tree-based genetic programming approach [23].

6   Conclusion

In this paper, we had the ambitious objective of presenting a new framework for evolutionary computation called EO, that would include all evolutionary computation paradigms as well as new ones, with novel data structures evolved, general or particular variation operators, and any population-level change operators.

EO has a practical implementation in the shape of the EO class library, which is public and freely available under the LGPL (FSF's Library, or lesser, general public license) from http://eodev.sourceforge.net . This library has already been applied to problems in which, traditionally, binary or oating point representations were used, using instead as evolving object the same data structure one want to obtains as a result, such as a neural net or a bidimensional fuzzy logic controller.

As possible lines of future work, we will try the implementation of   EOlib in dierent OO languages, such as Java, and its interoperability with each other. Another feature is an application generator, that will use high-level evolutionary computation languages such as EASEA [10], and an operating-system independent graphical user interface.

Acknowledgements.   This work has been supported in part by FEDER I+D project 1FD97-0439-TEL1, CICYT TIC99-0550, and INTAS 97-30950.

References

1. T. B ack. Self-adaptation in genetic algorithms. In F. J. Varela and P. Bourgine, editors, Procee dings of the First European Conference on Artificial Life. Toward a Practice of Autonomous Systems , pages 263–271, MIT Press, Cambridge, MA.
2. Th. B ack and H.-P. S chwefel. An overview of evolutionary algorithms for parameter optimization.   Evolutionary Computation , 1(1):1–23, 1993.
3. W. Banzhaf, P. Nordin, R.E. Keller, and F.D. Francone.   Genetic Programming An Introduction On the Automatic Evolution of Computer Programs and Its Applications . Morgan Kaufmann, 1998.
4. S. BenHamida and M. Schoenauer. An adaptive algorithm for constrained optimization problems. In M. Schoenauer et al., editor,   Procee dings of the 6th Conference on Parallel Problems Solving from Nature , pages 529–539. Springer-Verlag, LNCS 1917, 2000.
5. J. L. Bernier, C. Ilia Herr´ aiz, J. J. Merelo, S. Olmeda, and A. Prieto. Solving mastermind   using GAs and simulated annealing: a case of dynamic constraint optimization. In   Parallel Problem Solving from Nature IV , pages 554–563. Springer-Verlag, LNCS 1141, 1996.

6. J. G. Castellano, M. Garc´ a-Arenas, P. A. Castillo, J. Carpio, M. Cillero, J. J. Merelo, A. Prieto, V. Rivas and G. Romero. Objetos evolutivos paralelos. In XI Jornadas de Paralelismo , Universidad de Granada Depto. ATC, pages 247–252, 2000.

7. P. A. Castillo, J. Gonz´ alez, J. J. Merelo, A. Prieto, V. Rivas, and G. Romero. G-Prop-III: Global optimization of multilayer perceptrons using an evolutionary algorithm. In GECCO99 , 1999.

8. P.A. Castillo, J.J. Merelo, V. Rivas, G. Romero, and A. Prieto. Evolving Multilayer Perceptrons. Neural P rocessing Letters 12(2):115–127, 2000.

9. P. Collet, E. Lutton, F. Raynal, and M. Schoenauer. Polar ifs + individual gp = ecient inverse ifs problem solving. Genetic Programming and Evolvable Machines , 1(4), 2000.

10. P. Collet, E. Lutton, M. Schoenauer, and J. Louchet. Take it easea. In M. Schoenauer et al., editor, Procee dings of the 6th Conference on Parallel Problems Solving from Nature , pages 891–901. Springer Verlag, LNCS 1917, 2000.

11. Carlos Cotta, Enrique Alba, and Jos´ e M. Troya. Utilizing dynastically optimal forma recombination in hybrid genetic algorithms. In Thomas Back Agoston E. Eiben, Marc Schoenauer, editor, Parallel Problem Solving From Nature – PPSN V, pages 305-314. Springer Verlag, LNCS 1498, 1998.

12. K. Deb. Multi-Objective Optimization Using Evolutionary Algorithms . Chichester, UK: Wiley, 2001.

13. K. Deb, S. Agrawal, A. Pratab, and T. Meyarivan. A fast elitist non-dominated sorting genetic algorithm for multi-objective optimization: Nsga-ii. In M. Schoenauer et al., editor, Procee dings of the 6th Conference on Parallel Problems Solving from Nature , pages 849–858. Springer-Verlag, LNCS 1917, 2000.

14. A.E. Eiben, P.-E. Raue, and Z. Ruttkay. Genetic algorithms with multi-parent recombination. In Y. Davidor, H.-P. S chwefel, and R. Manner, editors, Procee dings of the 3rd Conference on Parallel Problems Solving from Nature , pages 78–87. Springer Verlag, LNCS 866, 1994.

15. D. B. Fogel, L. J. Fogel, and J. W. Atmar. Meta-evolutionary programming. In R. R. Chen, editor, Procee dings of 25th Asilomar Conference on Signals, Systems and Computers , pages 540–545, Pacific Grove, California, 1991.

16. B. Freisleben and P. Merz. A genetic local search algorithm for solving symmetric and asymmetric traveling salesman problems. In Oriceedubgs if tge 1996 IEEE International Conference on Evolutionary Computation , pages 616–621. IEEE Press, 1996.

17. J.J. Grefenstette. Optimization of control parameters for genetic algorithms. IEEE Transactions on Systems, Man and Cybernetics, SMC-16 , 1986.

18. F. Jouve H. Hamda, E. Lutton, M. Schoenauer, and M. Sebag. Compact unstructured representations in evolutionary topological optimum design. Intl J. of Applied Intelligence , 2001. To appear.

19. J org Heitk oter and David Beasley. The hitch-hiker's guide to evolutionary computation, (FA for comp.ai.genetic). Available from http://surf.de.uu.net/encore/www/ .

20. R. Hinterding and Z. Michalewicz. Your brain and my beauty. In D.B. Fogel, editor, Procee dings of the Fifth IEEE International Conference on Evolutionary Computation , IEEE Press, 1998.

21. L. Kallel and M. Schoenauer. Alternative random initialization in genetic algorithms. In Th. B ack, editor, Procee dings of the 7th International Conference on Genetic Algorithms , pages 268–275. Morgan Kaufmann, 1997.

22. M. Keijzer, V. Babovic, C. Ryan, M. O'Neill and M. Cattolico Adaptive Logic Programming. In GECCO01 , 2001.
23. M. Keijzer, C. Ryan, M. O'Neill, M. Cattolico and V. Babovic Ripple Crossover in Genetic Programming. In EuroGP 2001 , 2001.
24. C. Kane and M. Schoenauer. Topological optimum design using genetic algorithms. Control and Cybernetics , 25(5):1059–1088, 1996.
25. Teuvo Kohonen. Self-Organizing Maps . Springer, Berlin, Heidelberg, 1995.
26. J. Maynard-Smith. The theory of evolution . Penguin, 1975.
27. J. J. Merelo, J. Carpio, P. Castillo, V. M. Rivas, and G. Romero. Finding a needel in a haystack using hints and evolutionary computation: the case of genetic mastermind. In Late breaking papers at the GECCO99 , pages 184–192, 1999.
28. J. J. Merelo and D. Milone. Evolutionary algorithm for speech segmentation. Submitted , 2001.
29. Z. Michalewicz and M. Schoenauer. Evolutionary Algorithms for Constrained Parameter Optimization Problems. Evolutionary Computation , 4(1):1–32, 1996.
30. J. Paredis. Coevolutionary computation. Artificial Life , 2:355–375, 1995.
31. N. J. Radclie. Equivalence class analysis of genetic algorithms. Complex Systems , 5:183–20, 1991.
32. I. Rechenberg. Evolutionstrategie: Optimierung Technisher Systeme nach Prinzipien des Biologischen Evolution . Fromman-Hozlboog Verlag, Stuttgart, 1973.
33. V.M. Rivas, J. J. Merelo, I. Rojas, G. Romero, P.A. Castillo, and J. Carpio. Evolving 2-dimensional fuzzy logic controllers. Submitted.
34. V. Rivas, P. Castillo, and J. J. Merelo. Evolving RBF neural nets. In Procee dings IWANN'2001 , Springer-Verlag, LNCS, 2001. To appear.
35. G. Romero, M. Garc ía-Arenas, J. G. Castellano, P. A. Castillo, J. Carpio, J. J. Merelo, A. Prieto, and V. Rivas. Evolutionary computation visualization: Application to G-PROP. pages 902–912. Springer, LNCS 1917, 2000.
36. E. Ronald. When selection meets seduction. In L. J. Eshelman, editor, Procee dings of the 6th International Conference on Genetic Algorithms , pages 167–173. Morgan Kaufmann, 1995.
37. O. Roudenko, T. Bosio, R. Fontana, and M. Schoenauer. Optmization of car front crash members. In EA'01 , 2001. Submitted.
38. K. Abboud, and M. Schoenauer. Hybrid surrogate mutation: preliminary results. In EA'01 , 2001. Submitted.
39. A.V. Spirov, D.L. Timakin, J. Reinitz, and D Kosman. Experimental determination of drosophila embryonic coordinates by genetic algorithms, the simplex method, and their hybrid. In Procee dings of Second European Workshop On Evolutionary Computation In Image Analysis And Signal P rocessing , April 2000.
40. A.V. Spirov and J. Reinitz. Using of genetic algorithms in image processing for quantitative atlas of drosophila genes expression. Available from http://www.mssm.edu/molbio/hoxpro/atlas/atlas.html   .
41. A. Tang. Constructing GA applications using TOLKIEN. Technical report, Dept. Computer Science, Chinese University of Hong Kong, 1994.
42. M. Wall. Overview of GALib. http://lancet.mit.edu/ga , 1995.

Backwarding: An Overfitting Control for Genetic Programming in a Remote Sensing Application

Denis Robilliard and Cyril Fonlupt

Universit´e du Littoral-Cˆ ote d'Opale
LIL
BP 719
62228 Calais Cedex, France
robillia@lil.univ-littoral.fr
phone: +33-321 465 667

Abstract. Overfitting the training data is a common problem in supervised machine learning. When dealing with a remote sensing inverse
problem, the PAR, overfitting prevents GP evolved models to be successfully applied to real data. We propose to use a classic method of overfitting control by the way of a validation set. This allows to go backward
in the evolution process in order to retrieve previous, not yet overfitted
models. Although this "backwarding" method performs well on academic
benchmarks, there is not enough improvement to deal with the PAR. A
new backwarding criterion is then derived using real satellite data and
the knowledge of plausible physical bounds for the PAR coecient in
the geographical area that is monitored. This leads to satisfactory GP
models and drastically improved images.

1 Introduction

One central problem in supervised machine-learning is the overfitting problem.
The learning algorithm is usually evaluated on a training set, so one possible
pitfall is simply to learn by heart the examples (or training cases) from the
training set, and answer at random in any other cases. Thus the model derived
from the learning algorithm seems perfect as long as it is confronted to already
seen data, but lacks generalization. Of course most machine learning schemes
avoid the extreme solution we just described, but still tend to adapt so well to the
distinctive characteristics of the training set that they lost some generalization
ability. In the framework of Genetic Programming (GP), overfitting tends to
appear in later generations, when the error measured over the training set slowly
decreases as the search progresses, while the error measured over an independent
validation set typically decreases during the first generations, and then increases
while the learning set biases are more and more perfectly learned.

Overfitting has been addressed by many people, notably in the case of decision tree learning [1,2], with such techniques as post-pruning the tree, stopping
earlier the growth of the tree, or adding noise to the training cases. Overfitting
is also a signi6ant diculty for artifiial neural networks (ANN) and some techniques have been issued to decrease this factor like the weight decay method [3],

P. Collet et al. (Eds.): EA 2001, LNCS 2310, pp. 245–254, 2002.
ⓒ Springer-Verlag Berlin Heidelberg 2002

that tries to keep the weights of the ANN small. One of the standard methods for controlling overfitting is to provide a validation data set to the algorithm in addition to the training data set. As quoted by Mitchell [3], in the case of ANN, "two copies of the network weights are kept: one copy for training and a separate copy of the best-performing weights so far, measured by their error over the validation set. Once the trained weights reach a significantly higher error over the validation set than the stored weights, training is terminated and the stored weights are returned as the final hypothesis". Even if the overfitting issue is also problematic for the Genetic Programming paradigm, it has not yet been deeply studied. As explained by Banzhaf   et al [4] (page 230), "as GP is very computationally intensive, GP researchers frequently opt for no statistical validation and do not take into account the overfitting process".

In this paper we adapt the scheme that is quoted above, using a validation set as in [3]. We call this scheme "backwarding", since we do not use the solutions provided by the later generations of GP, but rather go backward in the evolution process until the point when overfitting is not yet too important. This scheme is tested on a set of inverse regression problems, resulting in a noticeable increase in precision. Then we turn to a real-world inverse problem, the Photosynthesis Available Radiation problem a.k.a. PAR problem and show that, in this case, both basic GP and a simple implementation of backwarding GP are unable to deal with the great variability of the PAR coecient: we failed in obtaining a simulated validation set representative enough for controlling overfitting. To avoid the need for a large simulated validation set, very costly in computer time, we propose to consider validation just like if we were facing a classification rather than a regression problem. We use a new validation criterion based on whether or not a set of satellite image pixels are out of plausible bounds for the PAR coecient, when inverted by our GP models. The lower and upper bounds are derived from expert knowledge on the geographical area that is monitored by the satellite. Backwarding can then reduce overfitting and the quality of satellite images is drastically improved. This article may be seen as a sequel to a previous paper published at the PPSN'2000 conference [5] where a closely related remote sensing problem, Ocean Color, was tackled.

In Section 2 of this paper, we remind the implementation of the backwarding method for the GP paradigm, and test results are given for some academic problems. Section 3 presents the Photosynthesis Available Radiation problem (PAR). Then Sect. 4 shows how the backwarding method was adapted and successfully applied to the PAR.

2   The Backwarding Mechanism

2.1   Algorithm Description

As explained in the introduction, overfitting can be very harmful for Genetic Programming. Nonetheless it does not seem to have been deeply studied in the GP literature [4]. In the dierent Kozaš books, the overfling factor has only been very quickly looked at. For instance, in [6], when trying to evolve a program to classify whether or not a segment of a protein sequence is a transmembrane domain, he faced the overfitting factor when evolving the programs. Actually, two fitness sets were used. A so-called   in-sample   set of fitness cases was used during the learning procedure while the true measure of performance of the evolved algorithm was performed on an   out-sample   set of fitness cases (not used during the learning phase). As explained by Koza, after some generations the evolved predicting programs are being more and more fitted to the idiosyncrasies of the particular in-sample fitness cases. However, even if the overfitting process was mentioned in this book, no solutions were proposed to deal with it.

In this paper, we propose to use a scheme, hereafter called   backwarding, inspired from a well known method in the machine learning community. At each generation of the GP algorithm, if there is a new best program with respect to the training set, this program is also compared to the last best one with respect to an independent validation set. A solution that improves on both sets is then stored and the last such solution is returned at the end of the GP process. The GP is stopped after a fixed number of generations as is usual. This implies that some computing time is wasted on calculating overfitted solutions, but the focus here is not on saving time by detecting when overfitting appears but rather on retrieving non overfitted solutions. The backwarding algorithm is summed up in Tab. 1.

Table 1.   Main steps of the backwarding GP algorithm.

```
run one generation of GP algorithm
best-training := best-of-generation on training set
best-validation := best-of-generation on training set
for all other generations of GP algorithm do
        compute new generation as usual
        if best-of-generation    best-training on training set then
            best-training := best-of-generation
            if best-training    best-validation on validation set then
                    best-validation := best-training
            endif
        endif
done
return best-validation program as GP solution
```

2.2   Preliminary Results

In this section, we show how the backwarding method can improve results on some inverse regression problems. We try to invert three functions: $f_1(x) = x^4 + x^3 + x^2 + x$, $f_2(x) = x^3$ $\cos(x)$ $\sin(x)$ $(\sin^2(x)$ $\cos(x)$ $1)$, and $f_3(x) =$

$x^2 + \exp(x) + \overline{2}$. Standard GP parameters (number of generations, genetic operator rates, ...) were used. The parameters detail and the data files are available at URL http://www-lil.univ-littoral.fr/robillia/Research/Backwarding/. Experiments were conducted with the lilGP library [1], slightly modified to add the backwarding algorithm. In Tab. 2 we show, for both basic and backwarding GP, the performance on the learning set, on the validation set and on an independent test set (bold figures). The figures gives the approximation error averaged over 10 runs, computed with Equ. 1. The fourth column gives the average size of the solutions, i.e. the number of program tree nodes.

$$\text{error} = \frac{1}{n} \sum_{i=1}^{n} C_{computed} \quad C_{expected} \tag{1}$$

Table 2.   A brief comparison of basic and backwarding GP on some test problems. First 3 columns are errors measured on 3 dierent data sets, the last column is the average size of solutions (rounded).

Problem	basic GP			
	learning	validation	test	size
f1	198.01	289.47	269.58	340
f2	74.46	100.91	124.05	232
f3	5.13	5.43	5.44	189

Problem	backwarding GP			
	learning	validation	test	size
f1	217.02	242.77	217.22	160
f2	78.21	91.91	106.92	213
f3	5.15	5.39	5.40	156

For all three problems, backwarding leads to more robust solutions, as can be seen on the independent test set. Notice that the backwarding solutions are smaller in size, which could be interpreted favorably in light of the "Occam's razor" principle. These good results were expected since this technique is derived from a well tested machine learning scheme. This method could help to deal with the "brittle" criticism that is often argued against GP models.

3   The Photosynthesis Available Radiation Problem

Remote sensing is a very active research area among biologists and physicists. One main goal of remote sensing applications is to monitor the evolution of ocean water characteristics. Notably, the primary production plays a key role for the evaluation of the global carbon cycle and is thus of great scientific concern, notably to understand the so-called greenhouse eect. The   photosynthesis available

---

[1]   URL http://isl.cps.msu.edu/GA/software/lil-gp

radiation also known as PAR is the number of photons available for photosynthesis in the visible (400   700nm) wavelength interval. The PAR values are often stated in photons .s $^{-1}$ m $^{-2}$.

Estimating the PAR is a step towards obtaining the primary production, but it is a dicult problem: the light coming from the sun and going through the earth atmosphere, is modified by solar radiance scattering and absorption from the air, from particle suspended in the air, and finally, from within the ocean, water molecules and dissolved and suspended particles [2] (for instance, the absorption and scattering properties of the water depend on the phytoplankton in the water). After light enters ocean, some of it is eventually scattered back up through the surface. This light is called the water-leaving radiance, and it can be detected from space. Generally, less than 10 of the total light detected by the satellite is water-leaving radiance. Radiative transfer analysis for water is also complicated by the fact that water optical properties, as well as those of biological constituents, are spectrally dependent. Waters are roughly categorized as:

- case I waters (also called K1): it corresponds to the open ocean (90 of the ocean can be viewed as K1), where phytoplankton dominates.
- case II waters (also called K2): it corresponds to coastal waters, where sediment and yellow substance may have to be taken into account, depending on the specific geographic area that is monitored. The K2 is known to be a much harder problem than K1 and is still mostly unexplored. As a matter of fact, the K2 is the most important, as about 50 of the world population lives next to the coastal waters. This setup is typical of the English Channel and the North Sea. The experiments presented in this paper are representative of the K2.

Solutions have been proposed to solve the direct problem, that is simulating the amount of radiations received by a satellite spectrometer using models of reectance derived either from empirical data [7,8], or from a radiative transfer code such as the OSOA model [9,10]. OSOA is based on the successive orders of scattering method. It makes use of the Mie theory, and takes into account the inuence of marine particles on the polarization of the water-leaving radiance. We are interested here in the inverse problem,   i.e. estimating the energy available for photosynthesis from the water-leaving radiance.

Formally, let L be the the the signal emitted by the sun. This signal is partly absorbed and partly reected by the sea water and its constituents. The energy level of the reected part of   L can be measured by a satellite spectrometer. This reected energy is measured on a set of specified   wavelengths. The set of monitored wavelengths of the "SeaWIFS" sensor is 410 nm, 443 nm, 490nm, 510 nm, 560 nm, and 665 nm. The amount of light available for photosynthesis is spectrally dependent. Moreover, the available energy transferred in the water column must be modeled, and we use the following formula:

$$K_d( ) = \frac{1}{z} \log(\frac{E_0(, z)}{E_0(, z_{surf})})$$

---

[2] from the MODIS web site

where z indicates the depth in the water column, $E_0(, z_{surf})$ is the available energy at the sea surface at   wavelength   and $E_0(, z)$ is the available energy at wavelength   at depth   z for photosynthesis.

In order to estimate the amount of light available for photosynthesis ($E_0(, z)$)) and to solve the PAR problem , one has to compute the attenuation coecient $K_d()$ and then get E_0 by applying the formula. This is far from being a trivial problem as the K_d coecient depends on the   wavelength as well as the various marine constituents concentrations. In this study, we focus on computing this coecient K_d from the measured reected energy. We face here a regression problem, searching a function   f such as:

$$K_d( ) = f(L_{410}, L_{443}, L_{490}, L_{510}, L_{560}, L_{665})$$

The training cases consist of the inputs (the set of reected energy in selected wavelengths) and the expected output (the attenuation coecient). Note that it could be possible but economically not feasible, to simultaneously record the values of the reectance measured on the satellite spectrometer and send boats to analyze the water column, and to get real data for learning. Thus, one main interest of having simulation codes for the direct problem is the ability to generate training and validation data sets.

4   Experimental Results

To our knowledge, this is the first application of genetic programming to the PAR problem. In the next subsection we present results obtained with basic GP. Although the GP model matches the learning data, it fails in inverting real satellite images, even when applying the backwarding method as it is explained above. However, it appears possible to consider the PAR as a classification rather than a regression problem, when dealing with the backwarding part of the algorithm. This, in Sect. 4.2, allows to select GP models that perform well with real data.

4.1   Basic GP and Simple Backwarding

In these firsts experiments, only a training set was used and then an additional validation set was added to allow backwarding. The parameters setting is given in Tab. 3. As the attenuation coecient is spectrally dependent and as the algorithm is aimed at using the SeaWIFs data, the attenuation coecient is computed for the 6 dierent   wavelengths.

These results are compared with a method based on ratios of radiance   wavelengths developed in Devred's PhD [11]. Tab. 4 sums up the results for the 6 dierent   wavelengths (error is relative RMS error, see Eq. 2). Due to their stochastic natures, all GP experiments were run 10 times and the best-of-all was chosen. Except for $K_d(560)$, the GP results are an improvement over the traditional method and these figures are considered to be very good by physicists.

Table 3.   Parameters setting: PAR problem

Objective	Compute attenuation coecient K_d for 412, 443, 490, 510, 560 and 670nm
Function set	+,-,,/
Fitness case	250 results of radiative transfer simulation
Population size	5000
Maximum number of generations	250
Crossover probability	85
Copy probability	10
Mutation probability	5
Selection	tournament of size 5
Maximum tree depth	10

Table 4.   GP scheme versus traditional method

Coe. (K_d)	rel. RMS (GP) ()	rel. RMS (Devred's method) ()
412nm	6.5	6.85
443nm	3.9	7.0
490nm	2.4	9.2
510nm	4.7	N/A
560nm	2.3	1.8
665nm	1.5	3.8

$$\text{relative RMS} = \frac{1}{n} \sum_{i=1}^{n} \left(\frac{C_{computed} - C_{expected}}{C_{expected}} \right)^2 \tag{2}$$

with n the size of the training set
and C the value to be approximated

However, when the evolved program was applied to real data obtained by the satellite sensors, most computed points on the image were out of range. Fig. 1 shows two images taken over the Channel in September 99, computed by the GP model: only a small part of these images can be computed and displayed (e.g. on the left image, 8175 points out of 15000 are outside of the plausible range). It is very plausible that we face an overfitting problem, since the training set is almost perfectly matched. Furthermore, even when using backwarding with a validation set of 250 new values, the results were slightly better, but did not significantly improve the quality of the images.

4.2   Adapting the Backwarding Method for the PAR

From the physicists and biologists experience, we learned that small modifications of the radiance can lead to very dierent values for the K_d especially with the 412 and 443 channels. Thus training and validation sets should be very large

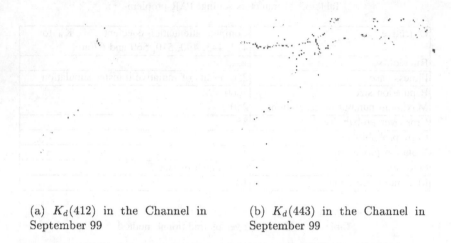

(a) $K_d(412)$ in the Channel in   (b) $K_d(443)$ in the Channel in
September 99   September 99

Fig. 1.   Two bad images, based on overfied GP models for the PAR coecient.

to be useful, in order to sample the diversity of values encountered with real data.

Unfortunately validation and learning data are computationally very expensive to produce. Instead of using a limited validation set, we propose to use a criterion derived from real data in the following way:

1. 10 images of the Channel were selected (note that in our case, an image consists in a set of 6 matrices, one for each   wavelength)
2. 20000 pixels were randomly chosen in these images. These pixels will make our validation set. We do not know of course the expected output values for these pixels, but according to the biologists these values can be bounded. For instance, in the Channel, $K_d(443)$ values may range between 0 and 1   .6. Our validation criterion will be the number of out of bounds points.
3. run the backwarding algorithm as introduced in Sect. 2. The learning set remains the same as in the previous experiment, but when it comes to the validation set, the evolved programs are tested with our new criterion: the lowest number of pixels out of bounds, the better the program.

The GP parameters setting for this new version of backwarding GP is the same that was presented in Tab. 3. The only dierence lies in the new validation method. As in the previous subsection, the GP algorithm was given 10 runs and the best-of-all was chosen. Tab. 5 sums up the results. The relative RMS error is not as good as in the previous section, but the model is now applicable to real data on which biologists and physicists are used to a 50 error, so these results are considered quite satisfactory at the moment.

The programs were applied to the same data as in the previous subsection and the resulting pictures are displayed in Fig. 2. The quality of the images is clearly improved, showing a spatially continuous structure over the sea (neighboring pixels are very similar). Land and sea features can be distinctly recognized.

Table 5.   Backwarding GP scheme

Coecient (K_d)	relative RMS (GP) ()	of badly classified points
412nm	15.49	313
443nm	8.5	66
490nm	13.36	30
510nm	17.97	3
560nm	22.02	8
665nm	4.14	43

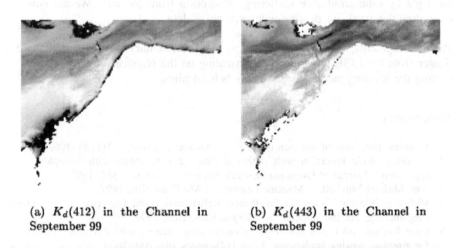

(a) $K_d(412)$ in the Channel in
September 99

(b) $K_d(443)$ in the Channel in
September 99

Fig. 2.   Same images displayed in Fig. 1 using our adapted backwarding validation method.

More examples and full-color images are available on-line at URL http://www-lil.univ-littoral.fr/fonlupt/Recherche/Par.

This GP approach was also compared with an algorithm designed by the NASA that provides numerical results for $K_d(490)$ (the way this algorithm works is not made public). When our images are compared with those from the NASA algorithm, a relatively high level of correlation of 0 .81 is reached. This confirms that our GP model gives meaningful results.

5   Conclusions and Future Works

We have presented in this paper a GP approach to tackle the photosynthesis available radiation problem (PAR). For synthetic data, we have shown that GP improves on traditional methods based on empirical ratios of radiance. Nonetheless, due to the variance and the complexity of the problem, overfitting greatly hamper the use of the evolved programs with real data. Some increases in the size of the data and/or validation set have not improved the results as much as was expected. Thus a new criterion for validation was derived from real data

and a knowledge of realistic bounds for the PAR coecient in the geographical area that is studied. This provides us with a new validation set that works
on a boolean/classification mode and the results obtained in the Channel strait
are considered to be good by the biologists and physicists (continuous spatial
structure and values of the attenuation coecient near the expected results).
Comparing with available NASA satellite product, shows a correlation coecient of 0.81 with our model.

This encouraging work will be extended in the near future. A dicult problem
encountered by the biologists is the application of their models on noisy data.
As explained in Sect. 3, remote sensing is very sensitive to the modification of
the light by solar irradiance scattering, absorption from the air... We are now
extending this work for dealing with such noisy data.

Acknowledgments.   We would like to thank Emmanuel Devred and Richard
Santer from the LISE laboratory, for providing us the physical models for generating the learning cases sets and many helpful hints.

References

1. J. uinlan. Induction of decision trees.   Machine Learning , 1(1):81–106, 1986.
2. J. uinlan. Rule induction with statistical data - a comparison with multiple
   regression. Journal of Operation Research Society , 38:347–352, 1987.
3. Tom Michael Mitchell.   Machine Learning . Mc Graw-Hill, 1997.
4. Wolfgang Banzhaf, Peter Nordin, Robert Keller, and Frank Francone.   Genetic
   Programming, An Introduction . Morgan Kaufmann, 1999.
5. Cyril Fonlupt and Denis Robilliard. Genetic programming with dynamic fitness
   for a remote sensing application. In   [12], pages 191–200, 2000.
6. John Koza. Genetic Programming II: Automatic Discovery of Reusable Programs .
   The MIT Press, 1994.
7. A. Morel. Optical modeling of the upper ocean in relation to its biogenous matter
   content (case I waters).   Journal of Geophysical Research , C9(93):10479–10768,
   1988.
8. A. Morel. Light and marine photosynthesis: a spectral model with geochemical
   and climatological implications.   Prog. Oceanogr. , 26:263–306, 1991.
9. Malik Chami. Développement d'un code de transfert radiatif pour le syst   eme océan-
   atmosph ere. Application au d´ etroit du Pas de Calais . PhD thesis, Universit´ e du
   Littoral - C^ote d'Opale, 1997. in French.
10. M. Chami, E. Dilligeard, and R. Santer. A radiative transfer model for the com-
    putation of radiance and polarization in an ocean-atmosphere system. polarization
    properties of suspended matter for remote sensing purposes. 2000. To appear in
    Applied Optics.
11. Emmanuel Devred.   Estimation du PAR (Photosynthetically Active Radiation) dans
    les eaux du cas II par t´ elédétection spatiale . PhD thesis, Universit´ e du Littoral -
    Côte d'Opale, France, 2001. In French.
12. Marc Schoenaueur, Kalyanmo Deb, G   unter Rudolph, Xin Yao, Evelyne Lutton,
    Juan Julian Merelo, and Hans-Paul S   chwefel, editors.   Parallel Problem Solving
    from Nature VI , volume 1917 of Lecture Notes in Computer Science , Paris, France,
    September 2000. Springer.

Avoiding the Bloat with Stochastic Grammar-Based Genetic Programming

Alain Ratle [1] and Michèle Sebag[2]

[1] LRMA- Institut Supérieur de l'Automobile et des Transports 58027 Nevers France
Alain.Ratle _isat@u-bourgogne.fr
[2] LMS CNRS UMR 76-49, Ecole Polytechnique 91128 Palaiseau France
Michele.Sebag@Polytechnique.fr

Abstract.   The application of Genetic Programming to the discovery of
empirical laws is often impaired by the huge size of the search space,
and consequently by the computer resources needed. In many cases, the
extreme demand for memory and CPU is due to the massive growth of
non-coding segments, the introns. The paper presents a new program
evolution framework which combines distribution-based evolution in the
PBIL spirit, with grammar-based genetic programming; the information
is stored as a probability distribution on the grammar rules, rather than
in a population. Experiments on a real-world like problem show that this
approach gives a practical solution to the problem of introns growth.

1   Introduction

This paper is concerned with the use of Genetic Programming (GP) [1,2] for the
automatic discovery of empirical laws. Although GP is widely used for symbolic
regression [3,4], it suers from two main limitations. One fst limitation is that
canonical GP oers no way to incorporate domain knowledge besides the set
of operators, despite the fact that the knowledge-based issues of Evolutionary
Computation are widely acknowledged [5,6].

In a previous work [7] was described a hybrid scheme combining GP and
context free grammars (CFGs). First investigated by Gruau [8] and Whigham
[9], CFG-based GP allows for expressing and enforcing syntactic constraints
on the GP solutions. We applied CFG-based GP to enforce the dimensional
consistency of empirical laws. Indeed, in virtually all physical applications, the
domain variables are labelled with their dimension (units of measurement), and
the solution law must be consistent with respect to these dimensions (seconds
and meters should not be added). Dimensional consistency allows for massive
contractions of the GP search space; it significantly increases the accuracy and
intelligibility of the empirical laws found.

A second limitation of GP is that it requires huge amounts of computational
resources, even when the search space is properly constrained. This is blamed
on the bloat phenomenon, resulting from the growth of non-coding branches
(introns) in the GP individuals [1,10]. The bloat phenomenon adversely aects
GP in two ways; on one hand, it might cause the early termination of the GP runs

P. Collet et al. (Eds.): EA 2001, LNCS 2310, pp. 255–266, 2002.
ⓒ Springer-Verlag Berlin Heidelberg 2002

due to the exhaustion of available memory; on the other hand, it significantly increases the fitness computation cost.

In this paper a new GP scheme addressing the bloat phenomenon is presented, which combines CFG-based GP and distribution-based evolution. In distribution-based evolution, an example of which is PBIL [11], the genetic pool is coded as a distribution on the search space; in each generation, the population is generated from the current distribution; and the distribution is updated from the best (and possibly the worst) individuals in the current population.

In this new scheme, termed SG-GP (for Stochastic Grammar-based GP), the distribution on the GP search space is represented as a stochastic grammar. It is shown experimentally that this scheme avoids the apparition of introns, which oers new hints into the bloat phenomenon.

The paper is organized as follows. Next section briey summarizes context-free grammars (CFGs) and CFG-based GP, in order for the paper to be self contained. The principle of Distribution-based evolution is presented in section 3, and related works are discussed [12]. Stochastic Grammar based GP is detailed in Section 4. An experimental validation of SG-GP on real-world problems is reported in Section 5, and the paper ends with some perspectives for further research.

2   CFG-Based GP

2.1   Context Free Grammars

A context free grammar describes the admissible constructs of a language by a 4-tuple S, N, T, P , where S is the start symbol, N the set of non-terminal symbols, T the set of terminal symbols, and P the production rules. Any expression is iteratively built up from the start symbol by rewriting non-terminal symbols into one of their derivations, as given by the production rules, until the expression contains terminals only. Fig. 1 shows the CFG describing the polynoms of variable X, to be compared with the standard GP description from the node set $N = +, $ and terminal set $T = X, R $:

$$
\begin{aligned}
N &= exp, op, var \\
T &= +, , X, R \qquad\qquad // \text{ R stands for any real-valued constant} \\
&\quad S \quad := \quad exp \quad ; \\
P &= \quad exp \quad := \quad op\ exp\ exp \qquad\qquad var \quad ; \\
&\quad op \quad := \quad + \quad ; \\
&\quad var \quad := \quad X\ R \ ;
\end{aligned}
$$

Fig. 1.   Context Free Grammar for polynoms of any degree of variable   X

Note that non-terminals and terminals have dierent meanings in GP and in CFGs. GP terminals (resp. non-terminals) stand for domain variables and constants (resp. operators). CFGs terminals comprise domain variables, constants, and operators.

2.2   CFG-Based GP

On one hand, CFGs allow one to express problem-specific constraints on the GP search space. On the other hand, the recursive application of derivation rules allows the build up of a derivation tree (Fig. 2), which can be thought of as an alternative representation for the expression tree.

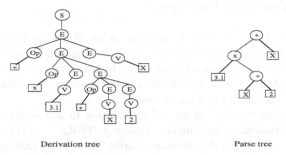

Derivation tree   Parse tree

Fig. 2.   Derivation tree and Corresponding Parse tree

Derivation trees can be manipulated using evolution operators. In order to en-sure that CFG-compliant ospring are produced from CFG-compliant parents, crossover is restricted to swapping subtrees built on the same non-terminal sym-bol; mutation replaces a subtree by a new derivation tree built on the same non-terminal symbol [8,9]. These restrictions are quite similar to that of Strongly Typed GP [13].

2.3   Dimensionally-Aware GP

As mentioned in the introduction, the discovery of empirical laws makes it desir-able to account for the units of domain variables. These units can be expressed wrt elementary units, and represented as vectors (e.g.   Newton = mass   length time ^{2} is represented as vector [1 , 1, 2]). Restricting ourselves to a finite number of compound units, we associate a non-terminal symbol to any com-pound unit allowed. The associated derivation rule describes all possible ways for generating an expression of the given unit. An automatic grammar genera-tor takes as input the elementary units and the set of compound units allowed, and produces the CFG describing all dimensionally consistent expressions in the search space [1]. Although the CFG size is exponential, enforcing these restric-tions linearly increases the crossover complexity in the worst case, and does not modify the mutation complexity.

Compared to most CFGs used in the GP literature [8], the dimensional-CFG is huge (several hundreds non-terminal symbols, several thousands of deriva-tions). The inefficiency of CFG-GP in this frame, already reported by [14], was blamed on the initialization operator. This drawback was addressed by a specific, constrained grammar-based initialization process, building a CFG-compliant   and suciently diverse   initial population. The core of the procedure is a two-step

---

[1] The production rule associated to the start symbol specifies the unit of the sought solution; it can also enforce the shape of the solution, according to the expert guess.

process: a) for any given non-terminal symbol, all derivations compatible with
the maximum tree-depth prescribed (ensuring that the final expression will have
admissible size) are determined; b) the non-terminal symbol at hand is rewritten
by uniformly selecting one compatible derivation (see [7] for more details).

3   Distribution-Based Evolution

Contrasting with genetic evolution, distribution-based evolution deals with a
high-level (intentional) description of the best individuals encountered so far, as
opposed to the (extensional) description given by the current population itself.
This intentional description is a probability distribution on the solution space,
which is updated according to a set of rules.

As far as we know, the first algorithm resorting to distribution-based evolution is Population-based Incremental Learning (PBIL)   [11], concerned with
optimization in $0, 1^n$. In this scheme, distribution M is represented as an element of $[0, 1]^n$, initialized to $M_0 = (.5, \dots, .5)$.
At generation t, M_t is used to generate the population from scratch, where the
probability for any individual X to have its i-th bit set to 1 is given as the i-th
component of M_t. The best individual X_{best} in the current population is used
to to update M_t by relaxation [2], with

$$M_{t+1} = (1 - \alpha)M_t + \alpha X_{best}$$

M_t is also randomly perturbed (mutated) to avoid premature convergence.

This scheme has been extended to accommodate dierent distribution models
and non-binary search spaces (see [15,16] among others).

Distribution-based evolution has been extended to GP through the   Probabilistic Incremental Program Evolution (PIPE)   system [12]. The distribution on
the GP search space is represented as a Probabilistic Prototype Tree (PPT); in
each PPT node stand the probabilities for selecting any variable and operator
in this node. After the current individuals have been constructed and evaluated,
the PPT is biased toward the current best and the best-so-far individuals. One
feature of the PIPE system is that the PPT grows deeper and wider along evolution, depending on the size of the best trees, since the probabilities of each
variable/operator have to be defined for each possible position in the tree.

Interestingly, one main dierence between PIPE and canonical GP is the
increased diversity of the individuals.

4   Stochastic Grammars-Based GP

4.1   Overview

Distribution-based evolution actually involves three components: the representation (model) for the distribution; the exploitation of the distribution in order to

---

[2] Other variants use the best two individuals, and possibly the worst one too, to update
the distribution.

generate the current population, which is analogous in spirit to the genetic initialization operator; the update mechanism, evolving the distribution according to the most remarkable individuals in the current population.

In CFG-GP, initialization proceeds by iteratively rewriting each non-terminal symbol; this is done by selecting a derivation in the production rule associated to the current non-terminal symbol (e.g.   exp   is either rewritten as a more complex expression,   op exp exp   , or a leaf   var   , Fig. 1). The selection is uniform (among the derivations compatible with the maximum tree size allowed, see Section 2.3). It comes naturally to encode the experience gained from the past generations, by setting selection probabilities on the derivations.

Distribution representation   . Finally, the distribution on the GP search space is represented as a stochastic grammar: each derivation d_i in a production rule is attached a weight w_i, and the chances for selecting derivation d_i are proportional to w_i.

Distribution exploitation   . Practically, the construction of the individuals from the current stochastic grammar is inspired from the CFG-GP initialization procedure. For each occurrence of a non-terminal symbol, all admissible derivations are determined from the maximum tree size allowed and the position of the current non-terminal symbol as in [7]; the selection of the derivation d_i is done with probability p_i, where

$$
p_i = \begin{cases} \dfrac{w_i}{\displaystyle\sum_{k \text{ admissible derivs.}} w_k} & \text{if } d_i \text{ is an admissible derivation} \\ 0 & \text{otherwise} \end{cases} \tag{1}
$$

This way, weights w_i need not be normalized.

Distribution update   . After all individuals in the current population have been evaluated, the probability distribution is updated from the N_b best and N_w worst individuals according to the following rules: for each derivation d_i,

- Let b denotes the number of individuals among the N_b best individuals that carry derivation d_i; weight w_i is multiplied by $(1 +)^b$;
- Let w denotes the number of individuals among the N_w worst individuals that carry derivation d_i; weight w_i is divided by $(1 +)^w$;
- Last, weight w_i is mutated with probability p_m; the mutation either multiplies or divides w_i by factor $(1 + _m)$.

All w_i are initialized to 1. Note that it does not make sense to have them normalized; they must be locally renormalized before use, depending on the current set of admissible derivations.

This distribution-based GP, termed SG-GP, involves five parameters besides the three standard GP parameters (Table 1).

4.2   Scalar and Vectorial SG-GP

In the above scheme, the genetic pool is represented by a vector   W, coding all derivation weights for all production rules. The storage of a variable length

Table 1.   Parameters of Stochastic Grammar-based Genetic Programming

Parameter	Definition
	Parameters specific to SG-GP
N_b	Number of best individuals for probability update
N_w	Number of worst individuals for probability update
	Learning rate
p_m	Probability of mutation
m	Amplitude of mutation
	Canonical GP parameters
P	Population size
G	Maximum number of generations
D_{max}	Maximum derivation depth

population is replaced by the storage of a single fixed size vector; this is in sharp contrast with canonical GP, and more generally, with all evolutionary schemes dealing with variable size individuals.

One limitation of this representation is that it induces a total order on the derivations in a given production rule. However, it might happen that derivation d_i is more appropriate than d_j in higher levels of the GP trees, whereas d_j is more appropriate in the bottom of the trees.

To take into account this eect, a distribution vector W_i is attached to the i-th level of the GP trees (i ranging from 1 to D_{max}). This scheme is referred to as Vectorial SG-GP , as opposed to the previous scheme referred to as   Scalar SG-GP .

The distribution update in Vectorial SG-GP is modified in a straightforward manner; the update of distribution W_i is only based on the derivations actually occurring at the i-th level among the best and worst individuals in the current population.

4.3   Experiment Goal

SG-GP oers a new perspective of the real causes for the apparition of introns, i.e. non coding segments in the GP solutions.

Factually, the proportion of introns in the GP material grows exponentially along evolution [17]. As already mentioned, the intron growth is undesirable as it drains out the memory resources, and increases the total fitness computation cost.

However, it has been observed that pruning the introns in each generation, significantly decreases the overall GP performances [1]. Supposedly, introns protect good building blocks from the destructive eects of crossover; as the useful part of the genome is condensed into a small part of the individual, the probability for a crossover to break down useful sections is reduced by the apparition of introns.

But intron growth might also be explained from the structure of the search space [10]. Consider all genotypes (GP trees) coding a given phenotype (program). There exists a lower bound on the genotype size (the size of the shortest tree coding the program); but there exists no upper bound on the genotype size (a long genotype can always be made longer by the addition of introns). Since there are many more long genotypes than short ones, longer genotypes will be selected more often than shorter genotypes (everything else being equal, i.e. assuming that the genotypes are equally fit) [3].

Last, intron growth might also be a mechanical eect of evolution. GP crossover facilitates the production of larger and larger trees: on one hand, the ospring average size is equal to the parent average size; on the other hand, short size ospring usually are poorly fit; these remarks together explain why the individual size increases along evolution.

But the information transmission in SG-GP radically diers from that in GP. As there exists no crossover in SG-GP, there should be no occasion for building longer individuals, and no necessity for protecting the individuals against destructive crossover.

Experiments with SG-GP are intended to assess the utility of introns. If the intron growth is beneficial   per se, then either SG-GP will show able to produce introns or the overall evolution results will be significantly degraded. If none of these eventualities is observed, this will suggest that the role of introns has been overestimated.

5   Experimental Validation

5.1   Test Problem

The application domain selected for this study is related to the identification of rheological models. These problems have important applications in the development of new materials, especially for polymers and composite materials [19]. The target empirical law corresponds to the Kelvin-Voigt model, which consists of a spring and a dashpot in parallel (Fig. 3). When a constant force is applied, the response (displacement-time relation) is

$$u(t) = \frac{F}{K} \left(1 - e^{-\frac{Kt}{C}} \right)$$

Fitness cases (examples) are generated using random values of the material parameters K and C and loading F. The physical units for the domain variables and for the solution are expressed with respect to the elementary   mass , time and length   units (Table 2). Compound units are restricted as the exponent for each elementary unit ranges in   2, 1, 0, 1, 2 . The dimensional grammar is generated as described in Section 2.3, with 125 non-terminal symbols and four operators (addition, multiplication, protected division and exponentiation). The grammar size is about 515 k.

---

[3] To resist the intron growth, a parsimony pressure might be added to the fitness function [18]; but the relative importance of the actual fitness and that of the parsimony term must be adjusted carefully. And the optimal trade-o might not be the same for the beginning and the end of evolution.

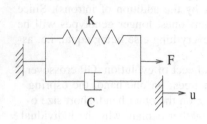

Fig. 3.   Kelvin-Voigt Model

Table 2.   Physical Units

Physical units			
uantity	mass	length	time
Variables			
E (Force)	+1	+1	−2
K (Elastic elements)	+1	0	−2
C (Viscous elements)	+1	0	−1
t (time)	0	0	+1
Solution			
u (displacement)	0	+1	0

5.2   Experimental Setting

SG-GP is compared [4] with standard elitist GP. The eciency of SG-GP is assessed with respect to the quality of solutions, and in terms of memory use. All results are averaged on 20 independent runs.

GP and SG-GP parameters are set according to a few preliminary experiments (Table 3). Canonical GP is known to work better with large populations and small number of generations [1]. uite the contrary, SG-GP works better with a small population size and many generations. In both cases, evolution is stopped after 2,000,000 fitness evaluations.

Table 3.   Optimization parameters

SG-GP		GP	
Parameter	Value	Parameter	Value
Population size	500	Population size	2000
Max. number of generations	4000	Max. number of generations	1000
N_b	2	P(crossover)	0.9
N_w	2	P(mutation)	0.5
	0.001	Tournament size	3
P_m	0.001		
m	0.01		

5.3   Experimental Results and Parametric Study

Fig. 4 shows the comparative behaviors of canonical GP and SG-GP on the test identification problem.

The inuence of the learning rate   is depicted on Fig. 5. The inuence of the mutation amplitude $_m$ is depicted in Fig. 6.a. Overall, better results are obtained with a low learning rate and a suciently large mutation amplitude; this can be interpreted as a pressure toward the preservation of diversity in the population.

---

[4] Due to space limitations, the reader interested in the comparison of CFG-GP with GP is referee to [7].

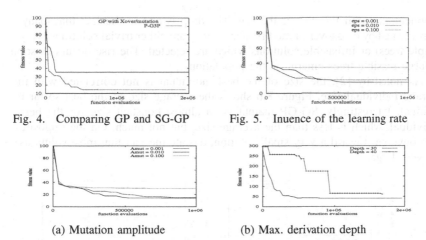

Fig. 4.   Comparing GP and SG-GP   Fig. 5.   Inuence of the learning rate

(a) Mutation amplitude   (b) Max. derivation depth

Fig. 6.   Parametric study of SG-GP

The maximum derivation depth allowed D_{max} is also a critical parameter. Too short, and the solution will be missed, too large, the search will take a prohibitively long time. Fig. 6.b shows the solutions obtained with maximum derivation depths of 30 and 40. As could have been expected, the solution is found faster for D_{max} = 30.

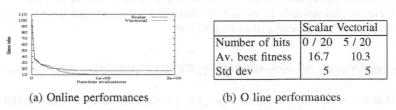

	Scalar	Vectorial
Number of hits	0 / 20	5 / 20
Av. best fitness	16.7	10.3
Std dev	5	5

(a) Online performances   (b) O line performances

Fig. 7.   Scalar vs. Vectorial SG-GP

The advantage of using a vectorial distribution model against a scalar one, is illustrated on Fig. 7.a, as Vectorial SG-GP significantly improves on Scalar SG-GP. Table 7.b points out that vectorial SG-GP finds the target law (up to algebraic simplifications) after 2,000,000 fitness evaluations on 5 out of 20 runs, while no perfect match could be obtained with scalar SG-GP.

CFG-GP results, not shown here for space limitations, show that even scalar SG-GP is more ecient than CFG-GP (see [7] for more details).

5.4   Resisting the Bloat

The most important experimental result is that SG-GP   does resist the bloat, as it maintains an almost constant number of nodes. The average results over all individuals and 20 runs is depicted on Fig. 9.a.

In comparison is shown the number of nodes in GP (averaged on all individuals, but plotted for three typical runs for the sake of clarity). The individual size first drops in the first few generations; and after a while, it suddenly rises exponentially until the end of the run. The drop is due to the fact that many trees created by crossover in the first generations are either trivial solutions (very simple trees) or infeasible solutions which are rejected. The rise occurs as soon as large feasible trees emerge in the population.

As noted by [2], the size of the best individual is not correlated with the average individual size. Figure 8.b shows the average size of the best-so-far individual. Interestingly, SG-GP maintains an almost constant size for the best individual, which is less than the average size, but not much. On the opposite, GP converges toward a very small solution, despite the fact that most solutions are very large.

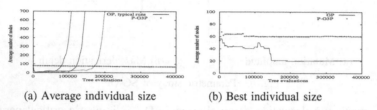

(a) Average individual size   (b) Best individual size

Fig. 8.   Solution size with GP and SG-GP, averaged on 20 runs

5.5   Identification and Generalization

As mentioned earlier on, SG-GP found the target law in 5 out of 20 runs (up to algebraic simplifications). In most other runs, SG-GP converges toward a local optimum, a simplified expression of which is:

$$x(t) = \frac{2\ t^2\ \frac{F}{K}}{\frac{C}{K}\ \frac{C}{K} + 2\ t^2} \tag{2}$$

This law is not on the path to the global optimum since the exponential is missing. However, the law closely fits the training examples, at least from an engineering point of view (Fig. 9.a).

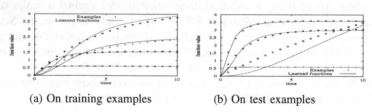

(a) On training examples   (b) On test examples

Fig. 9.   Correct Identification and Generalization with SG-GP

Even more important is the fact that SG-GP finds solutions which behaves well on test examples, i.e. examples generated after the target law, which have not been considered during evolution.

By construction, the target law perfectly fits the test examples. But the non-optimal law (Eq 2) also fits the test examples; the fit is quite perfect is three out of four cases, and quite acceptable, from an engineer's point of view, in the last case.

6   Conclusion

In this paper was presented a novel Genetic Programming scheme, combining grammar-based GP [8,9,7] and distribution-based evolution [11], termed SG-GP for Stochastic Grammar-based Genetic Programming. SG-GP diers from the PIPE system [12] as the distribution model used is based on stochastic grammars, which allows for overcoming one main limitation of GP, i.e. the bloat phenomenon.

Intron growth was suggested to be unavoidable for program induction methods with fitness-based selection [10].

This conjecture is infirmed by SG-GP results on a real-world like problem. Indeed, more intensive experiments are needed to see the limitations of the SG-GP scheme.

Still, SG-GP successfully resisted the intron growth on the problem considered, in the following sense.

First of all, SG-GP shows good identification abilities, as the target law was discovered in 5 out of 20 runs, while it was never discovered by canonical GP.

Second, SG-GP shows good generalization abilities; even in the cases where the target law was missed, the solutions found by SG-GP have good predictive accuracy on further examples (not considered during the learning task).

Last, these identification and generalization tasks are successfully completed by exploring individuals with constant size. No intron growth was observed; the overall memory requirements were lower by several orders of magnitude, than for canonical GP.

These results suggest that intron growth is not necessary to achieve ecient non parametric learning in a fitness-based context, but might rather be a side eect of crossover-based evolution.

Further research is concerned with examining the actual limitations of SG-GP through more intensive experimental validation. Eorts will be devoted to the parametric optimization problem (find the constants) coupled with non-parametric optimization.

References

1. J. R. Koza.   Genetic Programming: On the Programming of Computers by means of Natural Evolution . MIT Press, Massachusetts, 1992.
2. W. Banzhaf, P. Nordin, R.E. Keller, and F.D. Francone.   Genetic Programming An Introduction On the Automatic Evolution of Computer Programs and Its Applications . Morgan Kaufmann, 1998.
3. B. McKay, M.J. Willis, and G.W. Barton. Using a tree structures genetic algorithm to perform symbolic regression. In   IEEE Conference publications, n. 414   , pages 487–492, 1995.
4. J. Duy and J. Engle-Warnick.   Using symbolic regression to infer strategies from experimental data. In   Evolutionary Computation in Economics and Finance   . Springer Verlag, 1999.
5. N. J. Radclie. Equivalence class analysis of genetic algorithms.   Complex Systems , 5:183–20, 1991.

6. C. Z. Janikow. A knowledge-intensive genetic algorithm for supervised learning. Machine Learning , 13:189–228, 1993.
7. A. Ratle and M. Sebag. Genetic programming and domain knowledge: Beyond the limitations of grammar-guided machine discovery. In M. Schoenauer et al., editor, Proceed ings of the 6th Conference on Parallel Problems Solving from Nature , pages 211–220. Springer-Verlag, LNCS 1917, 2000.
8. F. Gruau. On using syntactic constraints with genetic programming. In P.J. Angeline and K.E. Kinnear Jr., editors, Advances in Genetic Programming II , pages 377–394. MIT Press, 1996.
9. P.A. Whigham. Inductive bias and genetic programming. In IEEE Conference publications, n. 414 , pages 461–466, 1995.
10. W. B. Langdon and R. Poli. Fitness causes bloat. In Soft Computing in Engineering Design and Manufacturing , pages 13–22. Springer Verlag, 1997.
11. S. Baluja and R. Caruana. Removing the genetics from the standard genetic algorithms. In A. Prieditis and S. Russel, editors, Proceed ings of ICML95 , pages 38–46. Morgan Kaufmann, 1995.
12. R. Salustowicz and J. Schmidhuber. Evolving structured programs with hierarchical instructions and skip nodes. In J. Shavlik, editor, Proceed ings of the 15th International Conference on Machine Learning , pages 488–496. Morgan Kaufmann, 1998.
13. David J. Montana. Strongly typed genetic programming. Evolutionary Computation , 3(2):199–230, 1995.
14. C. Ryan, J.J. Collins, and M. O'Neill. Grammatical evolution: Evolving programs for an arbitrary language. In W. Banzhaf, R. Poli, M. Schoenauer, and T.C. Fogarty, editors, Genetic Programming, First European Workshop, EuroGP98 , volume LNCS 1391, pages 83–96. Springer Verlag, 1998.
15. M. Sebag and A. Ducoulombier. Extending population-based incremental learning to continuous search spaces. In Th. B ack, G. Eiben, M. Schoenauer, and H.-P. Schwefel, editors, Proceed ings of the 5th Conference on Parallel Problems Solving from Nature , pages 418–427. Springer Verlag, 1998.
16. P. Larranaga and J. A. Lozano. Estimation of Distribution Algorithms. A New Tool for Evolutionary Computation . Kluwer Academic Publishers, 2001.
17. P. Nordin, W. Banzhaf, and F.D. Francone. Introns in nature and in simulated structure evolution. In D. Lundh, B. Olsson, and A. Narayanan, editors, Biocomputing and Emergent Computation , pages 22–35. World Scientific, 1997.
18. Byoung-Tak Zhang and Heinz M uhlenbein. Balancing accuracy and parsimony in genetic programming. Evolutionary Computation , 3(1):17–38, 1995.
19. I.M. Ward. Mechanical Properties of Solid Polymers . Wiley, Chichester, 1985.

Applying Boosting Techniques to Genetic Programming

Gregory Paris, Denis Robilliard, and Cyril Fonlupt

Universit´e du Littoral-Cˆote d'Opale
LIL
BP 719
62228 Calais Cedex, France
paris@lil.univ-littoral.fr
phone: +33-321 465 667

Abstract. This article deals with an improvement for genetic programming based on a technique originating from the machine learning field: boosting. In a fist part of this paper, we test the improvements oered by boosting on binary problems. Then we propose to deal with regression problems, and propose an algorithm, called GPboost, that keeps closer to the original idea of distribution in Adaboost than what has been done in previous implementation of boosting for genetic programming.

1 Introduction

The principle of Genetic Programming is the automatic generation of programs to solve problems, taking inspiration of the Darwinian model of evolution. Several steps of evolution are simulated, notably generating a population of programs (usually represented as trees), selecting the best programs for mating, implementing information interchange and stochastic variance for genetic crossover and mutation. Finally the best program is kept according to its performance. This quality, or fitness, of programs is measured by a fitness function usually based on the ability to retrieve data from a learning set. This allows us to consider genetic programming as a machine learning method, like decision trees for example.

The boosting algorithm is a scheme which improves machine learning methods without the need for more test cases. It rather operates by modifying the distribution of the examples in the learning set.

These two points lead us to apply the boosting technique to genetic programming. In this article, we test the improvement given by boosting, first on binary problems then on regression problems. We propose a dierent scheme from Iba's [Iba99] to apply Adaboost to genetic programming, especially on the interpretation of distribution. We also test several voting methods for models generated by boosting.

P. Collet et al. (Eds.): EA 2001, LNCS 2310, pp. 267–278, 2002.
c Springer-Verlag Berlin Heidelberg 2002

2   Boosting

2.1   Presentation

Boosting appeared at the beginning of the 90's, proposed by Schapire [Sch90] and Freund [Fre95]. It is dedicated to improving already known methods from the machine learning field, like decision trees, or rules systems (cf. "Machine Learning" by Tom Mitchell [Mit97]). It consists in giving several hypotheses on dierent distributions and then combining them to obtain a fial hypothesis. We can give an intuitive idea of the way boosting works, by saying the distribution is modified to put emphasis on hard cases in the learning set. When the base learning algorithm is a   weak learner   according to the PAC model [Val84], there is a theoretical proof that the error on the learning set of the final hypothesis is better than those of an "un-boosted" version of the algorithm. A weak learner is an algorithm able to provide a hypothesis whose precision is over 50 on binary problems with a positive confidence (i.e. intuitively the algorithm is better than random search). The principle of boosting even allows to improve indefinitely precision and confidence: "weak learning is strong" [Sch90].

2.2   The Adaboost Algorithm

Adaboost has been proposed by Freund and Schapire [FS96] for binary problems. Let $W(S,D)$ be a weak learner which can provide a hypothesis from a learning set S of m examples (x,y) obtained from the function $f : X \quad {1,+1}$ we are looking for. Examples in S take a distribution (or weight vector) D, such as $\sum_{i\,(x_i,y_i)\,\in\,S} D(i) = 1$. We can improve W with Adaboost:

Given: A learning set $S = (x_1,y_1),\ldots,(x_m,y_m)$; $x_i \in X$, $y_i \quad {1,+1}$;
          $W(S,D)$ a weak learner;
Adaboost algorithm:
Let D_1 be the distribution for iteration $t = 1$
$D_1(i)$ is the weight of example (x_i,y_i)
Initialize $D_1(i) := 1/m$ for all $(x_i,y_i) \quad S$
For $t = 1.. \, T$ do
        Run W using D_t
        Get weak hypothesis $h_t : X \qquad {1,+1}$
        Let $\alpha_t = \frac{1}{2} \ln \frac{1-\epsilon_t}{\epsilon_t}$
                with ϵ_t the error using distribution D_t
        Update distribution :
$$D_{t+1}(i) := \begin{cases} \frac{D_t(i)e^{-\alpha_t}}{Z_t} & \text{if } x_i \text{ is matched}, \\ \frac{D_t(i)e^{\alpha_t}}{Z_t} & \text{if } x_i \text{ is not matched}. \end{cases}$$
                where Z_t is a normalization factor so that D_{t+1} is a distribution
End For
Output :   Final hypothesis :
        $H(x) = \text{sign} (\sum_{1..T} \alpha_t \, h_t(x))$

Some further details on this algorithm:

- The algorithm loops on several distributions. For the first iteration D_1 is the uniform distribution on S.
- Parameter T is the number of boosting rounds, and thus it is the number of weak hypotheses that are combined to formulate the final hypothesis. T is a user defined parameter.
- The modification of examples weights is made according to how they are classified (right or wrong) by the weak hypothesis of the current round. Badly-classified examples have a higher weight in the next round.
- The value of the $_t$ coecient is computed in order to reduce training error as much as possible. It determines how weights vary from a round to an other and the importance to give to the hypothesis h_t in the final vote.
- To formulate the final hypothesis H, every hypothesis h_t votes with a certain confidence given by $_t$.

Other versions have been proposed for other problems, among them Ad-aboost.R by Drucker [Dru97] for regression problems.

2.3   Applying Boosting to Genetic Programming

There are many problems encountered when applying boosting to GP. First, genetic programming is not known as a weak learner, so the decrease of error is not guaranteed. Nevertheless, all tests have been successful. Secondly, a problem occurs with the interpretation of the distribution when dealing with regression problems. Iba [Iba99] keeps Drucker's proposition [Dru97], and picks up examples from the learning set using the distribution, before running the GP algorithm as usual, in order to process a round of boosted GP. This allows to keep the GP algorithm intact, but the precision given by weights is lost. We have chosen to try to keep this precision, and this has been achieved by taking into account the distribution properties inside the fitness function (see Sect. 4).

3   Boosting GP for Binary Problems

3.1   The Multiplexer Problem

Presentation.   The multiplexer can be seen as a black box which outputs one of its input data bits, depending on its input address bits. In the case of the 11-multiplexer, the output bit is selected among 8 input bits according to 3 address bits. This problem is considered as a dicult one for machine learning [Koz92].

Experiments and results.   We want to test the interest of boosting for GP. Every experiment compares the best function resulting from 10 runs of standard GP with the final hypothesis given by one run of 10 rounds of boosted GP. Parameters are the same in both experiments (cf. Tab. 1), and the comparison criterion is the number of matched cases. This experiment was repeated 5 times and results are summed up in Tab. 2.

Table 1.   Standard and boosted GP parameters for the 11-multiplexer problem.

	standard GP	boosted GP
Objective	Find a function for the multiplexer	
Terminal set	a_0, a_1, a_2 to select the input; $d_0, d_1, d_2, d_3, d_4, d_5, d_6, d_7$ for input value	
Operators	and, or, not, if	
Learning set	2048 (2^{11}) possible cases	
Fitness	Number of matched cases	$\sum D_i$ for matched cases * 2048
Population	500	
Generation number	200	
Rounds of boosting	/	10
Crossover probability	0.9	
Copy probability	0.1	
Selection Method	best (according to fitness)	
Maximum depth	15	
Creation method	half and half	

Note.   Weights are the same for the first round of boosting (1   / 2048 because there are 2048 cases). Hence the fitness function for this first round is the same for standard and boosted GP, and the first round of boosting is indeed a standard execution of GP.

Table 2.   Standard vs. boosted GP results for the 11-multiplexer problem.

function	max	min	average
standard GP	1472	1408	1440
boosted GP	1694	1597	1645

3.2   The Even-Parity Problem

Presentation.   The goal is to determine from several input bits if the number of bits with value 1 is even or not. We work on the even-6-parity problem meaning that there are 6 input bits, in a boolean classification framework so case 101100 should be associated with value false since there is an odd number of bits set to 1.

Experiments and results.   Five experiments have been processed like for the multiplexer problem, with same parameters for standard and boosted GP (cf. Tab. 3), and comparison criterion is again the number of matched cases (cf. Tab 4).

Table 3.   Standard and boosted GP parameters for the even-parity problem.

	standard GP	boosted GP
Objective	Find a function computing the even-6-parity	
Terminal set	$a_0, a_1, a_2, a_3, a_4, a_5$	
Operators	and, or, not, if	
Learning set	the 64 (2^6) possible cases with output value	
Fitness	number of matched cases	$\sum D_i$ for matched cases * $\overline{64}$
Population	100	
Generation	50	
Rounds of boosting	/	10
Crossover probability	0.8	
Copy probability	0.15	
Mutation probability	0.05	
Selection method	best (according to fitness)	
Max depth	15	
Creation method	half and half	

Table 4.   Standard vs. boosted GP results for the even-6-parity problem.

function	max	min	average
GP	46	44	44
GPboost	64	58	61

3.3 Conclusion

The good behavior of boosting techniques on binary problems was already demonstrated for many machine learning methods. On the two problems studied here, it is clear that boosting greatly improves GP. It is not demonstrated that GP is a weak learner, thus there is no proof that the error could be reduced indefinitely. Nonetheless good results were obtained, and GP seems to behave like a weak learner, at least on these two problems.

4 Boosting GP for Regression Problems

4.1 Presentation

We look at regression problems consisting in approximating a function $f : X$ R (with X a vector space of size n and R the set of real numbers), which is sampled in a discrete way to get a set of learning cases (x, y) X R . Genetic Programming performs well on regression problems, so, after the good results obtained on binary problems, it seems natural to apply boosting to regression.

Based on the work of Drucker [Dru97], Iba proposed in [Iba99] a version of Adaboost for GP and regression [1]. In Iba's work the fitness function remains the same as in standard GP, and the distribution serves at picking up examples to generate a new learning set for each boosting round as suggested by Freund and Schapire [FS97] when it is not possible to include the distribution in the algorithm. The probability for an example to be picked up is proportional to its weight, and any example can be picked up 0, 1 or several times, until enough examples have been retained to build the learning set. The standard GP algorithm is then run on the new learning set to compute the hypothesis associated to the current round of boosting. This scheme makes implementation easy but the precision of weights is somewhat lost in the process.

Another technique which shares some common features with boosting is the Stepwise Adaption of Weights, or SAW, see [JE01]. In this algorithm weights are modified according to the diculty to match examples. A main dierence with boosting comes from the fact that SAW works on only one function as result, when boosting generate a set of hypotheses that need to be combined.

In the next section we propose a boosting method that retains the precision of weights and operates on the whole set of examples for every round of boosting.

4.2   The GPboost Algorithm

GPboost may be seen as a template boosting algorithm for GP and regression problems, derived from Adaboost. In our algorithm we wish to take into consideration all examples with their weights. In order to achieve this goal, we need to build a weight based fitness function. GPboost is summed up with highlighted fitness function in Tab. 5.

To compute the final hypothesis F, T functions f_t have to be combined. The expression given here follows Iba's proposal of using the geometric median weighted by confidence coecients. In practice, to obtain the image of x, each function f_t gives a value $f_t(x)$. These values are sorted and the geometric median is taken to be $F(x)$.

In Tab. 5, we use as fitness function the sum of absolute dierences weighted by the distribution coecients. We multiply by m in order to get the same range of values for both standard and boosted GP, thus the first round of boosting is indeed a standard run of GP. We call GPboost a template algorithm, since a user may wish to change this fitness function to suit his need, for example using a RMS error-based one [2]. In this case, one should retain the base idea: using the distribution to weight the contribution of each learning case to the fitness evaluation. However it will not always be easy to keep the same range of fitness values as in a standard GP run.

---

[1] It seems that there is a typographical mistake in Iba's algorithm where update is $D_{t+1}(i) := \frac{D_t(i)^{1-L_i}}{Z_t}$, which is not conformant with Drucker's and adaboost algorithm.

[2] RMS error: $\sum_{i=1}^{N} f(x) - y^2/N$, with N the number of samples.

Table 5.   The GPboost algorithm.

Given: a learning set $S = (x_1, y_1), \ldots, (x_m, y_m)$; x_i X , y_i R
GP (S, D) a GP algorithm using a distribution   D on S.
GPboost Algorithm:
Let D_1 be the distribution for iteration $t = 1$
$D_1(i)$ is the weight of example (x_i, y_i)
Initialize $D_1(i) := 1 / m$ for all (x_i, y_i)   S
For $t = 1.. T$ do
    Run GP on D_t with fitness function:

$$\text{fit} = \sum_{i=1}^{m} (f(x_i) \quad y_i \quad D_t(i)) \quad m$$

    where f is a function in the GP population
The best-of-run function is denoted f_t
Compute loss for each example: $L_i = \frac{f_t(x_i) \quad y_i}{\max_{i=1 \ldots m} f_t(x_i) \quad y_i}$
Compute average loss: $L = \sum_{i=1}^{m} L_i D_i$
Let $_t = \frac{L}{1 \ L}$, the confidence given to function f_t
Update distribution:

$$D_{t+1}(i) := \frac{D_t(i) \quad {}_t^{1 \ L_i}}{Z_t}$$

    with Z_t a normalization factor so that D_{t+1} is a distribution
End for
Output: Final hypothesis:

$$F(x) = \min \quad y \ R \ : \quad \sum_{t:f_t(x) \ y} \log(1/ \ _t) \quad \frac{1}{2} \sum_{t=1}^{m} \log(1/ \ _t)$$

4.3   An Example Run of the GPboost Algorithm

In order to explain the behavior of our boosting algorithm, a simple example is now introduced. The parameters defining the run are given below:

Terminal	x
Operators	+, -, *, /
Learning set	5 samples from function $y = \frac{x^2}{2}$ $\{(-0.1, 0.005), (-0.5, 0.125), (1, 0.5), (0.2, 0.02), (0.7, 0.245)\}$
Fitness	$\sum_{i=1}^{5}(|f(x_i) - y_i| * D_i)/5$
Population	20
Generation	10
Rounds of boosting	5

Monitoring the evolution of weights.   The following table shows the weight evolution during the first three rounds of boosting:

Sample		Distribution		
x	y	D_1	D_2	D_3
-0.1	0.005	0.2	0.135645	0.091587
-0.5	0.125	0.2	0.155274	0.097203
1	0.5	0.2	0.433427	0.271331
0.2	0.02	0.2	0.140010	0.116996
0.7	0.245	0.2	0.135645	0.422883

Firstly, the weights of examples are the same, so this is equivalent to a standard run of GP. The first round of boosting results in best-of-run function $f_1(x) = x^4$, which is particularly bad on sample $(1, 0.5)$, hence the change in weights for D_2. The second round of boosting is then run, resulting in best-of-run function $f_2(x) = \frac{x^4}{1+x}$. Notice that this round provides a function which gives the exact value for $x = 1$, which was the learning case with the heaviest weight. Next rounds will have the same general behavior, putting the emphasis on learning cases with heaviest weights, but still accounting for all examples. Best-of-run functions and confidence in hypotheses for the whole run are given below:

hypothesis	confidence
$f_1(x) = x^4$	0.309380
$f_2(x) = \frac{x^4}{1+x}$	0.200801
$f_3(x) = x^4$	0.414084
$f_4(x) = \frac{x^4(x+2)}{(3x^3+x+1)(x+1)}$	0.730217
$f_5(x) = \frac{x^2}{1+x}$	0.158855

Computing the final hypothesis.   We now examplify how to compute the final hypothesis F, output of the algorithm. For any given value x, every function f_i proposes a value $f_i(x)$. We obtain $F(x)$ by computing the geometric median of the $f_i(x)$, weighted by their respective confidence coeficients. For intance let $x = 0.5$, we have: $f_1(x) = 0.0625$, $f_2(x) = 0.041667$, $f_3(x) = 0.0625$, $f_4(x) = 0.116305$, $f_5(x) = 0.166667$. The geometric median is $\frac{1}{2} \sum_{t=1}^{m} \log(1/\beta_t) = 2.907245$, and the sorting of values in ascending order gives (for $x = 0.5$): $f_2(x) \le f_1(x) \le f_3(x) \le f_4(x) \le f_5(x)$.

Then we begin to add terms $\log(1/\beta_t)$ in the same order than the sorted f_t:

$$\log(1/\beta_2) = 1.60544$$
$$1.60544 + \log(1/\beta_1) = 2.77862$$
$$2.77862 + \log(1/\beta_3) = 3.660312 \ge 2.907245$$

We can stop now since the third value is the first to be greater or equal to the geometric median, indicating that the final hypothesis is given by $f_3(x)$. So we have: $F(0.5) = f_3(0.5) = 0.0625$

4.4   Experiments and Results

We train our algorithm on the test function $f(x) = x^3 e^{-x} \cos x \sin x (\sin^2 x - 1)$, using 200 learning cases randomly chosen on $[-1; +1]$. Run parameters are given in Tab. 6. In his article, Iba notes that if the confidence values are all the same, the geometric and arithmetic medians are indeed the same for computing the final hypothesis. To explore this idea, we also compute the arithmetic median on two dierent conflence coecients: the GP fiuess function evaluation and the RMS error. So we have to compare the performances of five algorithms: standard GP, boosted GP with geometric median, arithmetic median, arithmetic fitness-based median and arithmetic RMS-based median. The results are summed up in Tab.7, computed on an independent test set, using the error measure: $err = \sum_{i=1}^{N} |f(x_i) - y_i| / N$, with N the number of cases in the test set.

Table 6.   Parameters for standard and boosted GP applied to regression problem.

	standard GP	boosted GP
Objective	Find the function matching points (x, y)	
Terminal set	x, random values $\in [-1, +1]$	
Operators	+, -, *, /, sin, cos, exp, log	
Learning set	200 points (x, y) randomly chosen with $x \in [-1, +1]$ and $y = x^3 * e^{-x} * \cos x * \sin x * (\sin^2 x * \cos x - 1)$	
Fitness	$\frac{\sum_{i=1}^{200} |f(x_i) - y_i|}{200}$	$\frac{\sum_{i=1}^{200} |f(x_i) - y_i| * D_i}{200}$
Population	500	
Generation	200	
Rounds of boosting	/	10
Crossover probability	0.85	
Copy Probability	0.05	
Mutation probability	0.1	
Selection method	tournament(size 7) for mutation and crossover, best for the copy	
Max depth	10	
Creation method	half and half	

Table 7.   Error measures for standard GP and GPboost with several medians.

Method	Best	Worst	Average
GP	0.0164	0.0325	0.0249
GPboost with :			
geometric median	0.0131	0.0276	0.0209
simple median	0.0131	0.0276	0.0209
arithmetic median on fitness	0.0131	0.0336	0.0211
arithmetic median on RMS	0.0131	0.0336	0.0211

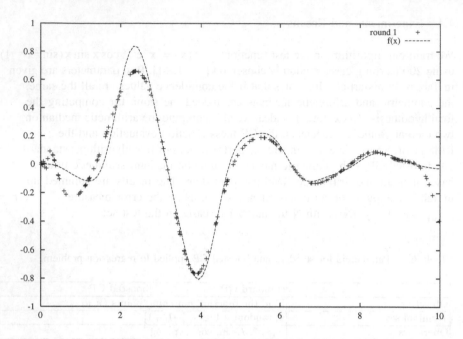

Fig. 1.   Target function and best-of function after the first round of boosting.

4.5   Results Analysis

Standard GP provides good results. However, with boosting results are improved.
We can see that standard GP encounters problems to approximate the function
on some sub-intervals, as is seen on Fig. 1. To introduce weights with boosting
helps to focus on these intervals for the next rounds. In Fig 2, we can see on the
plot where are the dicult areas, emphasized by larger weights.

The geometric median introduced by Drucker has the same error than the
standard median. We have verified that the choosen value (among every values proposed) by Drucker's method is the same as the value computed by the
arithmetic median. This is due to the fact that confidence values are very similar.

Using fitness and RMS error to measure the confidence did at times brought
dierent results than the standard geometric median. Although some predictions
were better, there is no improvement on average, within this experiment.

5   Conclusion and Future Works

All these results confirm the power of boosting and its applicability to genetic
programming. One advantage of boosting is that we simply have to take care of
the distribution without introducing deep modifications in the algorithm. Boosting may be seen as working around the algorithm, giving a better interpretation

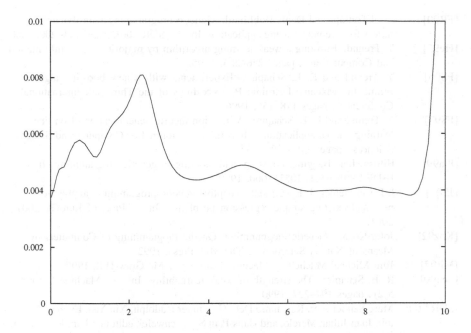

Fig. 2.   Weights values after the first update at end of round 1.

of its results. So it seems possible to combine it with other improves of genetic programming, in order to get even better results.

Some ideas have been presented for determining the final hypothesis (cf. Sect. 4.4), and we think this part deserves more work. Indeed, the loss function seems good for updating the weights, thanks to its good representation of relative errors between cases. Anyway, using this loss to compute the confidence in the hypotheses may be questionnable, since it does not really take care of the absolute error, which is intuitively related to the confidence from the user's point of view. Instead of one confidence value per hypothesis, it could be worth to use a confidence function varying according to   x.

Following these encouraging results on academic problems, we are now applying boosting to a real world problem: the ocean color (see [FR00]) which aims at quantifying ocean water components, such as chlorophyll-a, from remote sensing spectrometry data.

References

[BDE⁺99]   Wolfgang Banzhaf, Jason Daida, Agoston Eiben, Max Garzon, Vasant
            Honavar, Mark Jakiela, and Robert Smith, editors.   Procee dings of the Ge-
            netic and Evolutionary Computation Conference   , Orlando, Florida, USA,
            july 1999. Morgan-Kaufmann.
[Dru97]   H. Drucker. Improving regression using boosting techniques. In   Procee dings
            of International Conference on Machine Learning (ICML97)   , 1997.

[FR00]   Cyril Fonlupt and Denis Robilliard. Genetic programming with dynamic
           fitness for a remote sensing application. In   [SDR+ 00], pages 191–200, 2000.

[Fre95]   Y. Freund. Boosting a weak learning algorithm by majority.   Information
           and Computation , pages 256–285, 1995.

[FS96]   Y. Freund and R. E. Schapire. Experiments with a new boosting algo-
           rithm. In   Machine Learning: P rocee dings of the Thirteenth International
           Conference , pages 148–156, 1996.

[FS97]   Y. Freund and R. E. Schapire. A decision-theoric generalization of on-line
           learning and an application to boosting.   Journal of Computer and System
           Sciences , pages 119–139, 1997.

[Iba99]   Hitoshi Iba. Bagging, boosting, and bloating in genetic programming. In
           [BDE+ 99], pages 1053–1060, 1999.

[JE01]   J. I. van Hemert J. Eggermont. Adaptive genetic programming applied to
           new and existing simple regression problems. In   Proc. of EuroGP 2001 ,
           2001.

[Koz92]   John Koza.   Genetic Programming: On the Programming of Computers by
           Means of Natural Selection . The MIT Press, 1992.

[Mit97]   Tom Michael Mitchell.   Machine Learning . Mc Graw-Hill, 1997.

[Sch90]   R. E. Schapire. The strength of weak learnability. In   Machine Learning,
           5(2) , pages 197–227, 1990.

[SDR+ 00] Marc Schoenauer, Kalyanmo Deb, G   unter Rudolph, Xin Yao, Evelyne Lut-
           ton, Juan Julian Merelo, and Hans-Paul S   chwefel, editors.   Parallel Problem
           Solving from Nature VI , volume 1917 of Lecture Notes in Computer Sci-
           ence , Paris, France, September 2000. Springer.

[Val84]   L. G. Valiant. A theory of learnable.   Commun. ACM, 27(11) , pages 1134–
           1142, November 1984.

Dual Evolutionary Optimization

Rodolphe Le Riche [1] and Frédéric Guyon [2]

[1] Lab. de Mécanique de Rouen, UMR 6138, Saint Etienne du R   [ay], France
Rodolphe.Leriche@insa-rouen.fr
[2] Lab. de Bio-statistiques Bio-math´ ematiques, Univ. Paris 7, France
guyon@bach.urbb.jussieu.fr

Abstract.   The most general strategy for handling constraints in evolutionary optimization is through penalty functions. The choice of the
penalty function is critical to both success and eciency of the optimization. Many strategies have been proposed for formulating penalty
functions, most of which rely on arbitrary tuning of parameters. An new
insight on function penalization is proposed in this paper that relies on
the dual optimization problem. An evolutionary algorithm for approximately solving dual optimization problems is first presented. Next, an
ecient and exact penalty function without penalization parameter to be
tuned is proposed. Numerical tests are provided for continuous variables
and inequality constraints.

1   Introduction

Evolutionary optimization ([1]) stands today as one of the primary method for
tackling dicult optimization problems. Most practical applications of optimization are constrained problems. The issue of how to handle constraints in evolutionary optimization is therefore central and has received a lot of attention in
the last decade. The eciency of the method and its ability to generate optimal
feasible solution is at stake.

Four types of methods for handling constraints exist: penalization of infeasible solutions([2], [3], [4], [5], [6], [7]), projection of infeasible solutions onto
the feasible domain, co-evolution of populations which together solve the constrained optimization problem, and constraints representation building in the
course of the search. These approaches are related and have been coupled, like
co-evolution and penalty methods ([8] and [9]), or penalty and projection. A
review on constraints handling in evolutionary optimization can be found in
[10].

Among penalization strategies, one distinguishes static, dynamic and adaptive methods. Static penalties depend neither on the number of points sampled
during the search nor on their performance ([8]). Dynamic penalties are function
of the number of points sampled while adaptive penalties ([6], [7],[3],[2]) vary
with points evaluations. Mixed approaches exist, e.g. in [5].

Duality and related concepts such as Lagrange multipliers have yielded some
of the most ecient general purpose methods of mathematical programming for

P. Collet et al. (Eds.): EA 2001, LNCS 2310, pp. 281–294, 2002.

continuous, dierentiable and locally convex problems ([11]). Of particular practical importance are augmented Lagrangian functions and Lagrange multipliers updating techniques. On particular cases, it has been possible to formulate exact penalty functions ([12]).

Augmented Lagrangian functions and Lagrange multipliers updating have been applied to derive adaptive penalty functions in evolutionary algorithms. Bean and Hadj-Alouane ([6]) have proposed a penalty adaptation scheme which resembles Lagrange multipliers updating strategies. Kim and Myung ([2]) and Tahk and Sun ([9]) have used augmented Lagrangian penalty functions in evolutionary optimization, calculating Lagrange multipliers as a by-product of the search. In [9], a co-evolutionary algorithm simultaneously evolves a population of unknowns variables and a population of Lagrange multipliers.

The current work is also concerned with solving the dual optimization problem as a way to adapt a penalty function. Fundamentally, it diers from previous works in two aspects. First, an evolutionary algorithm is devised that explicitly solves the dual optimization problem. Second, the penalty function is not an augmented Lagrangian. Indeed, augmented Lagrangians were originally derived for mathematical programming. They are continuously dierentiable functions. They depend on the choice of a penalty parameter (the "augmented" term), which, if taken too small, leaves local optima. Continuous dierentiability is not needed in evolutionary optimization. The freedom gained in the formulation of the penalty permits removing the parameter and obtaining global optimality properties.

The text starts with a review of dual optimization principles. Then, linear and evolutionary algorithms are coupled to solve the dual optimization problem. Third, the non-equivalence of primal and dual problems is analyzed. It results in a new discontinuous exact penalty function that satisfies a minimal penalty rule. Finally, numerical tests are carried out where dual and primal problems are successively solved.

2   Dual Optimization

Dual optimization principles underlying the rest of the discussion are now reviewed.

2.1   Duality: Definitions and Fundamental Properties

The primal constrained optimization problem (P) is,

$$(P) \qquad \begin{array}{l} \min_{x \ S} \ f(x), \\ \text{such that} \ g(x) \quad 0. \end{array} \qquad (1)$$

where f, the objective function, and g, the constraint, are bounded functions (not necessarily continuous or dierentiable). We further assume that there is at least one feasible point in S, i.e., a point x such that $g(x) \quad 0$. The search is performed in the (primal) space of the design variables $x \ S$, which is a closed and bounded set. For the sake of simplicity, the number of constraints is limited to one. It should be noted that problems having $m \geq 1$ constraints can always be set in terms of a single constraint by taking the most critical constraint,

$$\min_{x \in S} f(x),$$
$$\text{such that } g(x) = \max_{i=1,m} g_i(x) \leq 0. \tag{2}$$

The set of solutions of (P) is denoted X^*, x^* is any element of X^*. The Lagrangian formulation (P_λ) of the primal problem is,

$$(P_\lambda) \qquad \min_{x \in S} L(x, \lambda), \tag{3}$$

where,

$$L(x, \lambda) = f(x) + \lambda\, g(x). \tag{4}$$

λ is a Lagrange multiplier. The set of solutions of (P_λ) is X_λ. We further assume that for each $\lambda \geq 0$, there exists at least a bounded solution $x_\lambda \in X_\lambda$. This assumption is fulfilled, for example, if f and g are continuous (Weierstrass Theorem, [11]).
The dual function is,

$$\partial(\lambda) = \min_{x \in S} L(x, \lambda), \tag{5}$$

and the dual problem is stated as,

$$(D) \qquad \max_{\lambda \geq 0} \partial(\lambda). \tag{6}$$

The dual search occurs in the Lagrange multipliers space. The solution of (D) are the Lagrange multipliers at the optimum, λ^*, and associated values of x are X_{λ^*}. An example of dual function is given in Fig. 1. When multiple constraints are handled through the maximization scheme of equation (2), λ^* is the optimal Lagrange multiplier of the most critical constraint.

The motivations for solving the dual optimization problem (D) are i) to directly solve the primal problem (P) when $X_{\lambda^*} = X^*$, ii) to calculate λ^*, which permits formulating exact penalty functions (cf. Section 4). At first glance, the dual problem seems much more complex than the primal problem since calculating the dual function involves solving an optimization problem, $\min_{x \in S} L(x, \lambda)$. However, favorable properties of the dual function added to the possibility of approximately solving (D) make duality a powerful approach for rational constraints handling.

Property 1 (Concavity of ∂). The dual function $\partial(\lambda)$ is concave in λ.

Property 2 (sub-gradient). For all $\lambda \geq 0$, let us denote $X_\lambda = \{ x_\lambda \in S / L(x_\lambda, \lambda) = \partial(\lambda) \}$. Then, for all $x_\lambda \in X_\lambda$, $g(x_\lambda)$ is a sub-gradient of ∂ at λ.

The two above properties, proofs of which can be found in [11], are valid under very general conditions (f and g bounded). They considerably simplify the resolution of (D) since ∂ is a concave function with a known sub-gradient. Lagrangian based penalty functions, such as the ones introduced in [6], [7], and [2], have penalty adaptation schemes where, schematically, the Lagrange multiplier is increased if the current best solution in terms of the penalized objective function is infeasible ($g(x_b) \geq 0$) and vice versa. Since $g(x_b)$ is an approximation of the sub-gradient of $\phi(\lambda)$, those penalty adaptation schemes are variations of a gradient based dual search. In the current work, an alternative strategy is taken to solve the dual problem (see Section 3 later).

2.2 Approximate Dual Problem

In terms of λ, (D) is easy to solve because it is a concave problem with a known subgradient. (D) has no local maximum and many algorithms exist to solve it (nondifferentiable optimization or linear programming, such as presented hereafter). The main difficulty remains the resolution of $\min_{x \in S} f(x) + \lambda g(x)$ at a given λ in the primal space. For this reason, the dual optimization problem is now approximated by restricting S to a discrete set of points T. It yields an approximate dual function,

$$\phi_T(\lambda) = \min_{x \in T} f(x) + \lambda g(x), \tag{7}$$

and an approximate dual problem,

$$(AD) \qquad \max_{\lambda \geq 0} \phi_T(\lambda), \tag{8}$$

where $T \subset S$ is a set of points of the primal space. By construction ϕ_T is concave and piecewise linear. (AD) can be formulated as a linear programming problem:

$$\max_{w, \lambda \geq 0} w,$$
$$(AD) \qquad \text{such that } f(x_i) + \lambda g(x_i) \geq w, \ x_i \in T, \tag{9}$$
$$\lambda \leq \lambda_{max},$$

where λ_{max} is an arbitrarily large upper bound on Lagrange multiplier meant to ensure the existence of a solution to (AD) even when all points in T are infeasible. The linear programming problem (AD) is efficiently solved by a simplex algorithm (cf. [13]). Let its solution be λ_T.

3 A Dual Evolutionary Algorithm

The dual evolutionary optimizer iterates between the primal problem (P_λ) and the approximate dual problem (AD). Based on a particular choice of λ, P_λ resolution by evolutionary optimization produces points to include in T. Based on T, (AD) resolution by the simplex algorithm yields a new Lagrange multiplier λ_T. Most dual optimization methods iterate between primal and dual spaces. Our algorithm bears particular resemblance to Dantzig's algorithm ([14]). The difference lies in the evolutionary primal optimization:

- It can visit different basins of attraction of the Lagrangian, $L(x, \lambda)$, during convergence, i.e., it can yield many judicious points to be included in T at each iteration.
- It can handle non-convex, discontinuous functions.

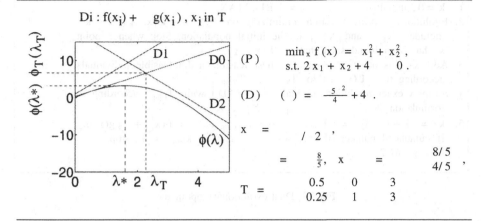

Fig. 1. Example of dual and approximate dual functions, problem with a saddle point.

A ow chart of a dual evolutionary optimizer is given in Fig. 2. X_k^f and X_k^i are the sets of feasible and infeasible active points of (AD) at iteration k, respectively. The evolutionary algorithm used is a steady-state algorithm with continuous mutation and crossover, and tournament selection ([1]). Evolutionary searches are stopped as soon as an improvement on L(x, $_k$) has been observed. This is an important implementation aspect as it saves much computational eort that would otherwise be spent minimizing the Lagrangian with $_k$ far from . As a side eect, this stopping criterion increases the number of resolutions of (AD). The cost of solving (AD) is however negligible because no evaluation of f or g is performed and the simplex algorithm is ecient. In all the tests performed (cf. section 5), the CPU time spent in (AD) is less than a percent of the total CPU time for T_k sets of up to 10000 elements. Further details on the simplex implementation, existence of X_k^f, and convergence rate of the method can be found in [13]. Important outputs of the algorithm are k_{final} and $X_{k_{final}}^f$.

k_{final} is an estimate of . $X_{k_{final}}^f$ is an estimate of the feasible points in X .

T_k gathers information from many potentially important points sampled by all evolutionary runs up to iteration k. Values of Lagrange multipliers are inferred from these points through an exact resolution of the approximate dual problem (AD). Such an approach is thought to be more ecient than gradient based dual searches which change $_k$ based on a local information, an approximation of g(x $_k$), x $_k$ X $_k$.

4   A Minimal, Exact, Penalty Function

The previous paragraph has introduced a coupled evolutionary / simplex algorithm for solving the dual problem (D). But the goal is to tackle the primal problem (P). In fact, problems having a saddle point at the optimum are readily

---

1. $k = 0$, initialize λ_0, $\mu_0 = $ DBL_MAX .
2. Evolutionary (primal) search minimizing on x $L(x, \lambda_k)$. If $k \neq 0$, include X^f_{k-1} and X^i_{k-1} in the initial population. Stop when a point x has been found such that $L(x, \lambda_k) \leq \mu_k$.
3. Add x plus other n_f and n_i best feasible and infeasible individuals according to $L(x, \lambda_k)$ to $T_k \to T_{temp}$.
4. Simplex exact resolution of dual $(\widetilde{AD})$ with T_{temp} according to formulation (9) $\to \lambda_{temp}$, X^f_k, X^i_k.
5. $k = k + 1$, $T_k = T_{temp}$, $\lambda_k = \lambda_{temp}$, $\mu_k = f(x^f_k) + \lambda_k g(x^f_k)$. If cumulated number of analyses $\geq N_{max}$, $k_{final} = k$, stop. Else go to 2.

Fig. 2.   Dual evolutionary optimizer

solved because in this case, $X^f = x^f$, x^f unique, $X^i = x^i$, x^i unique, and $x^f = x^i$ ([11]). In other terms, the dual and the primal problems are equivalent. The dual evolutionary algorithm provides $X^f_{k_{final}}$ (cf. Fig. 2), which includes an estimate of x^*. To sum up, the Lagrangian is a valid penalty function for problems having a saddle-point.

For problems without saddle point, solving $(\widetilde{D})$ does not directly provide a solution to (P), $X^f \neq X^i$. This can be seen on the example of Fig. 3 where $X^i = \{1.058, 4.58\}$, $x^f = 1.058$ and $X^f = \{4.5\}$. Problems without a saddle point require using another penalty function. Nevertheless, as will soon be seen, solving $(\widetilde{D})$ still generates information for properly penalizing the constraints : X^f contains at least one feasible element denoted x^f, $g(x^f) \leq 0$ (see [13]).

Let F_p denote any penalized objective function. The choice of the penalty function has a profound eﬀect on the evolutionary optimization eﬃciency. When too high a penalty is imposed on infeasible points, the population is prematurely pushed into the feasible domain, often far from optima x^*. Subsequent convergence to x^* can be extremely slow. In evolutionary terms, penalization makes F_p deceptive. Reciprocally, if too low a penalty is enforced, the algorithm converges into the infeasible domain. The optimal penalty function is problem dependent. However, several authors have described a reasonable heuristic, the minimal penalty rule, as a remedy against penalization induced deceptiveness (Davis [15], Richardson et al. [4], Smith and Tate [5]). It says: on the average, it is best to apply the smallest amount of penalty such that the algorithm converges to a feasible optimum, x^*. For calculation purposes, a more precise definition of "amount of penalty" is needed.

Definition 1 (Amount of penalty).   For optimization problems without a saddle point and such that there is an infeasible solution to the dual, x^i, the amount of penalty, r, is defined as,

$$r = F_p(x^i) - f(x^i) , \qquad (10)$$

where F_p is any penalized objective function.

A class of Lagrangian based exact penalty function is now introduced.

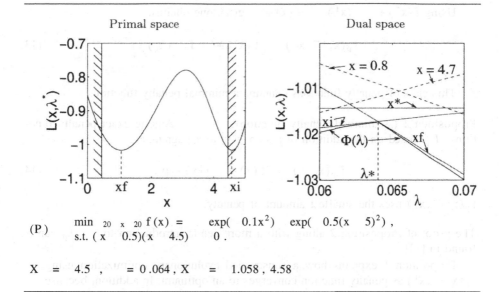

(P) $\min_{-20 \le x \le 20} f(x) = \exp(-0.1x^2) - \exp(-0.5(x-5)^2)$,
    s.t. $(x-0.5)(x-4.5) \le 0$.

$x^* = 4.5$, $\lambda^* = 0.064$, $X^* = 1.058, 4.58$

Fig. 3. Example of dual and approximate dual functions, problem without a saddle point.

Proposition 1 (A class of exact penalty functions).   For f and g bounded, let $f_p(x; \mu^+, x^f)$ be defined as,

$$f_p(x; \mu^+, x^f) = f(x) + H(g(x))[\mu^+ g(x) - \mu^+ g(x^f) - f(x^f) + f(x^-) + \epsilon],$$

(11)

where,

$$H(y) = 0 \text{ if } y \le 0, \ H(y) = 1 \text{ otherwise},$$

$$\mu^+ > \mu', \ x^f \in X / g(x^f) \le 0, \ \epsilon \ge 0,$$

x⁻ is the known feasible point with lowest f.

$f_p(x; \mu^+, x^f)$ has an absolute minimum at $x^- \in X^-$.

Proof: The result is obvious if $g(x) \le 0$. For all $x / g(x) \ge 0$, one shows that $f_p(x; \mu^+, x^f) \ge f_p(x^-; \mu^+, x^f) = f(x^-)$. Since $\mu^+ > \mu'$,

$$f_p(x; \mu^+, x^f) = f(x) + \mu^+ g(x) - \mu^+ g(x^f) - f(x^f) + f(x^-) + \epsilon$$
$$\ge f(x) + \mu' g(x) - \mu' g(x^f) - f(x^f) + f(x^-) + \epsilon.$$

(12)

Using $f(x^f) + \lambda g(x^f) \le f(x) + \lambda g(x)$, one obtains,

$$f_p(x; \lambda^+, x^f) \le f(x) + \lambda f(x). \tag{13}$$

This class of penalty functions contains a minimal penalty function.

Proposition 2 (A minimal penalty function). Among exact penalty functions, L_p, based on the addition of a step, ρp, to a Lagrangian,

$$L_p(x, \lambda) = f(x) + \lambda g(x) + \rho p, \tag{14}$$

$f_p(x; \lambda, x^f)$ uses the smallest amount of penalty.

The proof of Proposition 2 along with a more gentle introduction to f_p can be found in [13].

Proposition 1 explains how a constrained evolutionary optimization using $f_p(x; \lambda, x^f)$ as penalty function converges to an optimum. In addition, because it is a minimal penalty strategy (Proposition 2), it promotes fast convergence. An evolutionary optimizer for general constrained optimization problems is described in Fig. 4. Finally, we emphasize that no parameter of the penalty function is arbitrarily set since λ and x^f have a precise definition in terms of (λ D). ρ_0, N_{max}, n_f and n_i control the rate of convergence in the dual space. These parameters have little inuence compared to penalty parameters.

1. Run the dual evolutionary algorithm of Fig. 2
   $\lambda_{k_{final}}$, $x^f_{k_{final}}$, $x^i_{k_{final}}$.
2. Final evolutionary search minimizing on $x\ f_p(x; \lambda_{k_{final}}, x^f_{k_{final}})$. $x^f_{k_{final}}$
   and $x^i_{k_{final}}$ are included in the initial population.

Fig. 4.   Evolutionary optimization based on f_p.

5   Numerical Tests

Results on 4 test problems are presented. Each of them averages 50 independent runs. The mutation operator adds to the variables a Gaussian noise $N(0, \sigma)$, $\sigma^2 = (x_{max} - x_{min})^2/16$, $x_{min} = -20$, $x_{max} = 20$. The probabilities of crossover and mutation are $p_c = 0.7$ and $p_m = 0.4$, tournaments of size 2 select individuals. Population sizes, n_{pop}, and search lengths, N_{max}, are the same for the dual and primal evolutionary searches. The pairs (n_{pop}, N_{max}) are $(200, 10000)$, $(200, 10000)$, $(300, 100000)$ and $(300, 100000)$ for Tests 1 to 4, respectively. Other parameters of the algorithm are $\rho_0 = 20$, $n_f = n_i = 20$.

5.1   Comparison of Approaches

Dierent penalty approaches are compared to illustrate three claims. Firstly, the minimal penalty function promotes fast and reliable convergence as compared to arbitrarily tuned static penalty functions. Secondly, the dual evolutionary optimizer is a better strategy than an adaptive linear penalty. Thirdly, the dual evolutionary optimizer is not sensitive to its parameters setting. Those claims are checked using the hoop problem, which has an objective function composed of two linear functions and a narrow curved feasible domain :

$$\min_{x_1,x_2} f(x_1,x_2) \quad \text{such that} \quad \begin{aligned} R^2 \quad (x_1 \quad A)^2 \quad x_2^2 \quad 0, \\ (x_1 \quad A)^2 + x_2^2 \quad (R+E)^2 \quad 0, \\ 0 \quad x_1 \quad A, \quad 0 \quad x_2 \quad A, \end{aligned} \tag{15}$$

where,

$$f(x_1,x_2) = \begin{aligned} (x_1+x_2) \quad &\text{if} \quad x_1+x_2 \quad H \quad 0, \\ (x_1+x_2)+(\quad +1)H \quad &\text{otherwise}, \end{aligned} \tag{16}$$

and $A = 20$, $R = 18$, $E = 0.1$. The solution is $x = (1.9, 0)^T$. H is a parameter that controls the size of the basin of attraction of a local optimum. If $H = 5$, there is a local optimum at $(20, 18.1)^T$. If $H = 40$, there is no local optimum.

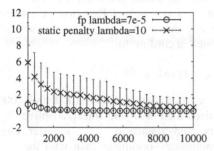

Fig. 5.   x   x   vs. nb. of analyses. Comparison of two amounts of static penalties, hoop problem.

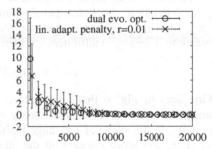

Fig. 6.   x   x   vs. nb. of analyses. Comparison of two adaptive approaches for estimating the penalty, hoop problem.

Fig. 5 illustrates the eect of the (static) amount of penalty on the convergence to x . The minimal penalty function $f_p(x; ,x^f)$ is compared to a linear (static, is fixed) penalty function,

$$F_p(x;) = f(x) + \max(0,g(x)), \tag{17}$$

where   is arbitrarily set to 10. Note that no dual search is performed on this plot. It is seen that the static penalty is much slower than the minimal penalty function. Another series of tests has been performed with   H = 5. The evolutionary algorithm using a static penalty function,   = 10, converges to the local (false) optimum 38 times out of 50, against 26 times when using the minimal penalty.

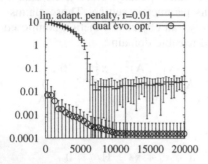

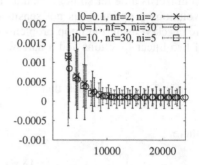

Fig. 7.   vs. nb. of analyses. Comparison of two adaptive approaches for estimating the penalty, hoop problem.

Fig. 8.   vs. nb. of analyses. Convergence of the dual evolutionary optimizer for dierent settings of $_0$, n_f and n_i, hoop problem.

In Fig. 6 and 7, the dual evolutionary optimizer of Fig. 4 is compared to an adaptive linear penalty algorithm. This last algorithm minimizes, at each primal iteration, the linear penalty function $F_p(x; _k)$ of Equation (17) using an evolutionary optimizer. Let use temporarily denote by $x(_k)$ the solution estimate. Lagrange multipliers are then updated according to,

$$_{k+1} = \max(0, (_k + rg(x(_k)))) . \tag{18}$$

One sees in Fig. 6 that convergence to   x   is faster, between 1000 and 7000 analyses, with the dual evolutionary optimizer than with the linear adaptive penalty. In the space of Lagrange multipliers, convergence to   is faster, more accurate and more stable with the dual evolutionary optimizer than with the linear adaptive penalty (cf. Fig. 7).

Fig. 8 shows how the dual evolutionary optimizer converges to   when its parameters setting ($_0$, n_f and n_i) changes. The method does not appear to be sensitive to parameters changes. The only visible feature is a slightly higher variance when $n_f + n_i$ decreases.

5.2   Convergence on Various Test Functions

The second test is the two humps function stated in Fig. 3. It has one variable and one constraint. It does not have a saddle point. The feasible solution of the dual problem, $x^f = 1.058$, is far from the optimum, x = 4.5. The algorithm rapidly converges to   , on the average after 5000 analyses (cf. Table 1). During the dual iterations, the evolutionary optimizer converges either to x^i or to x^f

(similarly for Tests 3 and 4 later). The final primal search using f_p robustly locates x . After 10000 analyses, the best search point according to f_p, x_b, is such that x_b x $= 4.6e$ 4 6.5e 4.

The third test has two variables and two constraints. The constraints are reduced to one constraint by considering only the most critical. It is formulated as ([10]),

$$\min_{x_1, x_2 \ [0.001, 20]} \frac{\sin((2\ x\ _1)^3)\sin(2\ x\ _2)}{x_1^3(x_1 + x_2)} ,$$
$$\text{such that } g(x_1, x_2) = \max(\ g_1(x_1, x_2), g_2(x_1, x_2))\quad 0 , \tag{19}$$
$$g_1(x_1, x_2) = x_1^2 \quad x_2 + 1 ,$$
$$g_2(x_1, x_2) = 1 \quad x_1 + (x_2 \quad 4)^2 .$$

Solutions of the primal and dual problems are :

$$x = \begin{matrix} 1.228 \\ 4.245 \end{matrix} , \quad = 87.348 , \ x^f = \begin{matrix} 1.604 \\ 4.155 \end{matrix} , \ x^i = \begin{matrix} 0.001 \\ 0.006 \end{matrix} .$$

Test 3 is known as a problem where fiding the right penalty is dicult ([10]). This can be understood by looking at a plot of the objective function on Fig. 9. There is a strong infeasible attractor (which, logically, is x^i) with a very low objective function, f 2000, near the origin. The optimum is far from that point and has a much higher objective function, f 0.1. It can only be seen on the right side zoom of Fig. 9 (where infeasible points have been removed). Feasible local optima are also visible on that plot. Since the infeasible attractor has such a low objective function, a large penalty is required. However, too large a penalty make the search likely to converge to a local feasible optimum. Test 3 has a large "duality gap", $f(x\ )$ $L(x^f, \ )$, and a large . Numerical experiments show that the dual evolutionary optimizer locates within 20 accuracy in 50000 analyses and 15 accuracy in 100000 analyses (cf. Table 1). Those results, which are substantially worse than those obtained on the other tests, illustrate how penalizing Test 3 is dicult. Nevertheless 15 accuracy in is sucient for f_p to robustly guide the search around x in the final primal search, x_b x $= 2.e$ 4 8.e 4.

The fourth and last test problem (from [10]) has 7 variables, 4 constraints,

$$\min_{20\ x_i\ 20} (x_1 \quad 10.)^2 + 5(x_2 \quad 12)^2 + x_3^4 + 3(x_4 \quad 11)^2 +$$
$$10x_5^6 + 7x_6^2 + x_7^4 \quad 4x_6 x_7 \quad 10x_6 \quad 8x_7 ,$$

such that,

$$127 + 2x_1^2 + 3x_2^4 + x_3 + 4x_4^2 + 5x_5 \quad 0 , \tag{20}$$
$$282 + 7x_1 + 3x_2 + 10x_3^2 + x_4 \quad x_5 \quad 0,$$
$$196 + 23x_1 + x_2^2 + 6x_6^2 \quad 8x_7 \quad 0,$$
$$4x_1^2 + x_2^2 \quad 3x_1 x_2 + 2x_3^2 + 5x_6 \quad 11x_7 \quad 0 .$$

Two of the constraints (the first and the last) are active at the optimum.
x $= (2.330499, \quad 1.951372, \quad 0.4775414, 4.365726, \quad 0.624487, \quad 1.038131,$

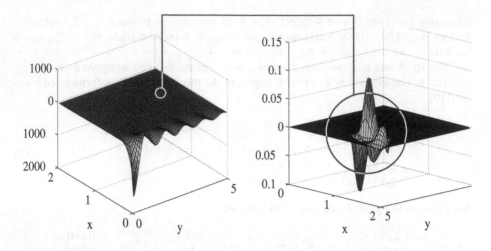

Fig. 9. Plots of the objective function of Test 3. On the right, only feasible points are drawn and the view point is changed so that feasible local minima are visible.

$1.594227)^T$, $\quad$ = 1.493. Although proof of existence of a saddle point is di-cult to establish formally for non-convex problems, Test 4 seems to have a saddle point because the numerically determined $\quad x^f$ and x^i are close to x . Constraints are handled by the max scheme of (2). Numerical experiments show that $\quad\quad$ is found after 50000 analyses (cf. Table 1). At the end of the last primal search,
x_b x $\quad$ = $2.7e$ 1 $1.5e$ 1.

These numerical experiments lead us to consider that solving the dual before the primal problem about doubles the price of the search. Convergence to $\quad\quad$ is achieved in 5000 analyses when N_{max} = 10000 analyses in Test 2, and in 50000 analyses when N_{max} = 100000 analyses in Test 4, but all N_{max} = 100000 analyses are necessary in Test 3. N_{max} is chosen as a typical search length with a static penalty function. Therefore, in this work, it is the length of the last primal search which minimizes f_p.

Table 1. Convergence to $\quad$: average $_k$ $\quad/\quad$ std. deviation, 50 runs.

No. analyses	500		5000		10000		50000		100000	
Test 2	0.02	0.02	2.e-5	4.e-5	1.e-5	2.e-5				
Test 3	0.71	0.15	0.41	0.15	0.32	0.15	0.20	0.11	0.15	0.07
Test 4	8.e3	6.e3	122	368	1.49	1.63	0.28	0.20	0.27	0.12

6   Concluding Remarks

A general method for handling inequality constraints in evolutionary optimization has been proposed. It is an adaptive penalty strategy based on duality. Beyond the ability, shared by all evolutionary methods, to tackle non-convex optimization problems, this approach has the following advantages: it does not require any penalty parameter to be tuned, the amount of penalty put on infeasible points is minimal, and it yields optimal Lagrange multipliers as a by-product of the search. Lagrange multipliers are important because they describe the sensitivity of the objective function at the solution to a change in constraints. In the current implementation of dual evolutionary optimization, solving for the Lagrange multipliers doubles the computational cost of a search using a traditional static penalty function.

References

1. Back, T.: Evolutionary Algorithms in Theory and Practice. Oxford Univ. Press, New York (1996)
2. Kim, J.-H., Myung, H.: Evolutionary Programming Techniques for Constrained Optimization. IEEE Trans. on Evolutionary Computation. July (1997) 129–140
3. Powell, D., Skolnick, M.M.: Using Genetic Algorithms in Engineering Design Optimization with Non-linear Constraints. In: Proc. of the Fourth International Conference on Genetic Algorithms. Morgan Kaufmann, San Mateo CA (1991) 424–431
4. Richardson, J.T., Palmer, M.R., Liepins, G. Hilliard, M.: Some Guidelines for Genetic Algorithms with Penalty Functions. In: Proc. of the Third International Conference on Genetic Algorithms. Morgan Kaufmann, San Mateo CA George Mason Univ., June 4-7 (1989) 191–197
5. Smith, A.E., Tate, D.M.: Genetic Optimization using a Penalty Function. In: Proc. of the Fourth International Conference on Genetic Algorithms. Morgan Kaufmann, San Mateo CA (1991) 499–505
6. Bean, J.C., Hadj-Alouane, A.B.: A Dual Genetic Algorithm for Bounded Integer Programs. Technical Report TR 92-53. Dept. of Industrial and Operations Eng., The Univ. of Michigan (1992)
7. Hadj-Alouane, A.B., Bean, J.C.: A Genetic Algorithm for the Multiple-Choice Integer Program. Technical Report TR 92-50. Dept. of Industrial and Operations Eng., The University of Michigan (1992)
8. Le Riche, R., Knopf-Lenoir, C., Haftka, R.T.: A Segregated Genetic Algorithm for Constrained Structural Optimization. In: Eschelman, L. (ed.): Proc. of the Sixth International Conference on Genetic Algorithms (ICGA95). Morgan Kaufman, San Francisco CA (1995) 558–565
9. Tahk, M.-J., Sun, B.-C.: Co-evolutionary Augmented Lagrangian Methods for Constrained Optimization. Submitted for publication in: IEEE Trans. on Evolutionary Computation. February (1999)
10. Michalewicz, Z., Schoenauer, M.: Evolutionary Algorithms for Constrained Parameter Optimization. Evolutionary Computation. Vol. 4   1 (1997) 1–32
11. Minoux, M.: Programmation Math´ ematique, Th´ eorie et Algorithmes. Vol. 1 and 2. Dunod, Paris (1983).
12. Howe, S.: New Conditions for Exactness of a Simple Penalty Function. SIAM Journal of Control. Vol. 11   2 (1973) 378–381

294 R. Le Riche and F. Guyon

13. Le Riche, R. Guyon, F.: Dual Evolutionary Optimization. Technical Report no. 01/2001. LMR, INSA de Rouen, France available at http://meca.insa-rouen.fr/rleriche   (2001)
14. Dantzig, G.B., Wolfe, P.: The Decomposition Algorithm for Linear Programming. Econometrica. Vol. 29   4 (1961) 767–778
15. Davis, L.: Genetic Algorithms and Simulated Annealing. Pitman, London (1987)

Using Evolutionary Algorithms Incorporating the Augmented Lagrangian Penalty Function to Solve Discrete and Continuous Constrained Non-linear Optimal Control Problems

Stephen Smith

School of Mathematics and Decision Sciences
Faculty of Informatics and Communications
Central ueensland University
Rockhampton
ueensland
Australia
s.smith@cqu.edu.au

Abstract. Constrained Optimal Control Problems are notoriously di-cult to solve accurately. Preliminary investigations show that Augmented Lagrangian Penalty functions can be combined with an Evolutionary Algorithm to solve these functional optimisation problems. Augmented Lagrangian Penalty functions are able to overcome the weaknesses of using absolute and quadratic penalty functions within the framework of an Evolutionary Algorithm.

1 Introduction

The vast majority of research into solving constrained optimisation problems using evolutionary algorithms has considered problems that involve the optimisation of a real valued function:

$$\text{Minimise} \qquad f(x) \tag{1}$$

subject to

$$g(x) \quad 0 \tag{2}$$
$$h(x) = 0 \tag{3}$$
$$x \quad . \tag{4}$$

The standard way to solve constrained problems using evolutionary algorithms is to use an evolutionary algorithm with continuous variables.

The constraints (2) and (3) are usually enforced by using either an absolute or a quadratic penalty function to incorporate the constraints into the fitness

P. Collet et al. (Eds.): EA 2001, LNCS 2310, pp. 295–308, 2002.

function. Equation (4) then defines the search space of an unconstrained problem, $x \in S \subseteq R^n$.

A larger class of problems are given by functional optimisation problems. Instead of searching R^n for an optimal point, functional optimisation involves searching for an optimal curve $c(t) \in R^n \forall t$ from within a given class of curves. This can be expressed as:

$$\text{Minimise} \quad f(x(t)) \tag{5}$$

subject to

$$g(x(t)) \leq 0 \tag{6}$$
$$h(x(t)) = 0 \tag{7}$$
$$x(t) \tag{8}$$
$$t_0 \leq t \leq t_1 . \tag{9}$$

Constrained Continuous Optimal Control Problems are a good example of a class of functional optimisation problems.

In this paper we investigate the eects of applying a third penalty technique, the Augmented Lagrangian penalty function instead of either the absolute or quadratic penalty function to optimal control problems.

2   The Problem

This paper considers how an augmented lagrangian penalty function can be incorporated into an evolutionary algorithm to solve optimal control problems of the following types:-

Discrete Problems

$$\text{Minimise} \quad (x_k) + \sum_{i=0}^{k-1} f_i^0(x_i, u_i) \tag{10}$$

subject to

$$x_{i+1} = f_i(x_i, u_i) \quad i = 0 \ldots k \tag{11}$$
$$x_0 = x0 \quad \text{(given)} \tag{12}$$
$$p(x_k) = 0 \tag{13}$$

where $x \in R^n$, $u \in R^m$ and p is vector function. u_i is usually assumed to be a constant over each interval. Thus overall u is a piecewise constant function in m dimensions.

Continuous Problems

$$\text{Minimise} \quad (x(t_1)) + \int_{t_0}^{t_1} f_0(x, u, t)\,dt \tag{14}$$

subject to

$$x = f(x, u, t) \tag{15}$$

$$x(t_0) = x_0 \quad \text{(given)} \tag{16}$$

$$g(x, u, t) \quad 0 \tag{17}$$

$$h(x, u, t) = 0 \tag{18}$$

$$p(x(t_1)) = 0 \tag{19}$$

$$P(x(t_1)) \quad 0 \tag{20}$$

where $x \quad R^n$, $u \quad R^m$ and g, h, p and P are vector functions. $u(t)$ is a curve in R^m t. In order to solve this problem numerically, $u(t)$ has to be approximated using some simpler class of curves. There are several ways of doing this. Firstly there are the collocation methods, in which u is approximated by a function of the form $_i$ $_i c_i(t)$ over the whole interval t_0 t t_1, where the $c_i(t)$ are a family of orthornormal functions, e.g. the chebychev polynomials. The problem then is to find the optimal combination of $_i$s. Another alternative is to discretise in the t dimension and approximate $u(t)$ with a piecewise continuous function. Typical examples include piecewise constant, piecewise linear, and linear or cubic splines.

In both case it should be noted that m n.

3   Traditional Approaches to the Problem

There are many dierent approaches that can be taken to solve the optimal control problems above. Most take the approach of assuming that the various components of the representation of the control curve u are the only unknowns. This then leads to an optimisation problem in R^{km} (the m components of the vector u at each of the k time steps). In these approaches (11) and (15) are treated as single large initial value problems (IVPs). Given a well defined u these equations can be used to explicitly determine the relevant values of x by 'integrating' through from start to finish. However, from an evolutionary algorithm point of view, this means that the constraints need not be in the search space. For example a constraint involving only x is implicitly defined without any explicit reference to the optimisation variables u.

An alternative approach is that of Mathematical Programming, in which the x values are also assumed to be unknowns. In the continuous case the relationship (15) can be re-expressed as a series of smaller problems over each of the discretisation ranges. On integrating these we get the relationships

$$x_{i+1}(t) = x_i(t) + \int_{t_i}^{t+1} f(x, u_i(t), t)dt \ . \tag{21}$$

Each of these integrations start at the point $x_i(t)$ and produce the point $x_{i+1}(t)$ which in general does not match the starting point of the next interval $x_{i+1}(t)$. Thus the resulting 'trajectory' is not continuous. Mathematical Programming then adds the extra constraints that

$$x_{i+1}(t) = x_{i+1}(t) \qquad i = 0 \ldots k \quad 1 \ . \tag{22}$$

The Mathematical Programming approach has increased the dimension of the optimisation problem from R^{km} to $R^{k(m+n)}$ and introduced an additional kn equality constraints. This means that the number of unknowns has at least doubled and a significant number of equality constraints have been added. Why do this? One reason is that it is sometimes very dicult to fid a curve $u(t)$ for which it is possible to integrate the system 'right through' without it 'blowing up'. Consider the problem $x = 10x + u$ with $x(0) = 100$. A control curve of $u(t) = 10$ will cause x to rapidly shoot o towards infiity. However, if this is discretised into 100 pieces then because each interval starts at a finite $x_i(t)$ none of the smaller IVPs can 'blow up'. Thus the error has been 'contained' within the discrepancies between $x_{i+1}(t)$ and $x_{i+1}(t)$. Another reason is that now all of the constraints lie within the search space. This then means that there is more chance of designing special operators based on the constraints [4].

When integrating the right hand side of (21), any of the standard methods for solving ordinary dierential equations may be used. If the problem is well behaved then the classical fourth order Runge-Kutta method is quite adequate. If the interval of integration, $[\ t_i, t_{i+1})$, is quite large then it is possible to take several small Runge-Kutta steps to maintain the necessary accuracy in the solution. Thus, while the continuous problem has been discretised, it is seen that the problem is still extremely complex.

The Mathematical Programming formulation of the problem can be used as an optimisation method in its own right or it may be used as a starter routine to find a suitable initial solution to feed into a more powerful technique which works on a formulation in which only the u's are treated as unknowns. This is a viable approach when, as above, it is dicult to determine an initial solution curve that can be integrated 'right through' or if the secondary method requires a starting curve in the neighbourhood of the optimal solution in order for it to converge, for example Newton type methods.

4   Penalty Techniques

A standard approach for handling constraints is to augment the function that is being optimised by adding in weighted terms that incorporate these constraints, thus converting the problem to an unconstrained one. Traditionally this can be done by using either a   barrier function or a penalty function [2].

Unfortunately barrier functions do not sit well in the evolutionary frame work since a mutation operator is quite likely to 'jump over' a constraint barrier.

The two most common penalty methods for handling constraints are the absolute and the quadratic penalty terms.

A third technique is the Augmented Lagrangian method.

4.1   Absolute Penalty Method

The absolute penalty function:

$$F = f(x) +   h_i(x)$$

is exact for a finite value of   . However, this approach produces a valley oor, in the search space, that is extremely narrow. (The profile of the valley has a "   " shape with very little room on the 'valley oor'.)

Our research has shown that this has the eect of crippling'an evolutionary algorithm once it has reached the valley oor. This premature curtailment of the evolutionary algorithm, which can easily be mistaken for premature convergence to a local minimum, is due to the fact that the ospring are almost always worse than the parent since they are located on the side of the valley, part way up the wall.

4.2   uadratic Penalty Method

The quadratic penalty method:

$$F = f(x) +   \frac{1}{2}h_i^2(x)$$

tends to produce a "   " shaped valley oor. Thus it presents a 'atter and wider' valley oor which the ospring can spread out over as they meander down towards the true solution. It can be shown that the quadratic penalty methods only converge to the true solution as the penalty multiplier term,   , tends to infinity. (The solution computed always lies in the infeasible region.) Numerically, as   tends to infinity the problem becomes more ill-conditioned, since the penalty term starts to dominate the computation.

4.3   Augmented Lagrangian Penalty Function

The augmented lagrangian penalty methods combine the best features of the previous two techniques. They have a suitably wide valley oor, yet will give the exact solution for a finite value of the multiplier term.

The most common form [2] of the augmented lagrangian penalty function is:

$$L_A(x,\ ,\ ,h\quad_i) = f(x) + h\ _i(x) + \frac{1}{2}h_i^2(x)$$

where   is the user supplied penalty multiplying factor, c.f. the   in the quadratic penalty function, and   is the Lagrange multiplier.   is automatically updated as the system approaches the optimal solution.

There are alternative formulations for this function, but this particular version shows the relationships it has with the clasical theory of Lagrange multipliers and how these have been   augmented   with a quadratic penalty term.

The augmented lagrangian method is known [2] to be able to force the   exact satisfaction of a constraint without the drawbacks of the other two methods.

Traditionally, all three formulations attempt to solve the original problem by solving a series of unconstrained subproblems.

A small (positive) value of   is chosen and the resulting unconstrained problem is solved. The value of   is then increased slightly and the new problem is solved using the final value of the previous problem as the starting value for the new problem. In an evolutionary algorithm context this is achieved by suitably increasing the value of   every G generations.

The only dierence for the augmented lagrangian penalty method is that the value of   is updated at the same time. (Initially,   = 0, thus to begin with the augmented lagrangian penalty method behaves like the quadratic penalty method.) There are many dierent formulae for updating   . It can be shown that the overall rate of convergence of a method using the augmented lagrangian penalty method depends on the rate of convergence of both   and x. Thus if one were using the augmented lagrangian penalty method in conjunction with Newton's method, then in order to retain the quadratic nature of the outer Newton method, it must be combined with a second order update method for the   s. However, since Evolutionary Algorithms are essentially zero order methods, any formula for updating the   s should be suitable. One such formula is:

$$_{j+1} = \ _j \quad h\ _i(x\ )$$

where x   should be the   exact minimum of the jth subproblem.

In traditional methods this is achieved by iterating each subproblem to convergence. However, in an evolutionary algorithm, there is no guarantee that the exact minimum of the previous 'subproblem' has been found when   and   are updated at the end of the   G generations. Running the EA until there is no improvement in the ftest individual of each subproblem is too inecient.

Any problems that may arise from using an inexact value for   x can usually be overcome by using a relaxation (damping) parameter,   , (c.f. the learning rate in neural networks). Thus the actual update rule used is

$$_{j+1} = \ _j \quad h\quad _i(x\ )$$

where   is typically 0.1 and   G is typically 10.

4.4   Application to Inequality Constraints

All three techniques can also be applied to the inequality constraints $g_i(x) \geq 0$, in a one sided way. This is done by defining a new function that only takes on positive values:

$$g^+(x) = \max(0, g(x))$$

g^+ only has an eect when a constraint is violated. For example, the augmented lagrangian penalty for an inequality constraint becomes:

$$L_A(x, , g_i) = f(x) + g_i^+(x) + \frac{1}{2}g_i^{+2}(x) . \tag{23}$$

This formulation means that there will be a discontinuity in the derivative of the fitness function at the constraint boundary, and the optimal solution if it lies on an inequality constraint. While this should have no eect on a standard evolutionary algorithm, since it does not use derivatives, the following alternative formula can be used in which the discontinuities in the second derivative occur at points away   from the optimal solution:

$$F = f(x) + \begin{cases} g(x) + g^2(x), & \text{if } g(x) \geq \frac{}{2}, \\ \frac{}{4}, & \text{otherwise} . \end{cases} \tag{24}$$

The formula for updating the value of   (with a damping factor added) is

$$_{j+1} = _j   \min(g, 0) . \tag{25}$$

Experimental work has shown that both (23) + (25) and (24) + (25) work well together for inequality constraints.

4.5   Other Lagrangian Approaches

There appears to be a limited amount of research being carried out into the use of Lagrangian methods within Evolutionary Algorithms. Two such research groups are:-

- H-J Kim and his students, in collaboration with D. Fogel, have been working on using the lagrangian function (not the augmented lagrangian penalty method) i n a 2 stage hybrid method [3]. The first stage uses an evolutionary Programming algorithm with a quadratic penalty method to determine a near optimal solution. The best individual is then used to seed a second evolutionary programming method based on a fitness function that uses Lagrange multipliers to incorporate the constraints.

– Helio Barbosa at LLNC in Brazil is approaching the problem from a co-evolutionary algorithm point of view [1]. The Augmented Lagrangian Penalty formulation can be transformed into a minimax problem and solved like a zero-sum dierential game. In this approach there are two competing populations, one that encodes the original set of unknown variables   x and tries to minimise the fitness function. The other population encode the Lagrange variables $_i$ and attempts to maximise the fitness function. In this context, the   variables are referred to as the   dual variables and the x variables as the primal variables.

Under certain conditions, on all the functions, it can be shown that the problem is well posed and that the optimal solution (x ,   ) satisfy the relationship:

$$L(x ,) L(x ,) L(x,)   x R^n, R \qquad (26)$$

where L is the fitness function incorporating the Augmented Lagrangian Penalty Function. This can also be formulated as:

$$\min_{x} \max L(x,) \qquad (27)$$

where the   s corrosponding to inequality constraints are restricted to non-negative values.

5   Examples

To demonstrate the potential of the augmented lagrangian penalty method we apply it to several simple examples.

5.1   An Example from Michalewicz

Michalewicz [4] gives an example of a discrete optimal control problem.

$$\text{Minimise} \quad \sum_{i=0}^{N 1} \overline{u_i} \qquad (28)$$

subject to

$$x_{i+1} = ax_i + u_i \qquad (29)$$

$$x_0 = 100 \qquad (30)$$

$$x_n = 100 . \qquad (31)$$

This is the harvest problem.   a is the growth rate of the population (of fish), x_i, and u_i is the quantity (of fish) harvested at the end of each of   n time periods.

The end condition ensures that the population is preserved for the beginning of the next cycle (year).

Michalewicz uses the example sizes of $n = 10, 20$ with $a = 1.1$. For ease of comparison we use the same sizes.

A simple worse case analysis can be used to calculate suitable upper and lower bounds on the variables. Studying $x_{i+1} = ax_i$ shows that x_i 800 n 21 (using the rule of 72). x must be positive for a meaningful real-life problem. A few trial runs, quickly show that a value of $= 100$ is a suitable upper bound on the u values. Again u is assumed to be positive, since the model has no 'cost' for adding to the (fish) population.

5.2 A Continuous Example

$$\text{Minimise} \quad \frac{1}{2} \int_0^1 x^2 + u^2 \, dt \tag{32}$$

subject to

$$x = x + u \tag{33}$$
$$x(0) = 1 \tag{34}$$
$$x(1) = 0 . \tag{35}$$

The curve $u(t)$ is approximated by a piecewise constant function consisting of k equal pieces, $u(t) = u_i$. This is solved using two dierent approaches. Firstly, equation (33) is used to 'integrate' the system through from $t = 0$ to $t = 1$ for a given control curve $u(t)$. This gives a problem with k unknowns, the values of the constant values of u and the single end point equality constraint (35). Secondly, both u and x are considered to be variables. This then gives a Mathematical Programming representation of the problem, having the extra characteristic constraints:

$$x_{i+1} = x_{i+1} \quad x_i + \int_{t_i}^{t_{i+1}} x + u_i \, dt \quad x(t_i) = x_i . \tag{36}$$

This has the eect of doubling the number of unknown variables and adding a further k equality constraints. The single large initial value problem (IVP), which may blow up, is replaced by k smaller ones, which hopefully will give bounded solutions. These then have to be 'joined together' by enforcing the equality constraints (36)

5.3 A Discrete Version of Example 2

To complete the picture, the discrete version of the same problem is solved using the same two techniques.

$$\text{Minimise} \quad \frac{1}{2N} \sum_{i=0}^{N-1} x_i^2 + u_i^2 \quad (37)$$

subject to

$$x_{i+1} = x_i + \frac{x_i + u_i}{N} \quad (38)$$

$$x_0 = 1 \quad (39)$$

$$x_n = 0 \ . \quad (40)$$

6   Results

In all of the examples it should be noted that the final end point condition, (31, 35, 40) had to be multiplied by a factor of 5. This was necessary to ensure that the end point condition dominated the performance index. There is a trade-o between reaching the desired target and increasing the overall cost. The penalty imposed for not reaching the target point must be greater than the 'saving' that is made by 'falling short' of the target. A small amount of trial and error is required to find a suitably small multiplier that ensures that the solution will eventually lock on to the target point to a high degree of accuracy, but is not too high that it does not allow the trajectory to move around in the early stages in order that it may settle down into its optimal shape.

All experiments were carried out using a custom-designed evolution program [4] which included a local hill-climber based on the simplex method.

The penalty multiplying factor in the augmented lagrangian penalty function was initially set to 0.01 and allowed to slowly grow to a maximum possible value of 50. The Evolutionary algorithm was run for 30000 generations with a population of 100. Following the advice of Larry Fogel [1]:

> We have to be thieves of the night.

we have begged, borrowed and probably stolen many ideas from many different people, especially in the area of operators.

Experiments have shown that when solving problems using an evolutionary algorithm and the Augmented Lagrangian Penalty method, it is better to approach a (hopefully global) minimum from below. That is we try to find a current best individual whose fitness function (including penalty terms) is   less than   the (unknown) constrained optimal value of the performance index. This is then steadily raised, by increasing the value of   until the constraints are satisfied and a (local) minima has been found. When testing on functions with known optimal values, it is seen that the best convergence is attained when the rate of increase of   is gradual enough to keep the value of the performance index below that of the optimal value, while the fitness function is kept just above.

---

[1] face to face conversation

If   is increased too quickly then the total penalty applied to the best individual can become large and this adversely eects the convergence to an optimal solution. Thus a careful eye should be kept on the dierence between the fness value and the actual value of the performance index for the current best individual. A large discrepancy could indicate that   is being incremented in too bigger steps or that there are not enough generations between increments. In this work   was incremented by 0 .00002 every 20 generations.

6.1   Harvest Problem

The algorithm has no problem with the smaller formulation of the Harvest Problem. The Algorithm consistently locks onto the optimal value. The error in hitting the target is down to machine accuracy. This is achieved with a maximum penalty value,   , of 50 and the final values of the lagrange multiplier,   , are given in the last column of the table.

n	fitness	end point error	
10	32.8209433819216	7.17648163117701E-14	2.038E-03
20	73.2376679269877	2.84217094304040E-12	1.281E-03

A typical set of u values for the case $n = 10$ is:

4.24097483205014E+00 5.13158445463120E+00 6.20920487958172E+00 7.51314003844254E+00
9.09091423494461E+00 1.10000186508910E+01 1.33099945385811E+01 1.61051036339984E+01
1.94872971069880E+01 2.35793320398069E+01

When attempting to solve the harder Mathematical Programming formulation, mixed results are obtained. For the smaller problem the method performs well, but it starts to struggle with the larger problem.

n	fitness	end pt error	max int error	
10	32.8190007266256	3.88097887160654E-14	1.44738726781668E-08	-2.291E-01
20	72.1541684845948	1.42247325030098E-15	1.47709364739512E-07	-2.307E-01

A typical set of x and u values for the case $n = 10$ is:

1.05237496697689E+02 1.10000000000000E+02 1.14898568254288E+02 1.19006792667003E+02
1.21979068519086E+02 1.23378736196669E+02 1.22664972839215E+02 1.19145664749898E+02
1.11948476932748E+02 1.00000000000000E+02
4.76250330215557E+00 5.76124629461049E+00 6.10143174568516E+00 7.38163235900985E+00
8.92840341461519E+00 1.07982390295868E+01 1.30516369771226E+01 1.57858053731660E+01
1.91117542921381E+01 2.31433246260194E+01

The algorithm manages to get close to the optimal fitness values, and manages to satisfy the end point condition very accurately. However, there are a few problems with the interior constraints. Most of them can be satisfied to 10-12 decimal places of accuracy, but 1 or 2 of the constraints are only satisfied to approximately 7-8 decimal places. While this is quite acceptable for engineering accuracy, it is hoped to improve on this in order to apply the technique to larger problems.

6.2   The Continuous Example

These problems were integrated through using the classical RK4 method using 16 and 32 intervals.

n	fitness	end point error	
16	0.302836008995567	1.14880327473088E-12	6.848E-02
32	0.298625337405764	1.87166254717042E-13	1.241E-01

A typical set of u values for the case n = 16 is:

```
-5.45203055037079E-01 -5.22653994565346E-01 -5.04070737981069E-01 -4.89487764639224E-01
-4.78766155765957E-01 -4.71760463373620E-01 -4.68452173359947E-01 -4.68852743768507E-01
-4.72922938498592E-01 -4.80711823034590E-01 -4.92273257876645E-01 -5.07683957404552E-01
-5.27097043377934E-01 -5.50633034914021E-01 -5.78480030915772E-01 -6.21701151700472E-01
```

When attempting to solve the harder Mathematical Programming formulation, mixed results are obtained. For the smaller problem the method performs well, but it again starts to struggle with the larger problem.

n	fitness	end pt error	max int error	
16	0.308120007266256	3.88097887160654E-13	1.44738726781668E-07	-2.291E-01
32	0.299597437247153	2.24606899966361E-12	4.27048582423933E-07	-2.746E-01

A typical set of x and u values for the case n = 16 is:

```
1.05237496697689E+02 1.10000000000000E+02 1.14898568254288E+02 1.19006792667003E+02
1.21979068519086E+02 1.23378736196669E+02 1.22664972839215E+02 1.19145664749898E+02
1.11948476932748E+02 1.00000000000000E+02
4.76250330215557E+00 5.76124629461049E+00 6.10143174568516E+00 7.38163235900985E+00
8.92840341461519E+00 1.07982390295868E+01 1.30516369771226E+01 1.57858053731660E+01
1.91117542921381E+01 2.31433246260194E+01
```

6.3   The Discrete Version of Example 2

To match the results given in the previous subsection, N was allowed to take on the values of 16 and 32. It should be noted that it is a well known fact that even in the limit as n   the discrete version of a problem does not have to converge to the same answer as the continuous version.

n	fitness	end point error	
16	0.294744848453616	2.70616862252382E-15	-1.192E-01
32	0.295152397265203	1.04083408558608E-16	-1.250E-01

A typical set of u values for the case n = 16 is:

```
-5.62113634179599E-01 -5.39440077768919E-01 -5.21252405787557E-01 -5.07403127603559E-01
-4.97790449487918E-01 -4.92333662384905E-01 -4.90958558399551E-01 -4.93686067463144E-01
-5.00524123536335E-01 -5.11549202754015E-01 -5.26812632011737E-01 -5.46483552460173E-01
-5.70701370866172E-01 -5.99673277302981E-01 -6.33646511585264E-01 -6.72892300908769E-01
```

When attempting to solve the harder Mathematical Programming formulation, again mixed results are obtained.

n	fitness	end point error	max interior error	
16	0.294747563550437	4.16529470969501E-11	3.17026726194124E-07	6.659E-01
32	0.297382009123979	1.25276185562329E-08	2.32234909760636E-06	7.891E-01

A typical set of x and u values for the case n = 16 is:

```
9.02175985896992E-01  8.12045661441916E-01  7.28852809980207E-01  6.51433542589089E-01
5.79833864575150E-01  5.12822436249790E-01  4.49959100998623E-01  3.91129401029641E-01
3.35621179258913E-01  2.82497003140196E-01  2.31862389229072E-01  1.83031856699407E-01
1.36024844063454E-01  8.98981682825151E-02  4.47692525037380E-02  8.33058941939003E-12

-5.65184225604561E-01  -5.39909191659732E-01  -5.19039917935492E-01  -5.09855461649805E-01
-4.94161227200457E-01  -4.92348864285245E-01  -4.92990809697763E-01  -4.91314945506498E-01
-4.97001164112640E-01  -5.14365625068361E-01  -5.27655345101623E-01  -5.49426104012754E-01
-5.69079189307642E-01  -6.02001876111981E-01  -6.32163765835295E-01  -6.71533678508132E-01
```

6.4   Comments

In all cases, the algorithm manages to get close to the optimal fitness values, and manages to satisfy the end point condition very accurately. However, there are a few problems with the interior constraints. Most of them can be satisfied to 10-12 decimal places of accuracy, but 1 or 2 of the constraints, (which ones varies from run to run), are only satisfied to approximately 7-8 decimal places. While this is quite acceptable for engineering accuracy, it is hoped to improve on this in order to apply the technique to larger problems. Research is currently being carried out to develop an automatic 'constraint polishing' algorithm to try and overcome this problem.

The overall evolutionary algorithm did have some trouble in consistently converging to the optimal values, but this is believed to be a function of the actual problem and that most evolutionary implementations would have diculty converging. This is partly due to the large range of values each variable can take on. This is compounded by the number of unknowns that are involved. Thus the curse of dimensionality strikes again. This can only be overcome by further research into (mutation) operators that are eective in high dimensions. Investigations along this line are being carried out.

Certainly the use of the Augmented Lagrangian Penalty function is assisting in overcoming this problem, because of the small coecients that are used. Having small values of   and   lead to a 'gently undulating' search space rather than a very 'angular' one for the absolute penalty method and a very 'high sided' one for the quadratic penalty method.

7   Conclusions

We have shown that the Augmented Lagrangian Penalty method can be used within the context of a Mathematical Programming formulation of both discrete and continuous constrained optimal control problems. This exploratory research has shown that results can be achieved that are comparable with those obtained when only the control curve,   u(t) is considered to be a variable, also using the augmented lagrangian Penalty method. This is quite significant given the   massive increase in the dimensionality of the search space and the number of equality constraints.

The harvest problem has shown that this approach to solving an optimisation problem using Mathematical Programming and Evolutionary algorithms is likely to struggle when the dependent variables x are able to range over large intervals, no matter what approach to incorporating the equality constraints is taken.

7.1 Future Work

This research has highlighted the need for further investigation of the high dimensionality control problems, especially when the x_i variables can take on a large range of values.

As an aside to this research, we believe that the area of Mathematical Programming should be investigated further as a potential source of meaningful problems that involve a large number of equality constraints. A suitable set of "test" cases, should be drawn up, published and maintained.

One of the principle reasons for this investigation into using Evolutionary Algorithms within a Mathematical Programming context is the fact that Evolutionary Algorithms have not done well in solving Continuous Optimal Control problems, when the optimal curve $u(t)$ has been known to have a medium to large oscillatory nature. They invariably get stuck in a local minima of a smooth curve. This is being actively researched at the moment, as it is perceived as a weakness in applying Evolutionary algorithms as a general purpose tool for solving a wide class of Optimal Control Problems.

References

1. Barbosa, H.J.C.: A Coevolutionary Genetic Algorithm for Constrained Optimization, Procedings of the 1999 Congress on Evolutionary Computation, Washington DC., pp 1605-1611.
2. Gill, P.E., Murray, W. and Wright, M.H.: Practical Optimization, Academic Press, (1981).
3. Myung, H., Kim, J-H.: Constrained Optimization Using Two-Phase Evolutionary Programming Proceedings IEEE International Conference on Evolutionary Computation, Nagoya, Japan. pp. 262-267.
4. Michalewicz, Z.: Genetic Algorithms + Data Structures = Evolution Programs Springer Verlag, Berlin, (1992).

Cooperative Coevolution for Learning Fuzzy Rule-Based Systems

Jorge Casillas [1], O. Cord´on[1], F. Herrera [1], and J.J. Merelo [2]

[1] Depto. Ciencias de la Computaci´ on e Inteligencia Artificial
Universidad de Granada, E-18071 Granada, Espa na
casillas,ocordon,herrera   @decsai.ugr.es   ,
[2] GeNeura Team, Depto. Arquitectura y Tecnolog´   a de Computadores
Universidad de Granada, E-18071 Granada, Espa na
jmerelo@geneura.ugr.es

Abstract.   In the last few years, the coevolutionary paradigm has shown an increasing interest thanks to its high ability to manage huge search spaces. Particularly, the cooperative interaction scheme is recommendable when the problem solution may be decomposable in subcomponents and there are strong interdependencies among them.

The paper introduces a novel application of these algorithms to the learning of fuzzy rule-based systems for system modeling. Traditionally, this process is performed by sequentially designing their dierent components. However, we propose to accomplish a simultaneous learning process with cooperative coevolution to properly consider the tight relation among the components, thus obtaining more accurate models.

1   Introduction

Fuzzy rule-based systems (FRBSs) constitute an extension of classical rule-based systems, because they deal with IF-THEN rules where antecedents and/or consequents are composed of fuzzy logic statements, instead of classical logic rules. This consideration presents two essential advantages: the key features of knowledge captured by fuzzy sets involve handling uncertainty and inference methods become more robust and exible with approximate reasoning methods of fuzzy logic. One of the most success applications of FRBSs is   system modeling [17], which in this field may be considered as an approach used to model a system making use of a descriptive language based on fuzzy logic with fuzzy predicates [23].

Several tasks have to be performed in order to design an FRBS for a concrete modeling application. One of the most important and dicult ones is to   derive an appropriate knowledge base (KB) about the problem being solved. The KB stores the available knowledge in the form of fuzzy IF-THEN rules. It consists of the rule base (RB), comprised of the collection of rules in their symbolic forms, and the data base (DB), which contains the linguistic term sets and the membership functions defining their meanings.

Partially supported by the Spanish CICYT, project PB98-1319

P. Collet et al. (Eds.): EA 2001, LNCS 2310, pp. 311–322, 2002.
c Springer-Verlag Berlin Heidelberg 2002

Numerous automatic methods based on ad hoc data-driven approaches [25] or on dierent techniques such as neural networks [15] or genetic algorithms (GAs) [4,20,21] have been developed to perform the derivation task. When only the derivation of the RB is addressed, methods generally operate in only one stage [24,25]. In this case, the DB is usually obtained from the expert information (if it is available) or by a normalization process.

However, methods that design both RB and DB are preferable since the automation is higher. In this case, we can distinguish between two dierent approaches:

- Simultaneous derivation : It relates to the process of directly obtaining the whole KB (RB and DB) from the available data in a simultaneous way [12, 13]. This task is usually known as learning p rocess.
- Sequential derivation : The task is divided into two or more stages, each of them performing a partial or complete derivation of the KB.
  Some methods learn the DB with a embedded approach [6,8] that may be used as one of the first stages.
  Generally, one of the last stages adjusts the previously learnt/obtained DB with slight modifications to increase the system performance [1,10,11]. This stage is known as tuning p rocess.
  In most cases, a sequential process by firstly learning the RB and then tuning the DB is considered [3].

When the RB and the DB are simultaneously derived, the strong dependency of both components is properly addressed. However, the derivation process becomes significantly more complex because the search space grows and the selection of an appropriate search technique is crucial.

Recently, the coevolutionary paradigm [16] has shown an increasing interest thanks to its high ability to manage with huge search spaces and decomposable problems. The direct decomposition of the KB derivation process (thus obtaining two interdependent components, learning of the RB and DB) makes coevolutionary algorithms with a cooperative approach [19] very useful for this purpose.

In this paper, we propose a KB derivation method within this novel evolutionary paradigm. Actually, a method has been already proposed by Pe   na-Reyes and Sipper with this cooperative coevolutionary philosophy [18]. However, opposite to it, our proposal performs a more sophisticated learning of the RB based on the Cooperative Rules (COR) methodology [2], whose good performance is related to the consideration of cooperation among rules. Once the rule antecedents (defining fuzzy subspaces) have been obtained, COR generates a candidate consequent set for each subspace and searches the consequents with the best global performance.

In the following sections, an introduction to coevolutionary algorithms, the proposed KB derivation method, some experimental results, conclusions, and further work are shown.

2   Coevolutionary Algorithms

Evolutionary algorithms (EAs) [14] are general-purpose global search algorithms that use principles inspired by natural population genetics. In a EA, each individual in the population represents a candidate solution to the problem and has an associated fitness to determine which individuals are used to form new ones in the process of competition. The new individuals are created using genetic operators such as crossover and mutation.

Within this field, a new paradigm has been recently proposed, coevolutionary algorithms [16]. They involve two or more species (populations) that permanently interact among them by a coupled fitness. Thereby, in spite of each species has its own coding scheme and reproduction operators, when an individual must be evaluated, its goodness will be calculated considering some individuals of the other species. This coevolution makes easier to find solutions to complex problems.

Dierent kinds of interactions may be considered among the species according to the dependencies existing among the solution subcomponents. Generally, we can mention two dierent kinds of interaction:

- Competitive coevolutionary algorithms   [22]: Those where each species competes with the remainder. In this case, increasing the fitness of an individual in a species implies decreasing the fitness of the ones other species, i.e., the success of somebody else entails the personal failure.
- Cooperative or symbiotic coevolutionary algorithms   [19]: Those where all the species cooperate to build the problem solution. In this case, the fitness of an individual depends on its ability to cooperate with individuals from other species.
  Figure 1 illustrates the cooperative approach. As shown, a set of selected individuals (called cooperators) is built in each species to represent it. Each individual is evaluated constructing solutions with it and cooperators of the remaining species.

Therefore, the use of cooperative coevolutionary algorithms is recommendable when the following issues arise [18]:

1. the search space is huge,
2. the problem may be decomposable in subcomponents,
3. dierent coding schemes are used, and
4. there is strong interdependencies among the subcomponents.

They also arise in problems where the training set is not known in advance, but created by the solution to the problem themselves, e.g., when collision avoidance behavior for two planes is being evolved simultaneously [7]. In that cases, training sets are created by the other planes which are being evolved.

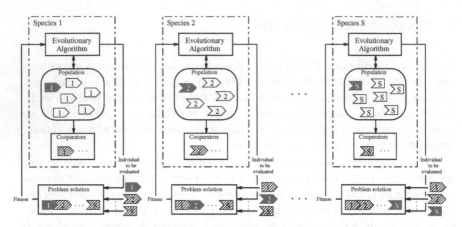

Fig. 1.   Cooperative coevolutionary scheme

3   A Cooperative Coevolutionary Algorithm for Jointly Learning Fuzzy Rule Bases and Membership Functions

Intuitively, we may decompose the problem of deriving a proper KB for an FRBS into two subtasks: to obtain fuzzy rule symbolic representations (learning the RB) and to define membership function shapes (learning the DB). Therefore, our coevolutionary algorithm consists of two species that cooperate to build the whole solution.

In the following subsections, a formulation for both learning tasks and the components of the cooperative coevolutionary algorithm are introduced.

3.1   The Knowledge Base Derivation Process

Learning Fuzzy Rule Bases

The RB learning task is based on the COR methodology [2]. Let E be the input-output data set, $e_l = (x_1^l, \ldots, x_n^l, y^l)$ on of its elements (example), and n be the number of input variables. Let A_i be the set of linguistic terms of the i-th input variable and B the set of linguistic terms of the output variable. Its operation mode is the following:

1. Define a set of fuzzy input subspaces, S_s s $1, \ldots, N_S$, with the antecedent combinations containing at least a positive example, i.e., $S_s = (A_1^s, \ldots, A_i^s, \ldots, A_n^s)$ A_1 $\ldots$ A_n such that $E_s =$ (with A_i^s being a label of the i-th input variable, E_s being the set of positive examples of the subspace S_s, and N_S the number of subspaces with positive examples). In this contribution, we will define the set of positive examples for the subspace S_s as follows:

$$E_s = e_l \quad E \quad i \quad 1, \ldots, n \; ,$$
$$A_{ij} \; A_i, \quad A_i^s(x_i^l) \quad A_{ij}(x_i^l) \; ,$$

with A_{ij} being a label of the i-th input variable and μ_T the membership function of the label T.

2. For each subspace S_s, obtain a set of candidate consequents (i.e., linguistic terms of the output variable) B^s to build the corresponding fuzzy rule.

In this contribution, we will define the set of candidate consequents for the subspace S_s as follows:

$$B^s = \{B_k \in B \mid e_{l^s} \in E_s \text{ where}$$
$$B_1 \in B, \; \mu_{B_k}(y^{l^s}) \geq \mu_{B_1}(y^{l^s})\} ,$$

with B_k being a label of the output variable.

3. Perform a combinatorial search among these sets looking for the combination of consequents (one for each subspace) with the best global accuracy.

For example, from the subspace $S_s = (\text{high}, \text{low})$ and the candidate consequent set in such a subspace $B^s = \{\text{small}, \text{medium}, \text{large}\}$, we will obtain the fuzzy rule:

$$R_s = \text{IF } X_1 \text{ is high and } X_2 \text{ is low THEN } Y \text{ is } B_s,$$

with $B_s \in B^s$ being the label selected by the combinatorial search to represent to the subspace S_s.

Learning Fuzzy Membership Functions

In our case, the derivation of the DB involves determining the shape of each membership function. These shapes will have a high influence in the FRBS performance. In this contribution, we will consider triangular-shaped membership functions as follows:

$$\mu_T(x) = \begin{cases} \frac{x-a}{b-a}, & \text{if } a \leq x \leq b \\ \frac{c-x}{c-b}, & \text{if } b \leq x \leq c , \\ 0, & \text{otherwise} \end{cases}$$

Therefore, different values of the parameters a, b, c will define different shapes of the membership function associated to the linguistic term T.

3.2   The Cooperative Coevolutionary Algorithm

Cooperative Interaction Scheme between Both Species

Let F_{ij} be the FRBS obtained by composing the subcomponents encoded in the chromosomes i and j of the species 1 (RBs) and 2 (membership functions), respectively. The objective will be to minimize the well-known mean square error (MSE):

$$MSE_{ij} = \frac{1}{2N} \sum_{l=1}^{N} \left(F_{ij}(x^l) - y^l \right)^2,$$

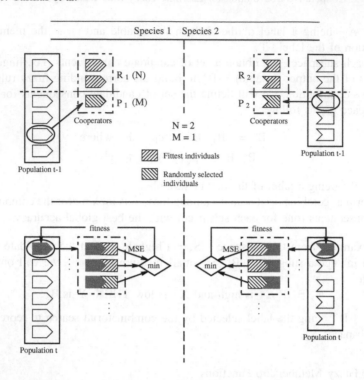

Fig. 2.   Interaction scheme considered in the learning method

with N being the data set size, $F_{ij}(x^l)$ being the output obtained from the designed FRBS when the l-th example is considered, and y^l being the known desired output.

Each individual of species 1 or 2 is evaluated with the corresponding   fitness function f_1 or f_2, which are defined as follows:

$$f_1(i) = \min_{\substack{j \ R_2 \ P_2}} MSE_{ij}$$

$$f_2(j) = \min_{\substack{i \ R_1 \ P_1}} MSE_{ij}$$

with i and j being individuals of species 1 and 2 respectively, R_1 and R_2 being the set of the fittest individuals in the previous generation of the species 1 and 2 respectively, and P_1 and P_2 being individual sets selected at random from the previous generation of the species 1 and 2 respectively. Figure 2 graphically shows the proposed interaction scheme.

Whilst the sets $R_{1\ 2}$ allow the best individuals to inuence in the process guiding the search towards good solutions, the sets $P_{1\ 2}$ introduce diversity in the search. The combined use of both kinds of sets makes the algorithm have a trade-o between exploitation ($R_{1\ 2}$) and exploration ($P_{1\ 2}$). The cardinalities of the sets $R_{1\ 2}$ and $P_{1\ 2}$ are previously defined by the designer.

A generational   [14] scheme is followed in both species. Baker's stochastic universal sampling procedure together with an elitist mechanism (that ensures to select the best individual of the previous generation) are used.

The specific operators used in every species are described in the following sections.

Species 1: Learning Fuzzy Rule Bases

An integer-valued vector (c) of size N_S (number of subspaces with positive examples) is employed as coding scheme . Each cell of the vector represents the index of the consequent used to build the rule in the corresponding subspace:

$$s \quad 1,\dots,N_S \; , \; c[s] = k_s \text{ s.t. } B_{k_s} \quad B^s.$$

The initial pool of this species is generated building the first individual as follows $s \quad 1,\dots,N_S$,

$$c_1[s] = \arg \max_{k_s \; B_{k_s} \; B^s} CV(R_{k_s}^s),$$

with

$$CV(R_{k_s}^s) =$$
$$\max_{e_{1s} \; E_s} Min \quad A_1^s(x_1^{l^s}),\dots, \quad A_n^s(x_n^{l^s}), \quad B_{k_s}(y^{l^s}) \; ,$$

and the remaining chromosomes generated at random:

$$p \quad 2,\dots, pool\_size \; , \quad s \quad 1,\dots,N_S \; ,$$
$$c_p[s] = \text{ some } k_s \text{ s.t. } B_{k_s} \quad B^s.$$

The standard two-point crossover operator is used. The mutation operator randomly selects a specific s $1,\dots,N_S$ where B^s 2, and changes at random $c[s] = k^s$ by $c[s] = k^s$ such that B_{k^s} B^s and $k^s = k^s$.

Species 2: Learning Fuzzy Membership Functions

As coding scheme , a 3-tuple of real values for each triangular membership function is used, thus being the DB encoded into a real-coded chromosome built by joining the membership functions involved in each variable fuzzy partition. A variation interval to every gene is associated to preserve meaningful fuzzy sets.

The initial population of this species is generated with a chromosome representing the original DB and the remaining chromosomes generated with the values at random within the corresponding variation interval.

The max-min-arithmetical crossover operator [10] is considered. If $C_v^t = (c_1,\dots,c_k,\dots,c_H)$ and $C_w^t = (c_1,\dots,c_k,\dots,c_H)$ are to be crossed, the following four ospring are generated:

$$C_1^{t+1} = aC_w^t + (1 \quad a)C_v^t, \qquad C_2^{t+1} = aC_v^t + (1 \quad a)C_w^t,$$
$$C_3^{t+1} \text{ with } c_{3,k}^{t+1} = \min \; c_k,c_k \; , \qquad C_4^{t+1} \text{ with } c_{4,k}^{t+1} = \max \; c_k,c_k \; .$$

The parameter a is defined by the designer. The resulting descendents are the two best of the four aforesaid ospring. As may be observed, its formulation avoids the violation of the restrictions imposed by the variation intervals.

With respect to the mutation operator   , it simply involves changing the value of the selected gene by other value obtained at random within the corresponding variation interval.

4   Experimental Results in the Electrical Maintenance Cost Estimating Problem

This experimental study will be devoted to analyze the behavior of the proposed derivation method jointly learning the RB following the COR methodology and the membership functions with cooperative coevolutionary algorithms (CORMF-CC). With this aim, we have chosen the problem of estimating the maintenance costs of the medium voltage electrical network in a town [5].

We will analyze the accuracy of the fuzzy models generated from the proposed process compared to the four following methods: the well-known ad hoc data-driven method proposed by Wang and Mendel (WM) [25]; a GA-based learning method following the COR methodology (COR-GA) [2]; and two sequential methods, WM+Tun and COR-GA+Tun, that firstly perform a learning of the RB with WM or COR-GA, respectively, and then adjust the membership functions with the tuning method proposed in [3].

With respect to the FRBS reasoning method used, we have selected the   minimum t-norm playing the role of the implication and conjunctive operators, and the center of gravity weighted by the matching strategy acting as the defuzzification operator.

4.1   Problem Description

Estimating the maintenance costs of the medium voltage electrical network in a town [5] is a complex but interesting problem. Since an actual measure is very dicult to obtain, the consideration of models becomes useful. These estimations allow electrical companies to justify their expenses. Moreover, the model must be able to explain how a specific value is computed for a certain town. Our objective will be to relate the maintenance costs of medium voltage line with the following four variables: sum of the lengths of all streets in the town , total area of the town, area that is occupied by buildings , and energy supply to the town . We will deal with estimations of minimum maintenance costs based on a model of the optimal electrical network for a town in a sample of 1,059 towns.

To develop the dierent experiments in this contribution, the sample has been randomly divided in two subsets, the training and test ones, with an 80-20 of the original size respectively. Thus, the training set contains 847 elements, whilst the test one is composed by 212 elements.   Five linguistic terms for each variable are considered.

4.2   Experimental Results and Analysis

The following values have been considered for the parameters of each method:

- COR-GA : 61 individuals, 50 generations, 0.6 as crossover probability, and 0.2 as mutation probability.
- Tuning stage of the WM+Tun and COR-GA+Tun methods   : 61 individuals, 300 generations, 0.6 as crossover probability, 0.2 as mutation probability, 0.35 for the weight factor in the max-min-arithmetical crossover, and 5 for the weight factor in the non-uniform mutation.
- CORMF-CC : 62 individuals (31 for each species), 300 generations, 0.6 and 0.2 for the crossover and mutation probabilities in both species respectively, 0.35 for the weight factor of the crossover operator in the species 2, the two fittest individuals ($R_{1\,2}$ = 2) and two random individuals ($P_{1\,2}$ =2) of each species are considered for the coupled fitness.

Ten dierent runs were performed for each probabilistic algorithm. The results obtained by the five methods analyzed are collected in Table 1, where MSE_{tra} and MSE_{tst} respectively stand for the error obtained over the training and test data sets. Arithmetic mean (x) and standard deviation ( ) values of the 10 linguistic models generated by each method are included. The best mean results are shown in boldface. A total of 66 fuzzy rules were obtained in all cases.

Table 1.   Results obtained in the electrical problem

Method	MSE_{tra}		MSE_{tst}	
	x		x	
WM	71,294	0	80,934	0
COR-GA	67,237	0	69,457	0
WM+Tun	24,667	1,350	34,143	2,452
COR-GA+Tun	24,255	1,349	31,393	2,831
CORMF-CC	15,435	1,094	22,573	1,557

In view of the obtained results, the CORMF-CC method shows the best performance combining both approximation (MSE_{tra}) and generalization (MSE_{tst}). Analyzing the two-stage methods (WM+Tun and COR-GA+Tun), we may observe how the tuning process significantly improve the accuracy degrees of the fuzzy models generated by the WM and COR-GA learning methods. However, when the derivation process is made in only one stage with the cooperative coevolutionary approach, the fuzzy model obtained overcomes the remainder thanks to the proper consideration of the dependency between the RB and the DB in the learning process. Moreover, the low standard deviations obtained show the robustness of the CORMF-CC algorithm.

Figure 3 illustrates the DB derived by the CORMF-CC method. Using the shown membership function shapes a good interpretability is kept up whilst the fuzzy model performance is improved.

(a) DB

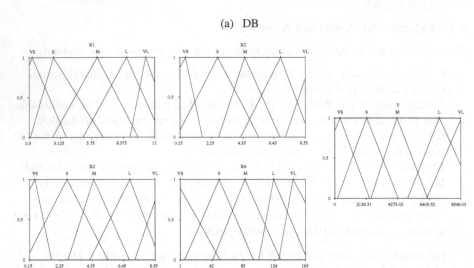

(b) RB

Rule	X_1	X_2	X_3	X_4	Y	Rule	X_1	X_2	X_3	X_4	Y	Rule	X_1	X_2	X_3	X_4	Y
R_1	VS	VS	VS	VS	VS	R_{23}	M	S	VS	S	S	R_{45}	L	L	L	S	L
R_2	VS	VS	VS	S	S	R_{24}	M	S	VS	M	S	R_{46}	L	L	L	M	L
R_3	VS	S	S	VS	S	R_{25}	M	M	M	S	M	R_{47}	L	L	L	L	VL
R_4	VS	S	S	S	S	R_{26}	M	M	M	M	L	R_{48}	L	L	M	S	M
R_5	VS	S	VS	VS	VS	R_{27}	M	M	S	S	S	R_{49}	L	L	M	M	M
R_6	VS	S	VS	S	S	R_{28}	M	M	S	M	M	R_{50}	L	L	M	L	L
R_7	S	VS	VS	VS	VS	R_{29}	M	M	S	VS	S	R_{51}	L	L	M	VS	M
R_8	S	VS	VS	S	S	R_{30}	M	L	L	S	M	R_{52}	L	VL	VL	S	VL
R_9	S	VS	S	VS	VS	R_{31}	M	L	L	M	L	R_{53}	L	VL	VL	M	VL
R_{10}	S	VS	S	S	S	R_{32}	M	L	M	S	M	R_{54}	L	VL	VL	L	VS
R_{11}	S	S	S	VS	S	R_{33}	M	L	M	M	M	R_{55}	L	VL	L	S	M
R_{12}	S	S	S	S	M	R_{34}	L	S	S	VS	S	R_{56}	L	VL	L	M	L
R_{13}	S	S	VS	VS	S	R_{35}	L	S	S	S	M	R_{57}	L	VL	L	L	VL
R_{14}	S	S	VS	S	S	R_{36}	L	S	S	M	M	R_{58}	VL	S	M	S	M
R_{15}	S	M	M	VS	M	R_{37}	L	S	S	L	M	R_{59}	VL	S	M	L	M
R_{16}	S	M	M	S	M	R_{38}	L	M	M	S	M	R_{60}	VL	S	M	VL	L
R_{17}	S	M	S	VS	S	R_{39}	L	M	M	M	M	R_{61}	VL	S	S	S	S
R_{18}	S	M	S	S	S	R_{40}	L	M	M	L	L	R_{62}	VL	S	S	L	M
R_{19}	M	S	S	VS	S	R_{41}	L	M	S	VS	S	R_{63}	VL	S	S	VL	L
R_{20}	M	S	S	S	S	R_{42}	L	M	S	S	M	R_{64}	VL	L	M	S	M
R_{21}	M	S	S	M	M	R_{43}	L	M	S	M	M	R_{65}	VL	L	M	L	L
R_{22}	M	S	VS	VS	VS	R_{44}	L	M	S	L	M	R_{66}	VL	L	M	VL	VL

Fig. 3.   KB derived by the CORMF-CC method, where   V S stands for very small , S for small , M for medium , L for large, and V L for very large

5   Concluding Remarks and Further Work

A KB derivation method that jointly learns the fuzzy rules and membership functions involved in an FRBS has been proposed. The fact of performing these tasks together allows the method to consider the tight relation between both components, thus obtaining better fuzzy models. However, this joint consideration becomes more dicult since the search space is signi6antly increased, thus being crucial the selection of a proper technique.

As David Goldberg stated, the integration of single methods into hybrid intelligent systems goes beyond simple combinations. For him, the future of Computational Intelligence " lies in the careful integration of the best constituent technologies " and subtle integration of the abstraction power of fuzzy systems and the innovating power of genetic systems requires a design sophistication that goes further than putting everything together [9].

In this contribution, this issue is addressed by using a cooperative coevolutionary approach with a sophisticated rule learning component based on the cooperation among the fuzzy rules derived. The good performance of the method compared with other classical hybridizations has been shown when solving a real-world problem. Nevertheless, the proposed modeling approach can be applied to other system modeling problems.

As further work, we propose to extend the components of the KB to be derived (number of labels, more exible fuzzy rules, etc.), to consider other metaheuristics to adapt each species, and to improve the interaction scheme for a better interdependency consideration and scalability to more than two species.

References

1. P. P. Bonissone, P. S. Khedkar, and Y. Chen. Genetic algorithms for automated tuning of fuzzy controllers: a transportation application. In   Procee dings of the 5th IEEE International Conference on Fuzzy Systems   , pages 674–680, New Orleans, LA, USA, 1996.
2. J. Casillas, O. Cord´ on, and F. Herrera. Dierent approaches to induce cooperation in fuzzy linguistic models under the COR methodology. In B. Bouchon-Meunier, J. Guti´ errez-R´os, L. Magdalena, and R. R. Yager, editors,   Techniques for constructing intelligent systems . Springer-Verlag, Heidelberg, Germany, 2001. In press.
3. O. Cord´on and F. Herrera. A three-stage evolutionary process for learning descriptive and approximate fuzzy logic controller knowledge bases from examples. International Journal of Approximate Reasoning   , 17(4):369–407, 1997.
4. O. Cord´on, F. Herrera, F. Homann, and L. Magdalena.   Genetic fuzzy systems: evolutionary tuning and learning of fuzzy knowledge bases   . World Scientific, Singapore, Singapore, 2001.
5. O. Cord´on, F. Herrera, and L. S´ anchez. Solving electrical distribution problems using hybrid evolutionary data analysis techniques.   Applied Intelligence , 10(1):5–24, 1999.
6. O. Cord´on, F. Herrera, and P. Villar. Generating the knowledge base of a fuzzy rule-based system by the genetic learning of the data base. To appear in   IEEE Transactions on Fuzzy Systems . Draft available at   http://decsai.ugr.es/   herrera/ .
7. N. Durand, J.-M. Alliot, and F. Medioni. Neural nets trained by genetic algorithms for collision avoidance.   Applied Intelligence , 13(3):205–213, 2000.

8. P. Glorennec. Constrained optimization of FIS using an evolutionary method. In F. Herrera and J. L. Verdegay, editors, Genetic algorithms and soft computing , pages 349–368. Physica-Verlag, Heidelberg, Germany, 1996.

9. D. E. Goldberg. A meditation on the computational intelligence and its future. Technical Report Illigal 2000019, Illinois Genetic Algorithms Laboratory, University of Illinois at Urbana-Champaign, Illinois, IL, USA, 2000. Available at http://www-illigal.ge.uiuc.edu/

10. F. Herrera, M. Lozano, and J. L. Verdegay. Tuning fuzzy controllers by genetic algorithms. International Journal of Approximate Reasoning , 12:299–315, 1995.

11. C. L. Karr. Genetic algorithms for fuzzy controllers. AI Expert , 6(2):26–33, 1991.

12. K. KrishnaKumar and A. Satyadas. GA-optimized fuzzy controller for spacecraft attitude control. In J. Periaux, G. Winter, M. Gal´ an, and P. Cuesta, editors, Genetic algorithms in engineering and computer science , pages 305–320. John Wiley Sons, New York, NY, USA, 1995.

13. L. Magdalena and F. Monasterio-Huelin. A fuzzy logic controller with learning through the evolution of its knowledge base. International Journal of Approximate Reasoning , 16(3):335–358, 1997.

14. Z. Michalewicz. Genetic algorithms + data structures = evolution programs . Springer-Verlag, Heidelberg, Germany, 3rd edition, 1996.

15. D. Nauck, F. Klawonn, and R. Kruse. Fundations of neuro-fuzzy systems . John Wiley Sons, New York, NY, USA, 1997.

16. J. Paredis. Coevolutionary computation. Artificial Life , 2:355–375, 1995.

17. W. Pedrycz, editor. Fuzzy modelling: paradigms and practice . Kluwer Academic, Norwell, MA, USA, 1996.

18. C. A. Pe na-Reyes and M. Sipper. Fuzzy CoCo: a cooperative coevolutionary approach to fuzzy modeling. To appear in IEEE Transactions on Fuzzy Systems . Draft version available at http://lslwww.epfl.ch/ penha/ .

19. M. A. Potter and K. A. De Jong. Cooperative coevolution: an architecture for evolving coadapted subcomponents. Evolutionary Computation , 8(1):1–29, 2000.

20. V. M. Rivas, J. J. Merelo, I. Rojas, G. Romero, P. A. Castillo, and J. Carpio. Evolving 2-dimensional fuzzy logic controllers. To appear in Fuzzy Sets and Systems , 2001.

21. I. Rojas, J. J. Merelo, J. L. Bernier, and A. Prieto. A new approach to fuzzy controller designing and coding via genetic algorithms. In Procee dings of the 6th IEEE International Conference on Fuzzy Systems , 1997.

22. C. D. Rosin and R. K. Belew. New methods for competitive coevolution. Evolutionary Computation , 5(1):1–29, 1997.

23. M. Sugeno and T. Yasukawa. A fuzzy-logic-based approach to qualitative modeling. IEEE Transactions on Fuzzy Systems , 1(1):7–31, 1993.

24. P. Thrift. Fuzzy logic synthesis with genetic algorithms. In R. K. Belew and L. B. Booker, editors, Procee dings of the 4th International Conference on Genetic Algorithms , pages 509–513, San Mateo, CA, USA, 1991. Morgan Kaufmann Publishers.

25. L.-X. Wang and J. M. Mendel. Generating fuzzy rules by learning from examples. IEEE Transactions on Systems, Man, and Cybernetics , 22(6):1414–1427, 1992.

Evolving Cooperative Ecosystems:
A Multi-agent Simulation of Deforestation
Activities

Ravi Srivastava and Amit Kaldate

University of Illinois at Urbana-Champaign
104 S. Mathews 117
Urbana IL - 61801 USA
Phone - (217) 333 2346, 244 8033
srivasta, kaldate   @uiuc.edu

Abstract.   Achieving cooperation among competing groups, particularly in the sphere of social and ecological resources, is an extremely
daunting realm of ecosystem management. This paper presents a multi-
agent model of the activities of two such competing groups   native
farmers and logging companies   using common pool natural resources,
namely virgin forests. In the model presented, native and logging agents
deal with conicting personal development and ecological conservation
objectives. The simulation results clearly depict that emergence of cooperative behavior among the agent groups ensures that indiscriminate
exploitation of vast amounts of natural resources is avoided at the cost of
only a relatively small compromise on development activities. Also, the
role of external (possibly government) agents is highlighted as eective
information exchange promoters.

1   Introduction

Achieving cooperation among stake-holders for sustainable development
throughout the world is widely recognized as a major challenge for the twenty-
first century (WCED, 1987). As resources become increasingly scarce, the need
to eciently manage them will become the paramount aim of all the nations
and societies in the world. In order to maintain resource utilization at sustainable levels, it is imperative that the stake-holders cooperate with each other,
share information and manage resources responsibly. In this paper, we present
modeling of stake-holder activities on a limited piece of land with indispens-
able natural resources. We especially look into the ways in which cooperation
among dierent agent groups can be generated to improve the overall goals of
development and sustainability. Also, the role of an external agent   like the government or Non-Governmental Organizations (NGOs)   is highlighted as being
an important factor in promotion of cooperative behavior through positive and
meaningful intervention.

The issues related to what are known as Common Pool Resources (CPR)
were first highlighted by Hardin (1968) as   Tragedy of Commons . The CPR are

P. Collet et al. (Eds.): EA 2001, LNCS 2310, pp. 323–337, 2002.

those resources that are subtractable and for which the exclusion of potential users or appropriators is dicult (Ostrom, Gardner, Walker, 1994). The universality of metaphors such as the tragedy of the commons was challenged by Ostrom (1990) by citing real world examples in which individuals appeared to organize their actions by establishing rules which facilitated a long term improvement in joint outcomes.

There has been an increasing use of agent-based modeling and simulation studies for social behavior (Gilbert Doran, 1994; Epstein Axtell, 1996; Kohler Gumerman, 2000). A multi-agent model allows significant analyses of social phenomenon and complex collective behavior. Franklin and Graesser (1996) define an autonomous agent as a system situated within and a part of an environment that senses its environment and acts on it, over time, in pursuit of its own agenda and so as to eect what it senses in the future. Thus, agents are particularly eective in modeling social behavior in an organizational setting, following the principles of the embodying environment. The primary properties of agents (Franklin Graesser, 1996) fall closely in line with the social agency modeling requirements. Agents' characteristics of being reactive, autonomous, goal-oriented, temporally continuous, communicative, adaptive, mobile, exible and character-possessive make them apt for social modeling studies. Axelrod (1997) contrasts agent based modeling with two standard methods of induction and deduction, emphasizing its emergence as a third way of doing science with thought experiments. The main purpose of agent-based modeling is to understand properties of complex social systems which are otherwise elusive. Computer simulations can have advantages over theoretical approaches when in addition to great model complexity and resistance to theoretical analysis, either no or a multitude of solutions exist for the problem under investigation (Simon, 1981). Computer models can also be used to monitor and analyze a system's behavior as it evolves over a period of time (Holland Miller, 1991).

Simulation models are increasingly used as decision support tools. In the case of natural resource management, any decision is seldom the result of one hypothetical decision-maker. Though some leaders may have limited power to inuence decision-making behavior of their groups, it is more often than not a matter of interactions between several stake-holders. In particular, agent-based modeling approaches allow the explicit representation of heterogeneous groups of agents and the analysis of their evolution at both individual and collective levels. Agent-based modeling has been applied for dierent applications to improve natural resources and environmental management   (Bousquet, Barreteau, Le Page, Mullon, Weber, 1999; Carpenter, Brock, Ghanson, 1999; Rouchier, Bousquet, Le Page, Bonnefoy, 2000; Sichman, Conte, Gilbert, 1998).   In addition to modeling the behavior and interactions among the agents, the role of external agents who act to intervene in the internecine or wasteful practices of competing agents is important. Such intervention strategies and management regimes for sustainable development have been discussed in some detail in recent research literature

(Christie A.T., 1997; Healey, 1998; Nielsen T., 1999). Sugden (1989) ana-lyzes the conditions under which collective rules regulating access to a natural resource can evolve and maintain themselves without conscious design, and without external enforcement

The approaches to evolve cooperation pass through many stages includ-ing mistrust, understanding of the potential benefits and finally mechanisms to arrive at some mutually beneficial arrangements (Caldart Ashford, 1998). But it is very dicult to explicitly determine the state a system is in and its rationale for doing so. Moreover it is not very clear that the system ar-rives at the improved stability as a result of negotiations. Such eorts to select types of negotiation structures and explicitly model them are still at an experimental stage. For natural resource management systems, interac-tions among stake-holders may take several channels such as the perception of the consequences of others' actions on the resource (reective agents, quite akin to rational beings of game theory), pairwise interactions or institution-alized collective frameworks (Rouchier, Barreteau, Bousquet, Proton, 1998). Tessier, Chaudron, and M uller (2001) provide an overview of dierent aspects of systems involving conicts and conict management strategies.

This paper is organized as follows. The next section discusses the problem at hand and its significance. It also defines the aims of the study. We then discuss the methodologies adopted for the problem solution and the performance yardsticks in section 3. This is followed by the discussion of actual modeling in section 4, Experimental setup details in section 5 and simulation results in section 6. We end with a note on model refinements and conclusions drawn from the study.

2   Problem Definition

To consider the cooperative behavior and its implications in a competitive ecosys-tem, we choose the concurrent usage of forest lands and forest woods by dierent groups, namely the shifting cultivation groups and loggers. In this section we will look at the problem as it is faced in vast regions of the tropical forest-lands, par-ticularly those in and around the South American Amazon forest belts, African Zaire basin etc.

2.1   Cursory Background

Over 90 of West Africa's original forests have been lost (FFI, 2001). More-over, around 77 of Africa's frontier forests are under moderate to high threat, among which, indiscriminate logging contributes 80 to the causes of con-cern. Shifting cultivation is almost an equally threatening cause of deforestation (Rowe, Sharma, Browder, 1992),(CGIAR, 1996). Native farmers derive their livelihood from sustenance cultivation for which they need to burn down the veg-etation and clear up a patch of land on a regular basis. Due to the quality and properties of the soil as well as repetitive and non-rotation farming, the cleared up lands soon become infertile forcing the farmers to shift to other regions and clear up a new forest area for cultivation. Apart from imparting some temporary richness to the soil (which is soon washed   away by rains), the burnt forest lands only contribute to ecological destruction, pollution and global warming not to mention the deforestation issue of course. And these are only a few of the drastic consequences. The burnt forests are valuable wood lost, which at other locations the loggers specifically require for their own industry. They fell canopies of high-rise trees, contributing to their share of the ecological damage.

2.2   Aims

An obvious strategy of constraining this fast depletion of forest lands, while allowing both logger as well as farmer groups to continue with their activities, is to develop cooperation between the two groups. An arrangement could be made, under which, the shifting farmers could move to areas where loggers have completed clearing the high-rise trees. The loggers on the other hand, too could plan their activities in a manner so as to carry on logging at places where the farmers could conveniently move in future.

In this study, we look at the evolution and benefits of such a cooperative behavior and the costs involved.

3   Methods Adopted

A multi-agent model is used for this simulation due to several advantages. The chief among these is that it helps us get to grips with the impact of individual and collective cognition in social systems and there is clear correspondence between the interactions in real world and modeled agents' world. The system agents are divided into three groups   the native cultivators, logging companies and (later) the intervening neutral agents.

However, the agents being considered in the ecosystem need to have a spatial location which also changes temporally. No two agents should occupy the same space at a given time. With this in mind, it is ensured through the system evolution rules that only one agent occupies a particular forest area at an instance. This gives our model the spatial adequacy benefits of lattice-base modeling which is also used extensively in ecosystem and other social systems' simulation studies. However, the decisions about agents' activities are not limited to neighborhood interactions alone, as is usually the case with lattice-based model.

3.1   Information Exchange

For information exchange between agents, an open channel model is used where each agent relays information about its location and other relevant attributes. Also, each agent randomly picks up one packet of information at each time step   packets correspond to the information relayed by one agent. However, not all packets have relevance for all agents. Thus, the information exchange language is coherent and the interaction is probabilistic, as would be in the real world. There is considerably accurate information available about native farmers these days, yet its availability to the most appropriate agencies at the appropriate times is rather less abundant.

3.2 Yardsticks

Development of the ecosystem is measured by the aggregate net income the groups manage to generate from their activities. The ecology management effectiveness is measured by aggregate of environmental conservation attributes of all agents. In addition, the overall environmental awareness of the groups is measured and is an indicator of eective intervention by government or other neutral agencies to improve information availability and resource accessibility. These features are explained in more detail in section 4 and are discussed in detail in appropriate model segments throughout the rest of this paper.

3.3 Plan of Simulation

To arrive at a meaningful comparison and role of cooperation with and without intervention, a three-stage simulation program is followed. First, a system is generated where there is no communication / cooperation between the two agent groups. The results of this system highlight the possibility and benefits that might accrue from cooperation among the agent groups and serve as a control case for comparisons. In the second stage, rules are developed for emergence of cooperation among the groups. Finally, the cooperation model is extended to the third stage where another agent category   the neutral government or action group agents   is introduced. The main role of these agents is to act as an information repository and disseminating body for native farmers so that the overall gains from the system are maximized while at the same time controlling the ecological impacts, which the government agents are committed to protect.

We now look at the details of the multi-agent model, the specific characteristics of each agent group and how they relate with the real-world situations.

4 System Model

This section describes the ecosystem model highlighting the assumptions made and the role of various attributes of the agents.

4.1 Assumptions

Some of the basic assumptions of our model are as follows:

- The ecosystem of forests is spread over a finite piece of land defined in two dimensions.

- Limiting boundary conditions are used (as against wrap-around) to keep the model closer to the real system.
- Each agent represents a   group of individuals in real life. Thus there is correspondence between the computer agents and real-life groups of a particular type, residing together. (It may be a colony of natives or a group of log-fellers operating together)
- Migration of agents within the system occurs after they have resided at the current location for a fixed period of time, which is dierent for native farmer and logger agents and is taken as a system input. This corresponds respectively to the time period after which the shifting cultivators are forced to move to new locations or when loggers have cleared a favorable land and need to shift to new forested areas. Government agents remain bound to a position, as would happen in real life (their spatial location is not significant in the model).
- For the purpose of simulations, each time step is assumed to be a representation of an actual time span of a few months. Specifically, an assumption of each time step representing a month would translate to a period of consideration equivalent to approximately 8 years, which is a reasonable time frame for studying system behavior.
- Interactions between the dierent agent groups is the crux of the model. Since the interaction within a particular agent group will not accrue any additional information to induce cooperation, these interactions are not considered important. Interaction occurs as detailed in section 3.1.
- Each agent group has a region within which it can move freely. This region is called region of inuence   . Since native farmers are only marginal-economy workers, they don't move beyond their region of inuence. However, native farmers can move to an area beyond their region of inuence with assistance from government agents (which is part of the third stage of simulation).
- Although not preferable, the logger agents can move beyond their region of inuence to fulfill the sustainable development goal. Moving beyond their region of inuence imposes a penalty on the loggers, which is reected in the decrement of their development attribute. However, meaningful usage of this freedom improves the overall system's environmental impact attribute.
- The model includes a group of government agents. These agents interact with native farmers and provide information about the location of the nearest logged land. Also, if this nearest logged area lies outside the region of inuence of native farmer, government agent provides assistance to translocate the farmers to this new location.
- Loggers submit to the government, information about the land they vacate after clearing trees. This is required only in case they fail to communicate the same to a feasible native group agent.
- Our model does not assume any additional outside resources and the improvement in the development and sustainability of the system is purely a result of cooperation and meaningful intervention.

- Without loss of generality, the initial system is generated randomly by considering logger and native group agents scattered around the overall forest area.

4.2   Attributes and Their Impacts

Group specific activities and the success with which the agents are able to execute them determines their individual character. It is also this list of character attributes which the agents selectively relay to the information channel. This list is meaningfully updated upon interaction with other agents, at each time step.

The attributes of agents, their initial values and their role in the ecological model are described below.

1. Type of group the agent belongs to (native farmers, loggers or intervening agency).
2. X, Y - the coordinates of agent's current location on the forest-land. Initially these coordinates are randomly assigned.
3. futureX, futureY   - the coordinate of agent's future location where s/he will be moving after the expiration of the stipulated stay period at one location. Initially, these future coordinates are randomly assigned within the region of inuence of the respective agents.
4. Sustenance level   - This attribute is applicable for native farmers and increases if the native agent moves to a region where logger agents have recently felled trees. This is reasoned on the basis that in such a case the native agent conserves the time and eort of clearing up a fresh forested area and the resources saved can be put to use in other productive activities. Also, if the sustenance level of a native agent grows beyond a pre-defined threshold (a system variable), the agent begins to contribute to the system development commercially (as described in development attribute below).
5. Income   - The income each agent gets from his/her activities is dierent for two groups of agents and is a function of the development level of the agent.
6. Environmental awareness   or concern - This indicates the level of environmental awareness or concern of each agent. It is significant to include this as an attribute of agents because if an agent has higher value for this attribute, which exceeds the threshold for its group, then it can override some of the more general rules and make moves to locations where the purpose of environmental protection is better served, even at a marginal cost of personal development. Each time a native agent interacts with a government agent and is suitably relocated to ecologically favorable spots, his/ her environmental concern and awareness increases.
7. Development   - This is the conicting attribute, which each agent tends to maximize in an isolated situation. Initially all native agents are assigned an identical value of zero while the Logger agents have a normally distributed random value in a pre-defined range. Based on their interactions with agents from other groups, their developmental value tends to increase or decrease. The development value is the overall measure of advancement or progress for individual agents.

8. Environmental Impact I   - In the context of our model, environmental impact I is ascribed to such activities which serve to maintain the forest cover and prevent burning down of trees. Thus it is defined as a binary variable. If the practice is sustainable, then it's value is +1 and in case of unsustainable practice, it is -1. More specifically, as a result of interaction or intervention, if the native farmer agent moves to the location which is already logged it will have a value of +1 and if it is moving to virgin forest land, then it has value -1. Similarly, for loggers, it has a value of +1 if the site vacated by logger agent will be occupied by a native farmer agent and -1 otherwise. Initially, due to the random nature of all future coordinates for all agents, this value is set to be -1 for all. Also, this gives a good measure of the non-cooperative scenario.

9. Environmental Impact II   - This is an attribute specific to Logging Industry agents. The loggers have +1 value if the new location they will be occupying is also planned to be occupied by a native agent further in the future. Thus it can be considered as a measure of improved planning of future activities. Initially, as for impact I, all impact II values are -1.

10. Time left   to change location - One of the major exibilities of model is that it allows each agent group to have its own time to stay at one location depending on the nature of its activities. This attribute indicates the time remaining before an agent can change its location and is a main determinant of the movements of agents after interactions. Initially all agents are assigned the maximum feasible stay time (at a particular location) as input to the system and this value decreases at each time step.

Having clearly distinguished the model assumptions and agent attributes, we now look at the experimental setup and simulation results.

5   Experimental Setup

All simulations were done for a finite piece of forest land assumed to be 1000x1000 square units. These land units serve as grid points where the agents can reside at any given point. Positions of agents on the land were denoted in Cartesian coordinate system and distances measured were Euclidean distances between two coordinates, as in the physical world. Specific rules were developed encompassing and reecting the assumptions and evolution criteria outlined above and each time step was considered to denote a period of one month.

Simulations were run for each scenario (non-cooperation, cooperation and intervention) for three dierent cases, as detailed in Results section below. The ecosystem inputs included   initial agent configuration, space restrictions, feasible time stay (at a fixed position) limits for agents, regions of inuence, environmental awareness sustenance thresholds,   and minimum desirable development gain at each time step .

These essential scenarios were simulated to highlight the main goal of our study. Each of these is discussed in detail in the remaining part of the paper.

6   Results and Discussion

Clearly, a cooperative system of agents evolves to develop improved sustainability (less negative environmental impact) at a small cost of individual developments. With intervention from a government or other external neutral agency both the environmental impact as well as the development scenario in the long run is improved.

6.1   Impact on Development

Figure 1 shows the results for development values for both native and logger agents along with the overall system's development values over a period of time. Results compare the three cases of non-cooperation, cooperation and cooperation with intervention from external (government) agents.

We see a slight drop in the development values of the loggers because under cooperation regime, they sometimes need to compromise on their individual interests in favor of the system's interest at large. However, for native agents, development improves under cooperation only, as otherwise they can hardly keep up with their sustenance requirements and do not contribute significantly to developmental goals. Also, the decrease in system's development due to cooperative behavior can be improved signißantly through eective intervention from government or NGOs.

Figure 1(b) shows that the model scales up to much higher ratios of logging agents in the overall systems. Figure 1(c) is the set of results when the logging agents decided to keep a high minimum development threshold for personal development (twice as much as in other cases) and did not let their development gains fall below the threshold while deciding to cooperate. Results are promising in as much as they show a similar trend as the other cases and indicate that the model is robust and does not crumple under high personal development pressures.

6.2   Impact on Sustenance Environmental Awareness

Figures 2 and 3 show the improvement in sustenance and ecological awareness levels of natives under inuence from cooperation and intervention. Under a cooperation regime, as mentioned earlier, the natives improve their sustenance levels when they manage to save time and resources otherwise spent on clearing virgin forests. They are also able to contribute to the system's development value. Also, an increase in the environmental awareness attribute implies ecologically better decision-making in the choice of new cultivation areas. Results show the crucial role of intervention in imbibing environmental awareness among native farmers in addition to the improvement in sustenance levels achieved through cooperation. The runs presented were taken for 30, 15 and 5 native farmer, logger and government agents respectively.

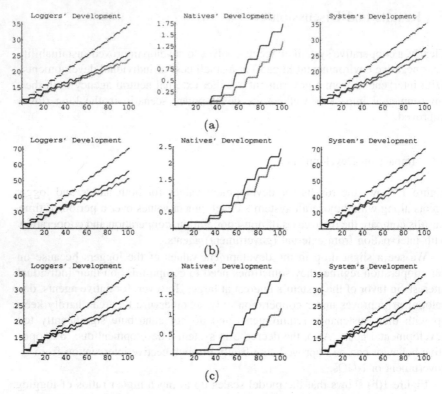

Fig. 1.   Comparison of Development values for loggers, natives and the overall Ecosystem vs. time steps; for Non-cooperation (dotted, red line), Cooperation w/o intervention (thick, green line) and Cooperation with Intervention (thin, blue line). (a) These runs were taken for an Ecosystem with 30 natives, 15 loggers and 5 government agents over 100 time steps. (b) These runs were taken for an Ecosystem with 35 natives, 30 loggers and 5 government agents over 100 time steps. (c) For these runs, the minimum development goal value for logger agents was increased two-fold. Agent population was maintained as in part a above.

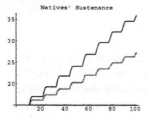

Fig. 2.   Comparison of sustenance values for natives vs. time steps; for Non-cooperation (dotted, red line), Cooperation w/o intervention (thick, green line) and Cooperation with Intervention (thin, blue line).

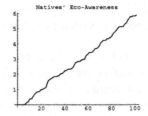

Fig. 3.   Comparison of ecological awareness values for natives vs. time steps; for Non-cooperation (dotted, red line), Cooperation w/o intervention (thick, green line) and Cooperation with Intervention (thin, blue line).

6.3   Ecological Impact

Figures 4(a) through (c) show the eects of the cooperation and intervention models on the ecological impact of logging and native settlement activities. Evidently, there is a remarkable improvement in the ecological system through cooperative behavior which can be further improved through intervention. Although initial cooperation reduces individual development values by a small amount, it is clear that intervention has a positive role in terms of development as well as ecological preservation. This highlights the importance of such initiative on the part of government or other neutral third parties.

Figure 4(b) shows the positive eect of the cooperation model when there are almost equal number of logger and native agents and the number of government agents is relatively less. This shows the scale-up behavior of the model which is quite promising. Figure 4(c) shows the ecological impact values for the case when logging agents decide to cooperate selectively, keeping a higher minimum threshold of developmental gains. Apart from the initial slowing down of the ecological remedy, the model proves robust for this agent behavior too.

Figure 5 shows the spatial representation of the ecosystem model and reinforces the results and conclusions presented in this section. It shows the simulated forest region after equal number of time steps under non-cooperation (a), simple cooperation (b) and cooperation with intervention (c) schemes. Clearly, the wasted forest-lands (white spots) are significantly more in the case of non-cooperation than for cooperation and information sharing.

The simulation study thus results in clear indication of the utility of cooperation among competing agents and highlights the role of simple meaningful intervention. The results also show that such cooperation does not hinder development drastically and in fact can prove to be supportive in the long run. Also, the model hints at better distribution of resources which helps in improving the life-styles of native shifting cultivators.

7   Future Work: Model Refinements

Several questions still remain to be answered and further new ones have sprung up during the course of the study itself. Possible directions of investigation and model refinements include the following:

- The behavior of logger agent can be modified to incorporate their environmental concern attribute, as is the case with native agents in this study.
- The current model assumes that the cost of relocating for natives and loggers is constant within the respective region of inuence and is also constant (though much higher) for all other regions. This cost could be made a function of the actual distance between the current position and the future position under consideration. This is partially implemented in the case of native and government agents' interaction where the native agent is relocated to the closest location available.

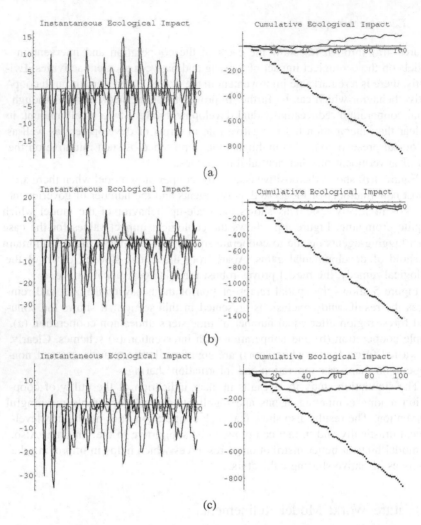

Fig. 4. Plot of the instantaneous and cumulative (over time) ecological impact parameter versus time steps; including Non-cooperation (dotted, red line), Cooperation (thick, green line) and cooperation with intervention (thin, blue line) cases. Results show a significant drop in negative ecological impact when agent cooperate. (a) These runs were taken for an Ecosystem with 30 natives, 15 loggers and 5 government agents over 100 time steps. (b) These runs were taken for an Ecosystem with 35 natives, 30 loggers and 5 government agents over 100 time steps. (c) These results correspond to the case when the minimum development value for logger agents was increased. Number of agents was kept same as in case a above.

– The attribute values and ratios assumed were based on cognitive reasoning and public domain information but not much statistical research was involved in their choice. This could be improved by studying a real-life scenario and any data available for such ecological systems.

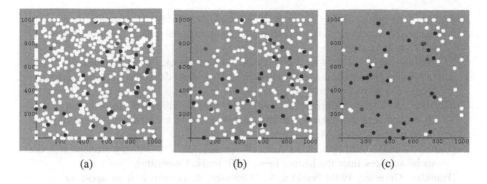

(a)   (b)   (c)

Fig. 5.   Spatial location plot for ecosystem agents. Dark (red) points represent native agents, deep gray (blue) points represent governments agents and the white spots depict the areas where logging activity was   not followed by native colony build-ups, thereby negatively aecting the ecosystem. (a) shows the conﬁguration at the end of 100 time steps under non-cooperation regime, while (b) and (c) show the configurations after equal number of time steps for simple cooperation and cooperation with intervention respectively.

– A preliminary investigation shows that the model is fairly robust. However, a thorough sensitivity analysis needs to be performed on the boundary values of the system parameters.

8   Conclusions

The work presented in this paper demonstrates the importance of evolving cooperative behavior to achieve sustainable development goals of society. The results also show clearly the trade-os between the developmental and environmental protection goals faced by societies. In this respect, it is important to note the increasing focus on cooperative activities in all human endeavors as the complexities and challenges of problems faced continue to increase. It also highlights the role government agencies can play by intervening as informative and guiding agents. This sharply contrasts with the traditional 'Command and Control' role executed by and expected of government agencies. The results lend insight to the system modeled and provide useful prediction of system's behavior which should directly lead to practical policy guidance.

References

[Axelrod, 1997] Axelrod, R. (1997).   The complexity of cooperation: Agent-based models of competition and collaboration   . Princeton University Press, Princeton NJ.

[Bousquet, Barreteau, Le Page, Mullon, Weber, 1999] Bousquet, F., Barreteau, O., Le Page, C., Mullon, C., Weber, J. (1999). An environmental modelling approach: the use of multi-agent simulations. In Blasco, F., Weill, A. (Eds.), Advances in Environmental and Ecological Modelling   (pp. 113–122). Paris: Elsevier.

[Caldart Ashford, 1998] Caldart, C., Ashford, N. (1998). Negotiation as a means of developing and implementing environmental policy. Working Paper, MIT, draft version.

[Carpenter, Brock, Ghanson, 1999] Carpenter, S., Brock, W., Ghanson, P. (1999). Ecological and social dynamics in simple models of ecosystem management. Conservation Ecology , 3 (2).

[Christie A.T., 1997] Christie, P., A.T., W. (1997). Trends in development of coastal area management in tropical countries: From central to community orientation. Coastal Management , 25 , 155–181.

[Epstein Axtell, 1996] Epstein, J., Axtell, R. (1996). Growing artificial societies: social sciences from the bottom up . MIT Press, Cambridge.

[Franklin Graesser, 1996] Franklin, S., Graesser, A. (1996). Is it an agent, or just a program?: A taxonomy for autonomous agents. Institute for Intelligent Systems, University of Memphis: Springer-Verlag.

[Gilbert Doran, 1994] Gilbert, N., Doran, J. (Eds.) (1994). Simulating societies: the computer simulation of social phenomenon . UCL Press, London.

[Hardin, 1968] Hardin, G. (1968). The tragedy of the commons. Science , 162 , 1243–1248.

[Healey, 1998] Healey, M. (1998). Paradigms, policies, and prognostications about the management of watershed ecosystems. In Naiman, R., Bilby, R. (Eds.), River Ecology and Management (pp. 662–682). Springer.

[Holland Miller, 1991] Holland, J., Miller, J. (1991). Artificial adaptive agents in economic theory. In American Economic Association Papers and P roceedings, Volume 81 (pp. 365–370).

[Kohler Gumerman, 2000] Kohler, T., Gumerman, G. (Eds.) (2000). Dynamics in human and primate societies: Agent-based modeling of social and spatial processes . Santa Fe Institute Studies in Sciences of Complexities: Oxford University Press.

[Nielsen T., 1999] Nielsen, J., T., V. (1999). User participation and institutional change in fisheries management: a viable alternative to the failures of 'top-down' driven control. Ocean and Coastal Management , 42 , 19–37.

[Ostrom, 1990] Ostrom, E. (1990). Governing the commons: The evolution of institutions for collective action . Cambridge, U.K.: Cambridge University Press.

[Ostrom, Gardner, Walker, 1994] Ostrom, E., Gardner, R., Walker, J. (1994). Rules, games, common pool resources . The University of Michigan Press.

[Rouchier, Barreteau, Bousquet, Proton, 1998] Rouchier, J., Barreteau, O., Bousquet, F., Proton, H. (1998). Evolution and co-evolution of individuals and groups in environment. IEEE Computer Society, Paris.

[Rouchier, Bousquet, Le Page, Bonnefoy, 2000] Rouchier, J., Bousquet, F., Le Page, C., Bonnefoy, J. (2000, July). Multi-agent modelling and renewable resource issues: the relevance of shared representations for interacting agents. In Moss, S., Davidson, P. (Eds.), Procee dings of the Second Workshop on Multi-Agent Based Simulation (MABS 2000) (pp. 181–198). Springer LNAI series.

[Rowe, Sharma, Browder, 1992] Rowe, R., Sharma, N., Browder, J. (1992). Deforestation: Problems, causes and concerns. In Sharma, N. (Ed.), Managing the World's Forests (pp. 34). Dubuque, Iowa: Kendall/Hunt Publishing Company.

[Sichman, Conte, Gilbert, 1998] Sichman, J., Conte, R., Gilbert, N. (1998). Multi-agent systems and agent based modelling . Springer.

[Simon, 1981] Simon, H. (1981). The sciences of the artificial . MIT Press.

[Sugden, 1989] Sugden, R. (1989). Spontaneous order.   Journal of Economic Perspectives , 3 (4), 85–97.

[Tessier, Chaudron, M   uller, 2001] Tessier, C., Chaudron, L., M   uller, H. (Eds.) (2001). Conicting agents : Conict management in multi-agent systems   . Kluwer Academic Publishers, Boston, MA.

The Impact of Environmental Structure on the Evolutionary Trajectories of a Foraging Agent

Ian R. Edmonds

School of Computing Information Systems and Mathematics
South Bank University, London SE1 0AA, United Kingdom
edmondi@sbu.ac.uk

Abstract.   A foraging agent using a sensorimotor controller is simulated in environments with varying ecological structure. The controller
is evolved in the dierent environments to produce a range of emergent behaviours, which are analysed and compared using data reduction
techniques: the behaviours are compared between environments and in
their evolutionary trajectories. The relationship between the evolutionary trajectories, the aordances in the dierent environments, and the
performance and onward evolution of controllers in their non-native environments is explored. The dierent environments have lead to agents
following dierent evolutionary trajectories and arriving at similar but
slightly dierent behaviours. These evolved controllers then evolve differently when challenged with a new environment.

1   Introduction

A foraging agent is simulated in several environments with varying ecological
structure. The agent uses a sensorimotor controller based on a neural network
with some limited memory, and is evolved in the dierent environments to produce a range of emergent behaviours. These behaviours are analysed and compared in order to explore the impact of environmental structure on the evolutionary trajectories and ongoing evolution of the controllers in non-native environments.

This paper extends the work reported in [1] which used a data reduction technique from the field of text based information retrieval, called Latent Semantic
Indexing (LSI) [2] [3], as a tool in understanding the complexity of the agents'
emergent behaviour. This introduction highlights some related work in 4 areas:
(i) techniques related to LSI which have been used to help in understanding the
complexity of emergent behaviour, (ii) how   shaping of an agent can be achieved
by providing particular sequences of experience, (iii) ways to measure or characterise environments and predict the impact they will have on agents in them,
and (iv) how dierent evolutionary histories can lead to dierential responses.

The further sections of this paper are: (2) gives an overview of the model:
the agent, controller, and environments, (3) gives a brief outline of results in the
previous work [1], (4) describes 3 new environments and shows the aordances
in 2 of these new environments produce dierent evolutionary trajectories, (5)

P. Collet et al. (Eds.): EA 2001, LNCS 2310, pp. 338–349, 2002.

compares agents' performances in their non-native environments and the results of evolving the agents in the 3rd new environment.

Part of the appeal of evolving sensorimotor controllers in ALife and situated robotics is that complex behaviour can emerge from the interaction of quite simple controllers and the agent environment (e.g. prey capture [4], obstacle avoidance and foraging [5], corridor navigation [6], garbage collection [7], [8], [9], rat navigation with hippocampal place cells [10]). However, trying to understand this emergent complexity can itself be challenging. A variety of techniques include those based on dynamical systems, e.g. used in [11] to uncover attractors in the phase portrait of an agent, used in [12] by developmental psychologists in tracking infant motor skills, and for a review see [13] which includes the use of Principal Components Analysis (PCA) as a data reduction technique. PCA is used in [14] to understand the evolutionary trajectories of nodes in a recurrent neural network under the evolutionary algorithm, SANE, and this use of PCA is similar to the use of LSI here, as a way of bridging the gap between high level, (distal) interpretation, and descriptions based on large volumes of detailed data.

Another issue of interest in ALife and situated robotics is the use of shaping and the way some researchers have provided sequences of particular experiences to their agents in order to help them evolve. In [4] the shaping follows two heuristics: (i) the density of relevant experiences is increased at early stages, and (ii) the diculty of the prey capture tasks is gradually increased. In [9] the density of certain obstacle experiences was artificially increased to aid the evolution, although other forms of shaping which were expected to be useful turned out to be a hindrance. This is related to another research theme that adopts the perspective of the environment, that of developing ways to measure or characterise environments and predict the impact they will have on agents within them. In [15] the relationship between an entropy measure of environmental structure and the adaptability of agents is explored, and in [16] the carrying capacity and complexity of an environment is related to the complexity of the behaviours that evolve.

Some recent work in the area of evolutionary psychology [17], [18], has shown how the evolutionary history of tungara frogs inuences their response to mating calls from closely related species. Dierent evolutionary trajectories of mating calls were assembled from a library of calls from: (i) related species, (ii) reconstructions of ancestral calls, and (iii) calls constructed by manipulating acoustic features (based on a PCA of the calls of the related species). The final call in each trajectory was always the tungara call, but they arrived there by dierent routes. Recurrent neural networks that were trained to simulate the frog responses were shown to be eective only if the training followed the trajectory of calls of their evolutionary ancestry. It is argued that this historical contingency is important in the shaping of cognitive function; the evolutionary trajectory constrains future adaptability.

2   An Overview of the Model

This is a very brief overview to provide a basis for understanding the results reported here. However, for a more detailed description of the model and the techniques and rationale behind the use of LSI see [1], [19].

The agent exists in a simulated world with plants that grow dierentially on a water resource gradient. The space is a 100 by 100 toroidal grid with 40 water sources: the water is diused using a standard lattice diusion formula leading to a water resource gradient. The first environment to be studied had the 40 water sources randomly distributed , e.g. see fig. 1 (a). Further environments will be discussed below. Plants can occupy cells in the grid, and are modelled using a lifecycle transition graph. They only grow in the wetter areas, and produce a structured and stable population of approximately 2,000, see fig. 1 (b) for plants in the random water environment (n.b. the water locations are random not the plants).

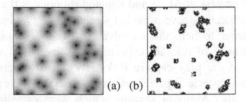

(a) (b)

Fig. 1.   The random water environment, (a) the water resource forms wet patches separated by arid areas, and (b) the plants grow in the wet patches.

The agent occupies one cell in the grid, and has 4 actions available to it: eat, move forward one step, and rotate by 90 degrees left or right. It has sensory input: of the water value at its current location and the gradient in its forward direction, and of the plants at its current location, directly in front, and the cells diagonally to the front left and front right. When a plant is eaten, it is removed from the world, and its biomass is absorbed by the agent.

In the previous work, three controller models were implemented to investigate how behaviour is eected by dierent memory structure: (i) a basic neural net, (ii) a neural net with memory, and (iii) a rule based controller. All 3 are essentially sensorimotor controllers, where (i) implements a reactive controller, (ii) provides internal state to the controller, and (iii) provides internal state in the sense of possible sequences of actions.

The basic neural networks contains feedforward nodes using a sigmoid activation function with one hidden layer. The input layer receives the 4 bit binary plant sense pattern and the 2 real water values. The 4 output nodes are treated as a stochastic output to select the agent's action in the given sensory state.

The neural net with memory is similar to the basic neural network, but with 5 additional memory input nodes fully connected to the hidden layer. Each memory is represented as a real valued time decaying integration of the stimulus, where the stimulus is each one of the actions performed and the amount of biomass eaten by the agent, and is akin to the battery levels used by robot controllers, e.g. [5], or in networks with recurrent loops e.g. [8].

The rule based controller iterates (at each step) through a sequence of 30 rules organised as a loop, and this allows for possible sequences of actions.

3   Previous Experiments in the Random Environment

A simulation consists of placing an agent in a random position in the world and having it make 3000 actions, during which time it may acquire some quantity of biomass. The agents were evolved during evolutionary runs of 1,000 generations, starting with 5 random parent genotypes. Ten mutants were produced from each parent to give a population of 50, and each mutant was tested in a number of simulations; the mean performance for each mutant was used as the fitness value. The best 5 mutants became the parents for the next generation.

In the previous work each of the 3 controller architectures were exposed to 10 evolutionary runs in the random environment and the results showed that the basic neural network performed worse, and that particular rule based controllers provided by far the best performers

The behaviours of the controllers were then described from 2 perspectives as in [8], (i) a distal perspective (i.e. a human observer), and (ii) using proximal descriptions based on large amounts of collected data and in this work using LSI as a data reduction technique. From the distal perspective, the behaviours were seen as falling within the classic foraging descriptions of exploitation and exploration [20]: the agents move quickly through the arid areas where there are no plants until they find a wetter patch with plants, at which point they start to eat and turn more. If they find themselves in an eaten out or barren wet patch, they may turn a little, but move on fairly soon.

Key dierences between the best performers of the 3 controllers were obvious, and most apparent in the exploitation behaviour. The basic neural network had a strategy of tending to head through the middle of a patch of plants and eating as it looped or moved with back and forth actions, see fig. 2 (a) and (b), leading to the patch of plants becoming broken up into isolated plants, becoming gradually more dicult to fid.

The neural network with memory controllers tended to eat by spiraling in to rectangular blocks, see fig. 2 (c) and (d), while the best rule based agents had the most eective strategy which involved eating by following the outer contours of the patch of plants, and gradually spiraling inwards, see fig. 2 (e) and (f).

In order to characterise the behaviours based on a proximal description, the agent's sense and action data was collected for every action during the simulations and was compiled into records of sequences of 15 actions, by moving a window over the actions: for the rest of the LSI analysis, each sequence of 15 actions will be called a behaviour.

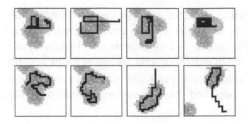

Fig. 2.   Paths of the agents over 60 actions in the random environment. The top row: (a) and (b) are the basic neural network eating by looping or a back and forth movement, (c) and (d) are the neural network with memory eating by spiraling inwards on a rectangular block. The bottom row are all the rule based controller: (e) and (f) eating along the contours of a plant cluster, (g) and (h) are finding a plant cluster using: in (g) a straight line approach, and in (h) a long legged zigzag.

The window size of 15 was selected based on a technique adopted from landscape ecology to identify scale factors. In [21] it is shown as being useful to identify fractal dimensions in beetle movement and plant distributions, and in [22] a related but more sophisticated technique is used to identify fractal dimensions in the movement of fish schools foraging on plankton swarms. In overview, a window is moved over the set of actions, and the Euclidean distance travelled in each window is extracted. The distribution of distances travelled gives a measure of walk curviness. This is repeated for dierent window sizes, and the distributions plotted against window size.

Data for behaviours of the 3 best performing controllers was analysed with LSI, and scatter plots in fig. 3, show the location of each behaviour in the space of the first 2 latent factors given by the PCA , with histograms showing the distribution of points in the vertical and horizontal dimensions. There are three

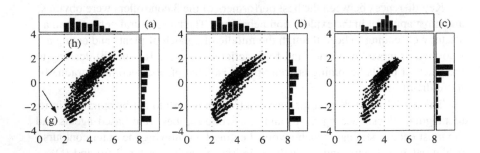

Fig. 3.   Scatter plots of the location of each behaviour in the space of the first 2 latent factors from the LSI analysis, with histograms showing the distribution of behaviours in the vertical and horizontal dimensions (agents in the random environment). (a) shows the basic neural network, (b) shows the neural network with memory, and (c) shows the rule based controller.

general points about the interpretation. Firstly, the meaning of the latent factors was determined, as in PCA, by looking at the factor loadings. These can be best understood by rotation to the axes indicated by the arrows (g) and (h) in fig. 3 (a), having the general meanings: in the direction (g)   away from the origin means an increase in the number of move actions, in the direction (h)   away from the origin means an increase in the numbers of plants seen and eat actions plus smaller contributions from turning actions, water values, and biomass eaten. Secondly, the banded nature of the points in the scatter plots perpendicular to axis (g) (i.e. the points form stipes at approximately 45 degrees) is a consequence of the discrete number of move actions in each behaviour: a behaviour with fewer moves will be in a band closer to the axis (h). Thirdly, there are 2 loose clusters: at around (4, 1) being associated with seeing plants and eating (i.e. exploitation), and another at (2.5, -3) being associated with maximum moving and seeing no plants (i.e. exploration); these clusters are most clearly seen in the horizontal and vertical distribution histograms in fig. 3 (c).

It can be seen, in fig. 3, that the basic neural network has the most dispersed footprint, indicating more varied behaviour, while the rule based controller has the tighter of the clustering around points (4, 1) and (2.5, -3), more eectively focussing its behaviours into exploitation and exploration. The neural network with memory lies between the other two.

Further analysis was conducted on the evolutionary trajectories of the be-haviours for the neural network with memory, fig. 4. The generations shown in fig. 4, and their approximate performances relative to the best are: g(1) 5, g(16) 25, g(40) 50, g(920) 100. The g(1) controller is quasi-random, and with the majority of behaviours lying close to the origin (0, 0) indicating little movement and seeing very few plants. The next snapshot, at g(16) shows a wider atter distribution, in the behaviour space - moving more and seeing more plants. At the g(40) snapshot, the distribution is beginning to focus in on the exploration cluster at (2.5, -3), but not yet into the exploitation cluster. By g(920), the 2 clusters are more distinct.

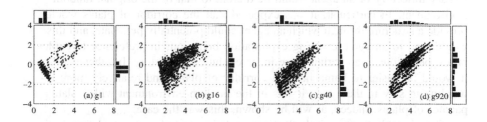

Fig. 4. Scatter plots of the location of each behaviour in the space of the first 2 latent factors from the LSI analysis, with histograms showing the distribution of behaviours in the vertical and horizontal dimensions, for successive generations of the neural network with memory (in the random environment): (a) is g(1), (b) is g(16), (c) is g(40), and (d) is g(920).

The trajectories of the other 2 controllers were fairly similar to the neural network with memory and indicate that the aordances in the random environment push a controller into evolving robust exploration behaviour, before exploitation will be honed.

4   Further Work with New Environments

Three further non-random environments were developed, the first two based on alternative layouts of the water sources, and the third one on changes to the plant structure; all three had the same number of water sources and parameter settings as the original environment. The first alternative, to be called the   percolating environment had the locations of water sources organised in a ribbon to provide an initial continuous corridor of plants (fig. 5 a), and the second, to be called the almost-percolating environment was like the percolating environment, but with some water sources moved out of the line so that the corridor of plants was broken up into shorter strands (fig. 5 b). The third, to be called the   depleted environment was based on the percolating environment, but with 80 of the plants randomly removed.

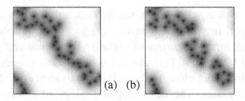

(a) (b)

Fig. 5. Two non-random water resource environments: (a) the percolating environment, (b) the almost percolating environment.

The intention was to create environments where the pressure on the evolutionary trajectory of an agent would be dierent to that of the original random environment. In particular, it was expected that the corridor of plants would provide a structure through which an agent could potentially percolate, and this would lead to an agent evolving exploitation behaviour in preference to exploration behaviour at an earlier stage in its evolutionary history. In other words, it would be easier to bump into the corridor of plants and relatively more rewarding to stay in them than it was in the random environment with isolated patches of plants. Indeed, as will be shown, this was the case.

The available biomass in both the percolating and almost percolating environments was approximately 40 more than in the random environment due to the localised focussing of the water resources creating more areas habitable by plants, while the depleted environment contained 20 of that available in the percolating environment. The depleted environment requires an agent to explore more, and provides less exploitable structure.

The neural network with memory controller was evolved in the percolating and almost percolating environments in the same manner as described for the previous work. The results are shown in fig. 6 along with the previous comparable results for the neural network with memory controller in the random environment. It can be seen that performances in the percolating and almost percolating environments are considerably higher, as would be expected, due to the increased amount of biomass available and the   easier to find structure of the plants.

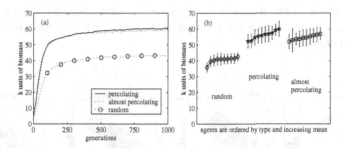

Fig. 6. The neural network with memory in 3 environments: (a) The means (using a moving average over the previous 20 generations) of the best performers of each of the 10 runs, (b) The means and standard deviations of the best overall performers from each run, ordered according to increasing mean.

LSI was applied to the newly evolved agents, and the LSI footprints of the three agents in their native environments are shown in fig. 7. The agent in the random environment (a), sees fewer plants than the other two, i.e. there are fewer points appearing further out from the origin than (6,2); it also does more exploration behaviour, i.e. more points around (2.5, -3).

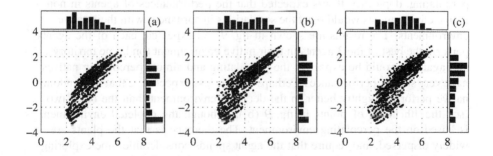

Fig. 7. Footprints of the neural network with memory agents in their native environments: (a) random, (b) percolating, (c) almost-percolating.

In comparing the evolutionary trajectories of the agents in the two new environment with that in the random environment, the expectations that the affordances would push the trajectory towards earlier exploitation rather than exploration are confirmed. Successive generations of agents in the percolating environment are shown in fig. 8, with similar stages of performance as were shown for the trajectory in the random environment in fig. 4, i.e. g(1) 5, g(8) 25, g(26) 50, g(984) 100. It can be seen that by the 50 performance stage, g(26), the points are clustering more on exploitation at around (4,1) and further out from the origin, than they were in the random environment, which at the 50 performance stage shows clustering on exploration. The evolutionary trajectory of agents in the almost percolating environment showed a similar trend to agents in the percolating environment with early clustering on exploitation rather than exploration.

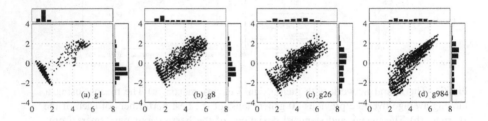

Fig. 8.   Footprints of successive generations of the neural network with memory agents in the percolating environment: (a) is g1, (b) is g8, (c) is g26, (d) is g984.

5   Comparing Agents in Non-native Environments

The best agent from evolution in each environment (3 in all: 1 each from random, percolating, almost percolating) was then exposed to sets of 50 simulation tests in environments other than the one it was evolved in (random, percolating, almost percolating, depleted). It was expected that the performances of agents in non-native environments would reect the similarities in aordances with their native environments. The results are shown in fig. 9. As expected, each of the agents perform the best of the 3 agents in their native environment (on home territory), and again as would be expected, the percolating and almost percolating natives have very similar performances. Perhaps what is less expected, is that the random native performs slightly better in the depleted environment than the other two (see the 4th block of points in fig. 9 (b)), although the depleted environment is based on the percolating environment. The dierence is that the plants are widely dispersed, and require that the agent spends considerable time exploring to find isolated plants: the random native appears to be slightly better at that exploration, perhaps due to its tendency to exploration that was seen in its early evolutionary trajectory (see fig. 4).

The next step was to allow the agents to evolve for a further 300 generations in the depleted environment, and the results are seen grouped in the 5th block of points in fig. 9 (b). It can be seen that the percolating and almost percolating natives have fie-tuned their performance more eectively than the random native. It is speculated that their evolutionary history of responding in the corridor of habitable area (with associated water sensory input) allows them to fine tune their networks more easily to exploring more in just this area, while the random native does not have this grounding to fall back on. This is an area for future work.

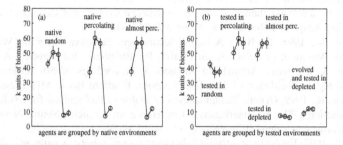

Fig. 9.   Performances of the neural network with memory as evolved in 3 environments (order: random, percolating, almost-percolating) tested in 5 environments (order: random, percolating, almost-percolating, depleted, onward evolution in depleted). The same results are shown in 2 arrangements to aid interpretation: (a) grouped by the agents' native environments, (b) grouped by the tested environment.

6   Conclusion

Dierent environments have been created with varying ecological structure but sharing an underlying set of parameters and processes. These dierent environments provide diering aordances which have lead to agents with the same sensorimotor controllers following dierent evolutionary trajectories and arriving at similar but slightly dierent behaviours. It has been shown that these evolved controllers will respond dierently to the opportunity to evolve further in a new environment containing characteristics of the original evolved agents' various native environments. The use of LSI has helped in uncovering the evolutionary trajectories and discovering meaning in the detailed data of sequences of actions.

Acknowledgements.   Many thanks to the anonymous referees for useful comments.

References

1. Edmonds, I. R., 2001, The Use of Latent Semantic Indexing to Identify Evolutionary Trajectories in Behaviour Space, in (eds) Kelemen, J., and Sosik, P., Advances in Artificial Life, 6th European Conference, ECAL 2001, LNCS; 2159, LNAI, Springer-Verlag
2. Deerwester, S., Dumais, S. T., Furnas, G. W., Landauer, T. K., and Harshman, R., 1990, Indexing by Latent Semantic Analysis, Journal of the American Society for Information Science, 41 (6), 391-407
3. Landauer, T. K., and Dumais, S. T., 1997, A Solution to Plato's Problem: The Latent Semantic Analysis Theory of Acquisition, Induction and Representation of Knowledge, Psychological Review, 104 (2), 211-240
4. Gomez, F., and Miikkulainen, R., 1997, Incremental Evolution of Complex General Behavior, Adaptive Behavior, vol. 5, no. 3/4, 317-342
5. Seth, A. K., 1998, Evolving Action Selection and Selective Attention Without Actions, Attention, or Selection, in Pfeifer, R., Blumberg, B., Meyer, J-A., and Wilson, S. W., (eds), Animals to Animats 5, Proceedings of 5th International Conference on Simulation of Adaptive Behavior, Bradford Book, MIT Press
6. Shipman, R., 1999, Genetic Redundancy: Desireable or Problematic for Evolutionary Adaption?, The 4th International Conference on Artificial Neural Networks and Genetic Algorithms (ICANNGA '99), April 1999
7. Nolfi, S., 1997, Evolving non-trivial behaviours on real robots: A garbage collecting robot, Robotics and Autonomous Systems, 22, 187-198
8. Nolfi, S., 1997, Using Emergent Modularity to Develop Control Systems for Mobile Robots, Adaptive Behaviour, vol. 5, no. 3/4, 343-363.
9. Calabretta, R., Nolfi, S., Parisi, D., and Wagner, G. P., 1998, Emergence of Functional Modularity in Robots, in Pfeifer, R., Blumberg, B., Meyer, J-A., and Wilson, S. W., (eds), Animals to Animats 5, Proceedings of 5th International Conference on Simulation of Adaptive Behavior, Bradford Book, MIT Press.
10. Foster, D. J., Morris, R. G. M., and Dayan, P., 2000, Models of Hippocampally Dependent Navigation, Using The Temporal Dierence Learning Rule, Hippocampus, vol. 10, issue 1
11. Husbands, P., Harvey, I., and Cli, D., 1995, Circle in the round: State space attractors for evolved sighted robots, Robotics and Autonomous Systems, 15, 83-106
12. Thelen, E., 1995, Motor Development, American Psychologist, Feb 95, 79-95
13. Beer, R. D., 2000, Dynamical approaches to cognitive science, Trends in Cognitive Sciences, vol 4, no 3, 91-99
14. Moriarty, D. E., and Miikkulainen, R., 1998, Forming Neural Networks Through Ecient and Adapted Coevolution, Evolutionary Computation, 5(4), 373-399
15. Fletcher, J. A., and Zwick, M., 1996, Dependence of Adaptability on Environmental Structure in a Simple Evolutionary Model, Adaptive Behavior, vol 4, 3/4, 283-315
16. Menczer, F., and Belew, R. K., 1996, From Complex Environments to Complex Behaviors, Adaptive Behavior, vol 4, 3/4, 317-363
17. Phelps, S. M. and Ryan, M. J., 2000, History inuences signal recognition: neural network models of tungara frogs, Proc. Royal Society London B, 267, 1633-1639
18. Ryan, M. J., Phelps, S. M., and Rand, A. S., 2001, How evolutionary history shapes recognition mechanisms, Trends in Cognitive Sciences, vol 5, 4, 143-148
19. Edmonds, I. R., 2001, Tracking the Evolution of a Foraging Agent, Technical Report, SBU-CISM-01-07, South Bank University, London

20. Gelenbe, E., Schmajuk, N., Staddon, J., and Rief, J., 1997, Autonomous search by robots and animals: A survey, Robotics and Autonomous Systems, 22, 23-34
21. Milne, B. T., 1991, Lessons from Applying Fractal Models to Landscape Patterns, in Turner, M. G., and Gardner, R. H., (eds), uantitative Methods in Landscape Ecology, Springer-Verlag, 199-235
22. Tikhonov, D. A., Enderlein, J., Malchow, H., and Medvinsky, A. B., 2001, Chaos and fractals in fish school motion, Chaos, Solitons and Fractals 12, 277-288

Learning as a Consequence of Selection

Samuel Delepoulle [1,2] , Philippe Preux [2], and Jean-Claude Darcheville [1]

[1] Unit´e de Recherche sur l' Évolution des Comportements et des Apprentissages (URECA), UPRES-EA 1059, Universit´ e de Lille 3, B.P. 149, 59653 Villeneuve d'Ascq Cedex, France, lastname@univ-lille3.fr
[2] Laboratoire d'Informatique du Littoral (LIL), Universit´ e du Littoral Cˆ ote d'Opale, UPRES-JE 2335, B.P. 719, 62228 Calais Cedex, France, lastname@lil.univ-littoral.fr

Abstract. Since the end of the XIX [th] century, the inuence of learning on natural selection has been considered. More recently, this inuence has been investigated using computer simulations. However, it has not yet been shown how the ability of learning can be the product of natural selection. This point is precisely the subject of this paper.

1 Introduction

Since it has been proposed independently by Lloyd Morgan [12], Osborn [13] and Baldwin [2], it is known that the activity of organisms during their lifetime can bring long term modifications to their genomes, and therefore plays a role in natural selection. The Baldwin eect has fist been experimented in the 1950s on Drosophila by Waddington [23,24]. It is now widely recognized that genetic evolution and learning are deeply intertwined processes. Today, computer simulations provide a tool to investigate the interaction between natural evolution and learning. Even if the complexity of the agents that are simulated is rather crude with regards to living organisms, and even though the natural processes are much simplified when simulated, it has been argued that this type of work is useful [10]. Using simulations, Hinton and Nowlan [9] were the firsts to show that learning can guide and speed-up evolution. More generally, we refer to [22] for a recent up-to-date review regarding the interaction between evolution and learning. Among other points that have yet to be explored, the ability to learn should be explained by way of natural selection if we want to remain within a strict selectionist point of view of evolution. Indeed, if natural selection is invoked as the basic process of evolution, it has to create the structures that evolve, as well as all the other processes. Among these, is the ability of learning. At the most basic level, learning is the ability for an animal to modify its behavior according to the stimuli it receives from its environment. This has been modeled by Thorndike as the law of eect, which says that the frequency of emission of a certain behavior increases when it has been followed by favorable

Samuel Delepoulle acknowledges the support of the "Conseil R´ egional Nord-Pas de Calais, France", under contract n 97 53 0283

P. Collet et al. (Eds.): EA 2001, LNCS 2310, pp. 350–361, 2002.
c Springer-Verlag Berlin Heidelberg 2002

consequences in the past [20,21]. Subsequently, the law of eect has been studied and experimented in numerous works and by numerous researchers [4,11]. Skinner proposed the principle of selection of behavior by its consequences [16,18] which is basically the same thing as the law of eect, even though the theoretical framework has evolved since Thorndike [5]. The emphasis we put on the law of eect clearly distinguishes our work from others, such as [1]. In our study, agents are not supervised (at least, not in a strong sense such as involving backpropagation mechanisms or so); they behave and they eventually receive stimuli on their input sensors, and emit behaviors, getting neither reward nor even a value measuring any goodness of the emitted behavior; their lifetime activity selects them for providing osprings to the next generation.

In the sequel, we first set up the stage by describing the model we use for agents, and processes of natural selection and learning. Then, we present the tasks the agents are facing, that is their environment. Afterwards, we present the result of the simulations. We finish with a discussion of this work.

2   The Model

In this section, we describe the agents that evolve, the processes of evolution and lifetime behavior. Natural selection is simulated using a genetic algorithm.

2.1   Evolved Structures

The agents that are evolving are made of a set of   N input sensors (IS) to let them perceive their environment, a set of   N behavior units (BU) to let them act onto their world, and a neural network which controls their activity and let them adapt to their environment during their lifetime (see Fig. 1). Agents are not located spatially in their world; they merely interact with each others. The neural network of an agent is made of   C layers, each of N neurons. The IS's receive binary stimuli from the environment. Let us call "unit" either an input sensor, a behavior unit, or a neuron. Then, each neuron receives the output of the N units of the previous layer (input connections) and the output of the   N units of the next layer (re-entrance connections); that is, each neuron receives 2N inputs. Owing to these connections, the neural network of an agent perceives its own behaviors since the BU's feed back the output layer of the network. Each BU is connected to one neuron of the last layer of the network in a one-to-one relationship. At each time step, only one BU is active, the one associated with the neuron having the highest potential, in a winner takes all fashion, ties being broken at random between neurons having the same potential. In this paper,   C has always been set to 3, and   N to 10. So, there is an input layer of neurons, an output layer, and a layer of hidden neurons. The characteristics of the neural network (that is, the characteristics of the neurons as well as those of the connections) are encoded in a genome. The response of each neuron is characterized by a boolean value which indicates whether the neuron is active or not, and by

6 real numbers: $, , , a, b$ [100, 100],   [1, 1]. These 6 parameters characterize the response of the neuron with regards to surrounding neurons and its own past activity. As far as any neuron is connected to each neuron of the two surrounding layers, each neuron is also characterized by 2   N weights. Each weight is characterized by a quadruplet (V, E_a, E_b, E_{ab}), where V is its initial value, while E_a, E_b and E_{ab} control how its value changes during learning. The value of these 4 parameters lies in [1, 1]. Finally, the whole network is characterized by two numbers A_c and A_p which are discussed below (see Sect. 2.3). To sum-up, the genome of an agent encodes   C   N neurons, each one constituted with one bit of activity, 6 real numbers, and 2   N weights, each weight being itself made of 4 real numbers, which totals in $CN(8N + 6) + 2$ numbers (plus one bit).

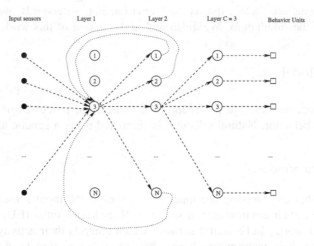

Fig. 1. The internal structure of the agents being evolved. For the sake of clarity, not all units and not all connections are represented. On the neuron 3 of the first layer, we can see 4 out of the N input connections coming from the input sensors, and 3 out of the N re-entrance connections coming from the neurons of the second layer. Refer to the text for more details.

2.2   Evolution

The evolution process is simulated using a genetic algorithm acting on the previously described genomes. Basically, the algorithm that is performed is:

Initialize a population of agents
While   stopping criterion is not fulfilled   do
    // lifetime
    For   i   [1, life_duration] Do
        For j   [1, number _of_agents] Do

```
        Activate
        Learn
     Done
  Done
  Evaluate the fitness of the agents
  // evolution
  Constitute the population of osprings using genetic operators
Done
```

// Activate	// Learn
Begin Activate	Begin Learn
Activate input neurons	For i [1, A_p] Do
For i [1, A_c] Do	Select one connection at random
Select one neuron at random	Update its weight
Update its potential	Done
Done	End Learn
Observe response	
End Activate	

This algorithm is described in the following paragraphs.

Genetic operators.   To constitute the population of osprings, we use 6 operators: one recombination and 5 kinds of mutation. Each mutation acts at a certain level of the genome: weights, neurons, and their expression, and how the network learns. Basically, one point-crossover is used on two parents to produce one ospring. The crossover can only cut between two dierent neurons. The resulting genome inherits the two parameters A_c and A_p from one of its two parents drawn at random. With regards to mutation, the first one acts on a weight. It consists in choosing at random a weight in the genome and modify its value of at most 10. This yields a mutation that has only a slight eect. The probability to apply this mutation is noted $_w$; it can be rather high as long as its eects are not very disruptive. The second mutation acts on a neuron and it consists in resetting at random all the characteristics of a neuron. The probability to apply this mutation is noted $_n$; obviously, its eects on the activity of the network can be rather important. Thus, we use a rather low value for $_n$. The third mutation concerns the expression of a neuron and merely acts on the activity bit of a neuron in the genome. It is applied with probability $_e$. Toggling this bit can have important consequences on the activity of the network. When inactive, a neuron can "travel" along generations without being noticed, and it can undergo mutations which do not modify the fitness of the agent (neutral mutations). When made active again, it can greatly modify the activity of the network and, thus, the fitness of the agent to which it belongs. The two last mutations concern the parameters A_c and A_p. With probability $_p$, each of these two variables can be modified independently. Their mutation changes slightly their value (10 units).

2.3   The Lifetime of an Agent

In this section, we describe how an agent learns during its lifetime. Before that, we describe how the neural network reacts to stimuli to produce its behavior, that is the procedure Activate of the algorithm.

Activate.   To come close to a concurrent activity of neurons, the neurons of an agent are activated as follows. Iteratively, A_c neurons are drawn at random, letting it possible that a neuron is drawn several times during a single invocation of "Activate". Let $l \in [1, C]$ and $n \in [1, N]$ drawn at random be the layer and the number in the layer of a neuron to be activated. This neuron should have its expression gene turned on. Let us note $A_t(l, n)$ the potential of this neuron at time t. $A_{t+1}(l, n)$ can be written as a function of its current potential $A_t(l, n)$ and the weighted sum of its inputs $Se_t(l, n)$, the weighted sum of its re-entrance $Sr_t(l, n)$, and h_t a random noise uniformly drawn in $[-1, 1]$. Then, the "Update its potential" step of the algorithm is as follows:

$$A_{t+1}(l, n) = f(\ _{ln}.Se_t(l, n) + \ _{ln}.Sr_t(l, n) + .A_t(l, n) + \ _{ln}.h_t) ,$$

where

$$Se_t(l, n) = \sum_{k=1}^{N} V_t^e(k.ln) \quad A_t(l-1, k)$$
$$Sr_t(l, n) = \sum_{k=1}^{N} V_t^r(k.ln) \quad A_t(l+1, k) \quad ,$$

where $V_t^e(k.ln)$ is the weight at time t of the kth input connection of neuron (l, n), and $V_t^r(k.ln)$ is the weight at time t of the kth re-entrance connection of the same neuron. The function $f(x)$ is linear by parts. It is determined according to the value of a_{ln} and b_{ln}:

– if $a_{ln} = b_{ln}$, then $g(x) = 2(x - a_{ln})(a_{ln} - b_{ln}) - 1$
  and
$$f(x) = -1 \quad \text{if } g(x) \quad -1$$
$$f(x) = g(x) \text{ if } -1 \quad g(x) \quad +1 \quad ,$$
$$f(x) = +1 \quad \text{if } g(x) \quad +1$$

– if $a_{ln} = b_{ln}$, then
$$f(x) = -1 \text{ if } x \quad a_{ln}$$
$$f(x) = +1 \text{ if } x \quad a_{ln} \quad .$$

The potential of all neurons that are not updated remains unchanged. Finally, as long as a neuron has its expression gene turned o, its potential remains 0.

Learning.   This paragraph describes the "Learn" action in the algorithm. Learning is not determined by genes but by the variables (V, E_a, E_b, E_{ab})) that are genetically encoded and produce the way the network is activated. Hence, learning is not strongly genetically predetermined but remains under the inuence of the environment to a large extent. We call "Learning" the activation of the network according to the stimuli it receives from its environment. It consists in a modification of the weights of the network. This modification is not made deterministically but at random: a connection of the network is drawn at random, and its weight is modified according to the neurons to which it is connected. This modification is made iteratively A_p times. Again, a connection can be selected more than once in a single invocation of "Learn".

More precisely, let $l \in [1, C], n, k \in [1, N]$, and $r \in \{false, true\}$ drawn at random.

If $r = false$, the weight of the kth input connection is updated according to:

$$V_{t+1}^e (k.ln) = V_t^e (k.ln) + E_a \cdot A_t(l, n) + E_b \cdot A_t(l-1, k) + E_{ab} \cdot A_t(l, n) \cdot A_t(l-1, k),$$

If $r = true$, the weight of a the kth re-entrance connection is updated according to:

$$V_{t+1}^r (k.ln) = V_t^r (k.ln) + E_a \cdot A_t(l, n) + E_b \cdot A_t(l+1, k) + E_{ab} \cdot A_t(l, n) \cdot A_t(l+1, k).$$

The description of the model is now finished.

3   Simulation

The agents have been the subject of three tasks. In each case, the fitness function is directly related to the behavior of the agents.

3.1   The Tasks

We describe three conditions under which the agents have been evolved, namely a discrimination task, a task known in the psychological literature as "mutual fate control", and a derived task we call "mutual fate control with selection of behavior".

Discrimination Task   aims at selecting those networks that are able to learn to emit behaviors the emission of which have been followed by favorable consequences in the past. This task consists in discriminating two stimuli S1 and S2. In the presence of S1, the behavior of the agent must be B1; in the presence of S2, it should be B2. S1 and S2 are input on two dierent ISŝ. B1 and B2 correspond to the activation of two dierent BUŝ. When the required behavior is emitted, a stimulus is put on a certain IS (distinct from the IS's that receive S1 and S2): this stimulation acts as the positive consequences (a reward). To obtain networks that are not only able to emit B1 when facing S1, and B2 with S2, the task is made of a series of epochs. In each epoch, one combination is rewarded: either S1-B1, S2-B2, or S1-B2, S2-B1. The agent receives no signal so as to know which of the two combinations is the rewarded one at a given moment; at the beginning of each epoch, one combination is selected at random with probability 0.5 to be the one which will be rewarded during the epoch.

Owing to this, agents should have adaptive capabilities, that is, they should be able to learn the good combination of stimulus-behavior, and they should be able to learn to adapt their behavior along their "lifetime" according to the stimuli they receive from their environment. Each epoch is made of 1 000 stimuli. As the agents have no means to measure time, this task is not markovian. Each stimulus is emitted with probability 0 .5 at each time step. The population is made of 10 agents. The fitness function is defined as the cumulated number of rewards that are received along the 10 epochs. One generation is made of 10 epochs. The maximal fitness is then 10 000. Initially, the population is drawn at random. The two worst individuals are removed from the population at the end of each generation. They are replaced by two osprings of the two flest agents of the current population obtained by way of the genetic operators.   w is set to 0 .05, $_n$ is set to 0 .01, and $_e$ is set to 0 .05. $_p$ is set to 1 .0. Initially, the values of A_p and A_c are drawn uniformly at random in [1 , 1000].

Mutual fate control   (MFC) is drawn from the field of social psychology and deals with a situation of cooperation. It was introduced in 1957 [15]. The idea is to confront two agents A and B which have two possible behaviors, B1 and B2. Their behaviors have only consequences for their party as follows:

 – if the behavior of A is B1, then B receives positive consequences,
 – if the behavior of A is B2, then B receives negative consequences,
 – if the behavior of B is B1, then A receives positive consequences,
 – if the behavior of B is B2, then A receives negative consequences.

So, an agent does not receive the consequences of its own behaviors but the consequence of the behaviors emitted by its party (hence, one controls its party's fate). This situation leads to complex dynamics which have been discussed in [6,7]. As in the previous task, the goal is to select adaptive agents, not agents that are only able to answer systematically by behavior B1. Thus, we perform 10 epochs and for half of the epochs, the consequences are reversed: if A (resp. B) emits B2, then B (resp. A) receives positive consequences, while if A (resp. B) emits B1, then B (resp. A) receives negative consequences. Once again, the agents receive no information with regards to the fact that a new epoch begins and in which among the two cases it falls in. For this task, the population is initialized at random. The selection step as well as the probability to apply the operators are identical to those of the first task. During 10000 iterations, two agents are facing each other. We do not select these two agents at random because this would lead to diculties with regards to our goal. Indeed, whatever an agent does, its fitness is fully determined by its party. Hence, an agent that would provide its party lots of rewards while receiving no reinforcer from its party would be rated very bad, while the second would be rated very high: this would be totally unfair, and would not match our goal: we want agents which provide rewards to their party to be fit, while those which do not provide rewards to their party fipoorly. To avoid this eect, we select one genome in the population and generate two clones out of it which then face each other in the task. As long as agents learn during their lifetime, this is not the same as if one agent was facing itself; actually, we can see the experiment as two twins facing each other. Moreover, we can add that it would also be interesting to study the evolution of the population where two dierent agents would face each other in the task. This has to be done and, it is not so clear whether, finally, the evolution of the fness of the population would be very dierent.

Mutual fate control with selection of behavior   is strictly identical to the previous one except with regards to how the initial population is set. Instead of being random, the population is initialized with agents that have been able to pass a test procedure. This test procedure is made in such a way that, to pass it, the agent should be able to perform a very basic learning. This procedure consists in systematically rewarding a certain behavior. If the frequency of emission of this behavior is higher than for other behaviors, the test is passed by the agent. 813 agents have been drawn at random to obtain a population of 10 agents that pass the test 10 times in a row. Once constituted, the population undergoes MFC.

3.2   Results

The agents and the processes that have been described have been implemented in Java to perform the simulations. This section presents the results of these simulations, task by task.

Discrimination task.   Fig. 2(a) plots against generations the average performance of the population of agents in the discrimination task. Basically, after a rapid increase of performance, it levels for a while, then, it increases again to reach a much higher level (approximately 8500). The first level corresponds to a population where agents are able to receive the reinforcer one time out of two: the ordinate is 5000 whereas the maximum that can be obtained is 10000 (see Sect. 3.1). Then, in this population, the ability to discriminate appears later: after 200 generations, approximately 90 of the agents are able to perform the discrimination task.

This simulation shows that selection is able to produce the ability to discriminate, that is to learn its behavior from the consequences its emission receives from the environment. It should be emphasized that the environment is dynamic: the reinforcers are not received deterministically after the emission of a behavior; nothing in the environment helps the agents know in which condition they are. The ability to emit a behavior that have brought favorable consequences in the past is the core of the law of eect and, subsequently, of the principle of selection of behaviors by their consequences. Thus, we have shown that this can be produced by natural selection. Based on that observation, the next two simulations show that the ability of learning confers a great advantage.

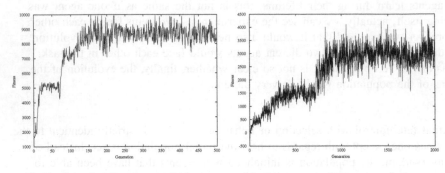

(a) Average performance of the agents facing the discrimination task against time. The performance is measured as the number of reinforcers that are received. The maximum is 10000.

(b) Average performance of the agents facing MFC against time. The performance is measured as the number of reinforcers that are received. The maximum is 10000. The initial population is made of agents which genome is drawn at random.

Fig. 2.   Evolution of performance of agents along time on the first two tasks.

Mutual fate control.   Fig. 2(b) plots the evolution of the average performance of agents in the MFC. Clearly, genetic selection leads to agents that have an increasing ability to control their party. This evolution is made of dierent phasis. Sharp increases in the performance happens from time to time, separated by periods that are rather steady. It is noticeable that after 2000 generations, the performance of the agents is still rather low: it receives only 35 of the rewards they could receive.

Mutual fate control with selection of behavior.   When the initial population of agents that face MFC is composed of agents that have passed the test procedure, the evolution of performance of the population is very dierent. Fig. 3 shows this dierence where A is the evolution of a random population while B is the evolution of the agents that have passed the test. At the beginning, the performance of the two populations are rather identical. But after very few generations, population B shows a much better performance than population A. After 200 generations, population B obtains 85-90 of the available reward. The discrepancy to 100 is due to the fact that at the beginning of each epoch, the agents have to adapt their behavior.

4   Conclusion and Discussion

In this paper, using computer simulations, we have shown that a population of agents can acquire the ability to learn during their lifetime by way of natural selection, giving way to the possibility of the Baldwin eect. Learning means that

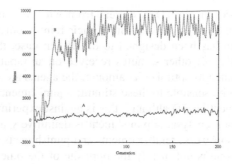

Fig. 3. Performance of two populations of agents facing the MFC against time: A is a population of agents initially drawn at random (that is, this is the same population as that of Fig. 2(b)); B is a population which is initially constituted of agents that pass a simple test of learning ability. This clearly advantages very much the population which performs well after 200 generations. As far as the environment is changing without any notice to the agents, 85-90 successes is a very good performance.

a certain structure in the agents is able to learn favorable associations between stimuli and the behavior to emit. Learning does not mean acquiring a stimulus-response reex: the environment is dynamic so that the agents have to be able to adapt their behavior to changing environments during their lifetime. Technically, this is known as operant learning, or instrumental conditioning. This work has thus to be considered as a step further the following known facts that have been shown using evolutionary algorithms: natural selection can produce fitter and fitter individuals along generations in static and in dynamic environments; natural selection can produce neural networks that act as control architecture of animats; the Baldwin eect can speed-up evolution. These points have already been raised and studied from the point of view of engineers to obtain good solutions for optimization problems, either numerical, or symbolic: hybrid algorithms have been a favorite theme of the EA literature for many years. We have also shown that once agents are able to select their behavior according to its consequences, the interaction between agents shows complex dynamics [7], and that complex behaviors can be acquired [14,8]. Clearly, an other originality of our work is that we put a strong emphasis on the interaction between two adaptive agents, not merely on the evolution and adaptation of a single agent in its environment. We think that this is an important point to obtain models and simulations that can bring something to the scientists who study life. Finally, we can also emphasise that the agents are facing non markovian environment.

At this point, we would like to relate this work to reinforcement learning. Clearly, the law of eect has inspired temporal dicrence (TD) methods as it has been argued in dierent places (see for example [3,6,19]). However, the formalization of the law of eect under an algorithmic form is far from straightforward. For sure, TD is appealing but there remains many unclear points, such as the definition of states and actions, the evolution law of -values or connection weights, and the role of time in the dynamics of behavior [7]. Furthermore, the status of the reinforcement in TD is dierent from what we consider here. In TD, the agents are optimizing the amount of their reinforcements because this is what the algorithm has been designed to do. In our case, the reinforcement is a mere stimulus like any other stimulus received on an input sensor, but, the agent is not programmed to optimize its amount, the agents are selected according to their ability to be sensible to these stimuli; a punishment is considered as a negative reinforcement by TD although this is against experimental evidences: according to behavioral analysis, a punishment is definitively not a negative reinforcement [17]. In our work, reinforcements and punishments are considered as dierent stimuli, thus one is not merely the opposite of the other: they are two dierent things. Finally, in the present work, our goal is to show that the law of eect can be the product of natural selection; we could have tried to evolve structures (using genetic programming for example) leading to TD; instead of that, we have favored an other approach which also leads to a reinforcement architecture based on neural networks. This approach has been chosen as being more "natural" to us, and in which we have tried to minimize the number of hypothesis as well as their remoteness from natural structures: neurons do exist (they can be seen, touched and even operated) while things like states are more elusive.

References

[1] D. Ackley and M. Littman. Interactions between learning and evolution. In Christopher Langton, Charles Taylor, J. Doyne Farmer, and Steen Rasmussen, editors, Artificial Life II , Santa Fe Institute Studies in the Sciences of Complexity, pages 487–509. Addison-Wesley Publishing Company, 1992.

[2] J.M. Baldwin. A new factor in evolution.   The american naturalist , 30, 1896.

[3] A.G. Barto. Reinforcement learning and adaptive critic methods. In D.A. White and D.A. Sofge, editors,   Handbook of intellkigent control: neural, fuzzy, and adaptive approach , pages 469–491. Van Nostrand Reinhold, 1992.

[4] C. Catania. Thorndikeš legacy: learning, selection, and the law of eect.   Journal of the experimental analysis of behavior , 72:425–428, 1999.

[5] P. Chance. Thorndike's puzzle boxes and the origins of the experimental analysis of behavior. Journal of the Experimental Analysis of Behavior , 72(3):433–440, 1999.

[6] S. Delepoulle.   Coopération entre agents adaptatifs ; ´ etude de la sélection des comportements sociaux, exp´ erimentations et simulations   . PhD thesis, Universit´ e de Lille 3, URECA, Villeneuve d'Ascq, October 2000. Th   ese de doctorat de Psychologie.

[7] S. Delepoulle, Ph. Preux, and J-Cl. Darcheville. Dynamique de l'interaction. In B. Chaib-Dra and P. Enjalbert, editors,   Proc. Mod eles Formels de l'Interaction, Toulouse , pages 141–150, 2001.

[8] S. Delepoulle, Ph. Preux, and J-Cl. Darcheville. Selection of behavior in social situations application to the development of coordinated movements. In   Applications of Evolutionary Computing   , volume 2037 of Lecture Notes in Computer Science , pages 384–393. Springer-Verlag, April 2001.

[9] G.E. Hinton and S.J. Nowlan. How learning can guide evolution. 1:495–502, 1987.

[10] J. Maynard-Smith. When learning guides evolution.   Nature , 329:761–762, October 1987.

[11] D. McFarland. Animal Behavior. Psychology, Ethology and Evolution . Longman Science and Technology, 1998.

[12] C. Lloyd Morgan. On modification and variation.   Science , 4:733–740, 1896.

[13] H.F. Osborn. Ontogenetic and phylogenetic variation.   Science , 4:786–789, 1896.

[14] Ph. Preux, S. Delepoulle, and J-Cl. Darcheville. Selection of behaviors by their consequences in the human baby, software agents, and robots. In   Proc. Computational Biology, Genome Information Systems and Technology , March 2001.

[15] J.B. Sidowski, B. Wycko, and L. Tabory. The inuence of reinforcement and punishment in a minimal social situation.   Journal of Abnormal Social Psychology , 52:115–119, 1956.

[16] B.F. Skinner. The behavior of organisms . Appleton-Century Crofts, 1938.

[17] B.F. Skinner. Science and human behavior . MacMillan, 1958.

[18] B.F. Skinner. Selection by consequences.   Science , 213:501–514, 1981.

[19] R.S. Sutton and A.G. Barto.   Reinforcement learning: an introduction . MIT Press, 1998.

[20] E.L. Thorndike. Animal intelligence: An experimental study of the associative process in animals. Psychology Monographs , 2, 1898.

[21] E.L. Thorndike. Animal Intelligence: Experimental Studies . Mac Millan, 1911.

[22] J. Urzelai. Evolutionary Adaptive Robots: artificial evolution of adaptation mechanisms for autonomous systems . PhD thesis, EPFL, Lausanne, Suisse, 2000.

[23] C. Waddington. Genetic assimilation for acquired character.   Evolution , 7:118–126, 1953.

[24] C. Waddington. Genetic assimilation of the   bithorax phenotype. Evolution , 10:1–13, 1956.

Coevolution and Evolving Parallel Cellular Automata-Based Scheduling Algorithms

Franciszek Seredy´nski[1] and Alber t Y. Zomaya [2]

[1] The University of Podlasie
Computer Science Department
Sienkiewicza 51, 08-110 Siedlce, Poland
sered@ipipan.waw.pl
[2] Parallel Computing Research Laboratory
Department of Electrical and Electronic Engineering
The University of Western Australia
Nedlands, Perth, Western Australia 6907, Australia
zomaya@ee.uwa.edu.au

Abstract. The paper reports new results on developing parallel algorithms for multiprocessor scheduling with use of cellular automata (CAs). The simpliest case when a multiprocessor system is limited to two processors, but without of any limitations on a size and parameters of parallel programs is considered. An approach called a selected neighborhood is used to design a structure of CAs for a given program graph. Coevolutionary genetic algorithm (GA) to discover rules of parallel CAs, suitable for solving the scheduling problem is proposed. Sequential and parallel scheduling algorithms discovered in the context of CAs - based scheduling system are compared.

1 Introduction

Multiprocessor scheduling even limi ted to the simplies t case considered in the paper when we have to do with the two processor sys tem bu t any parallel program is known to be as an NP-comple te problem. The prevaling majori ty of known scheduling algori thms are sequen tial algori thms, and a new perspec tive direc tion in this area is developing parallel scheduling algori thms [1].

A great hope today to solve problems like this are na turally inspired nonstandard compu tational techniques [3] such as molecular compu tation, gene tic algori thms, compu tation in cellular au tomata and quan tum compu ting.

In this paper, we review and ex tend the recen tly proposed technique for scheduling, based on applying cellular au tomata [5,6]. CA is a highly parallel and dis tribu ted system consis ting of single cells which behave according to local rules, and their in terac tion resul ts in a global behavior o f the system. Recen t resul ts [2,8] show that such CA sys tems, combined wi th evolu tionary techniques for discovering local rules, can be eec tively used to solve complex problems such as classifica tion, synchroniza tion or cryp tography.

The remainder o f the paper is organized as follows. The nex t section discusses the scheduling problem in the context of the CA solu tion and overviews curren tly

P. Collet et al. (Eds.): EA 2001, LNCS 2310, pp. 362–373, 2002.

results in this area. Sec tion 3 describes the coevolu tionary gene tic algori thm-based engine for discovering CA scheduling rules. Sec tion 4 presen ts experimen tal results concerning the discovery wi th coevolu tionary GA scheduling rules   for parallel CAs. Sec tion 5 contains conclusions.

2   Multiprocessor Scheduling with Cellular Automata

2.1   Scheduling Problem

Both a mul tiprocessor sys tem and a parallel program are represen ted by cor-responding graphs. A mul tiprocessor sys tem is represen ted by an undirec ted unweigh ted graph $G_s = (V_s, E_s)$ called a system graph . V_s is the set of N_s nodes represen ting processors and E_s is the set of edges represen ting bidirec tional channels be tween processors. A parallel program is represen ted by a weigh ted direc ted acyclic graph $G_p = V_p, E_p$, called a prece dence task graph or a pro-gram graph . V_p is the set of N_p nodes o f the graph represen ting elemen tary tasks. The weigh t b_k of the node k describes the processing time needed to execu te a task k on any processor o f the system. E_p is the set of edges of the precedence task graph describing the communica tion pa tt erns be tween the tasks. The weigh t a_{kl} of the edge (k,1) describes a communica tion time between the pair o f tasks k and 1 when they are loca ted in neighboring processors. Figure 1 (upper) shows examples o f the program graph and   the system graph. The program represen ted by the graph consis ts of 4 tasks with $b_0 = 1, b_1 = b_3 = 2, b_2 = 4$ (numbers on the left side of nodes), and $a_{01} = a_{02} = a_{13} = 1$ (numbers on the left side of edges). The sys tem graph represen ts a mul tiprocessor sys tem consis ting of two processors P0 and   P1 .

The purpose o f scheduling is to distribu te the tasks among the processors in such a way that t he precedence cons traints are preserved, and   the response time T (the total execu tion time) is minimized. The response time T depends on the allocation of tasks in the mul tiprocessor topology and on scheduling policy applied in individual processors:

$$T = f \text{ (allocation, scheduling \_policy)}. \tag{1}$$

We assume that t he scheduling policy is a user-defined paramc tcr, the same for all processors, bu t t he alloca tion is a subjec t t o change by a scheduling al-gorithm. We assume that, for each node k of the precedence task graph, se ts of predecessors(k) , brothers(k) (i.e. nodes having a t least one common predeces-sor), and successors(k) are defined. We also assume that, for each node k of the precedence task graph, parame ters such as static and dynamic level and co-level can be defined.

2.2   A Concep t of Cellular Au tom a ta-Based Scheduler

An idea o f CA-based scheduler is presen ted in Figure 1. Wi th each task of a program graph from Figure 1 (upper) an elemen tary cell o f CAs is associa ted. An ini tial state of the CAs correspond to an ini tial alloca tion of tasks in the

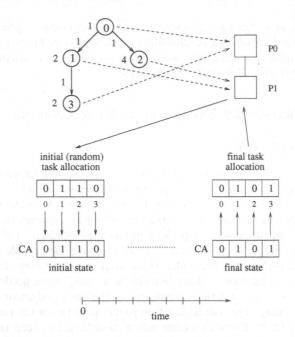

Fig. 1. An idea of CA-based scheduler: an example of a program graph and a system graph (upper), corresponding CA-based scheduler (lower)

two-processor sys tem (Figure 1 (lower-le ft)). Nex t, the CAs s tart t o evolve in time according to some rule. Changing s tates of the CA cells corresponds to changing the alloca tion of tasks in the system graph, wha t results in changing the response time T. A final s tate of the CAs correspond to a final alloca tion of tasks in the system (Figure 1 (lower-righ t)).

To cons truct t he CA-based scheduler one mus t solve several problems:(a) what is the topological s tructure of proposed CAs: linear as shown in Figure 1 (lower-le ft) or nonlinear, rela ted in some way to the topological s tructure of a program graph, (b) wha t kind of a local neighborhood o f a program graph is the most appropria te to design corresponding CAs, and (c) how   to find in a huge space of CA rules, the rule capable to solve the scheduling problem.

In the approach we adop t, developed in our previous works (see [5,6]) the struc ture of the CA is nonlinear and corresponds   to the topology of the program graph. Elemen tary cells are associa ted with tasks of a program graph and a some neighborhood o f a central task is crea ted. The cen tral cell takes only values 0 or 1, wha t results in considering the scheduling problem only   for the 2-processor topology: the state 0 or 1 of a cell means that a corresponding task is alloca ted either in the processor P0 or P1, respec tively.

The scheduler opera tes in two modes: a mode o f learning CA rules and a mode of normal opera ting. The purpose o f the learning mode is to discover eec tive rules for scheduling. Searching rules is conduc ted with use of GA [2]. For this purpose an ini tial random popula tion of rules is crea ted. Also a se t of random initial allocations of a program graph in to the system graph is crea ted. States of the CAs are ini tiated according to a given alloca tion of a program graph. The CAs equipped wi th a rule from the popula tion of rules s tart t o run during predefined number o f time steps. Changing s tates of the CAs correspond to changing an alloca tion of task of a program graph. The response   time T for a final allocation is evalua ted. This procedure o f evalua tion of a rule is repea ted for each allocation from the set of initial alloca tions. Even tually, a fi tness value T   for the rule is evalua ted as the sum o f final values o f T corresponding to each ini tial allocation of tasks. A fter evalua tion in a similar way o f all rules from the population, gene tic opera tors are applied.   Selection with elitism trans fers some percen t of the bes t rules, called elite to the new popula tion which will processed in the next genera tion. The remaining par t of the popula tion is crea ted by crossover between members o f the elite, and nex t mutation is applied to new members (children) o f the popula tion. Also a new se t of initial alloca tions is randomly crea ted at t he beginning o f each genera tions. Evolu tionary process is con tinued a predefined number o f genera tions, and when i t is comple ted discovered rules are s tored.

In the mode o f normal opcra ting, when a program graph is randomly allocated, the CA is ini tiated and equipped wi th a rule taken from the set of discovered rules. We expec t in this mode, that for any ini tial alloca tion of tasks of a given program graph,   the CAs will be able to find in a fini te number o f time steps, an alloca tion of tasks, providing an op timal or subop timal value o f T.

2.3   Selec ted Neighborhood

A neighborhood o f a central task consis ts of three subneighborhoods and includes this task. Each subneighborhood o f a cell associa ted with a task k is crea ted only by two selected represen tatives of a set of predecessors, bro thers and successors, respec tively. The represen tatives are selec ted on the basis o f respectively maximal and minimal values o f some att ribu tes of tasks in the given set. If corresponding tasks of a subneighborhood are missing in a program graph, dummy task are in troduced. In a given run o f the scheduling algori thm, one a t-tribu te for each se t of predecessors, bro thers and successor is selec ted. So, the selected neighborhood o f a given cell associa ted with a cen tral task consis ts of 7 cells, and includes this cell.

There is some scheme to calcula te a state of a neighborhood o f a given central task. The scheme assumes   that a state of each subneighborhood is calcula ted first. A state of a subneighborhood   takes one o f five values (see, [5,6]) corresponding to allocations of two tasks of the subneighborhood. A total number o f states of a neighborhood can be calcula ted as 2555 and is equal   to 250. A leng th of a rule (a transi tion function) is 250 bi ts. A space o f solutions of the problem is defined by a number 2 ^{250} of possible transi tion functions. GA wi th a popula tion of rules is used to discover an appropria te rule for CA to solve a scheduling problem.

Figure 2 shows a neighborhood o f the task 0 from Figure 1. Because the task 0 does not have any predecessor and bro thers the dummy tasks are crea ted to fulfill the requiremen t of a regular neighborhood consis ting of 7 tasks. The se t

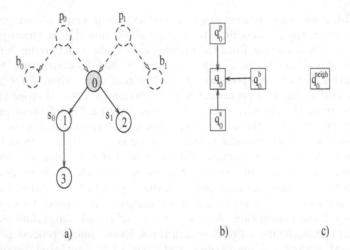

Fig. 2. Selected neighborhood: creating a neighborhood for the task 0 from Figure 1
(a), a state of the cell 0 depends on states of subneighborhoods created by predecessors,
brothers and successors (b), a state of a neighborhood of the cell 0 is evaluated (c)

of successors is represen ted by tasks 1 and 2, and the task 3 does no t belongs to
the neighborhood o f the task 0. A s tate of a neighborhood o f each cell associa ted
with a task is defined on the base o f states of corresponding subneighborhoods.

2.4   Previous Resul   ts

Results of experimen tal study of the version o f the CA-based scheduler running
under DOS have shown [5,6] that GA was able to discover eec tive rules for
scheduling o f a number o f program graphs known   from litera ture. However, discovered rules were working in a de terminis tic sequen tial mode of CA, i.e. only
one cell could upda te its state in time, and the order o f upda ting was predefined
by numbering o f tasks in a program graph. I t means that one of the most interesting features of CAs - their massive parallelism - was no t explored. Also the
frequency o f successful runs o f the scheduler resul ting in discovery o f an op timal
rule was low and the convergence o f the scheduler in learning mode was limi ted
to the case when the set of initial alloca tions of tasks was small.

For all these reasons an a tt empt t o develop a new enhanced Windows'98
version o f the scheduler was under taken. The main   feature of the scheduler is
a new, much more power ful coevolu tionary GA-based engine   for discovery CA
rules and some visualiza tion tools enabling tracing the work of the scheduler.

3   Coevolutionary Genetic Algorithm for Discovery CA
Rules

One of the most promising lines o f research in the area o f parallel evolu tionary compu ting (EC) is a developmen t of coevolu tionary algorithms . The idea

of coevolutionary algorithms comes from the biological observation of natural selection which shows that coevolving some number o f species defined as collections of phenotypically similar individuals is more realis tic than simply evolving a population containing representatives of one species. So, ins tead of evolving a population of similar individuals represen ting a global solution, it is more appropriate to coevolve subpopula tions of individuals represen ting specific par ts of the global solution.

Among a recen tly proposed coevolu tionary algori thms is the coevolutionary GA [4] described in the context of the constraint satisfaction problem and the neural ne twork optimization problem as a low level parallel EA based on a predator-prey paradigm. The algori thm is described below wi th use of a parallel processing language OCCAM-like no tation. In par ticular, sequen tial and parallel processes are specified by SE and PAR cons tructors, respec tively. Commen ts concerning the algori thm follow the symbols .

Coevolu tionary GA:
chromosome 1 : global structure representing a solution x of a problem
chromosome 2 : additional structure representing constraints y of a problem
optimization criterion : global function f (x, y)
population 1 : main subpopulation $P^1()$
population 2 : additional subpopulation $P^2()$
population structure : two interacting subpopulations
t = 0
SE
    initialize $P^1(t)$ and $P^2(t)$
    WHILE   termination _condition NOT TRUE
        SE
            t = t + 1
            SE   i = 1 FOR n_encounters
                SE
                    PAR j = 1 FOR 2   running coevolving subpopulations
                        SE
                            select individuals $I_k^j(t)$ from $P^j(t)$
                            confront selected individuals
                            evalua te result (fitness of individuals) o f confrontation
                select a pair o f paren ts in both $P^1(t)$ and $P^2(t)$
                crossover over pairs o f paren ts
                mu tate in children
                replace paren ts in $P^1(t)$ and $P^2(t)$
problem _solution = the best individual x from the subpopulation $P^1(t)$

The algori thm opera tes on the main subpopula tion $P^1()$ con taining individuals x, and an addi tional subpopula tion $P^2()$ con taining individuals y coding some cons train ts, condi tions or simply test points concerning a solu tion x. Both, or only one subpopula tion evolve to optimize a global func tion f (x, y).

A single ac t of coevolution is based on independen t selection of individuals x and y from subpopula tions, to encoun ter them and evalua te their f (x, y). The manner o f assigning a fi tness to the individuals s tems from the preda tor-prey

rela tion: a success o f one individual should be a   failure o f the second one. During one genera tion individuals are con fron ted a predefined number   n_encounters
times. At t he end of the evolu tion process, the best individual from $P^1()$ is
considered as a solu tion of a problem.

In the case of the CA-based scheduler the main popula tion of the coevolutionary GA con tains the N^{main} CA rules, and the addi tional popula tion con tains
the N^{test} tests - the ini tial alloca tions of a program graph. During a given generation each individual from the main popula tion is tested, as previously, on each
individual o f the addi tional popula tion. The same gene tic opera tors as described
earlier are applied to the main popula tion. The addi tional popula tion is ini tially
randomly crea ted, bu t now the set of tests in nex t genera tions will be con trolled
by its own GA.

As a fitness func tion of an individual- test of the addi tional popula tion we
choose the value o f T_{test} , which is the average o f final values o f T obtained by
all rules o f the main popula tion on this test. Gene tic opera tors of tournament
selection with elitism, crossover and mu tation are applied to individuals o f the
popula tion.

4   Experimental Results: out   tree 31 Case Study

The main purpose o f performed experimen ts was to study the inuence o f the
coevolu tionary GA on discovery scheduling rules   for parallel CAs. For this purpose a program graph called a binary   out   tree 31 was selected. The program
graph consis ts of 31 tasks with compu tational and communica tion costs equal to
1. Figure 3 presen ts a simpler varian t of the graph, a program graph   out   tree 15
consisting of 15 tasks only.

In all conduc ted experimen ts the following parame ters were used. The size
of a popula tion of rules N^{main} was equal to 100, with the size of elite equal to
20. Not only elite, but all individuals from the popula tion could take par t in
mating with probabili ty of crossover p_c^{main} = 0.9, and probabili ty of mutation
p_m^{main} = 0.1. Selected neighborhood was crea ted using level as at tribu tes of
task-predecessors and task-bro thers, and dynamic level as an a tt ribu te of task-
successors. To calcula te T for a given alloca tion of tasks a scheduling_policy of
the type: a task with the highes t value of a dynamic level-firs t, was applied.

CA was allowed to run 25 time steps, and the value o f T corresponding to a
final alloca tion of tasks was calcula ted as the average on the base o f 3 last steps
of CAs. The size o f a popula tion of tests N^{test} was equal 30. When coevolu tion
was turned on, the following gene tic opera tors were applied to the popula tion
of tests: tournamen t selection, and eli te with size equal to 2, crossover wi th
p_c^{test} = 0.9 and mu tation with p_m^{test} = 0.005. The evolu tionary process was
observed during 500 genera tions.

Experimen t 1:   discovery o f rules for de terminis tic sequen tial CAs

In the experimen t repor ted in this section it is assumed that CAs work sequen tially and de terminis tically. At a given momen t of time only one cell upda tes
its state. An order o f upda ting states of cells is defined by their order number

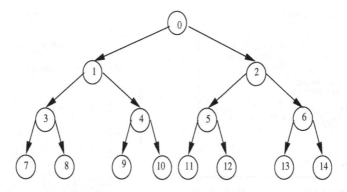

Fig. 3.   Program graph   out   tree 15

corresponding to tasks (a number in circles-nodes o f the program graph, see Figure 3) in a precedence   task graph. A single s tep of CAs is comple ted in N_p (N_p - a number o f tasks of a program graph) momen ts of time. A run o f CAs consists of a predefined number   G = 25 o f steps, with the tot al time steps equal to G N_p = 25   31 = 775.

Figure 4 shows resul ts of a typical experimen t with evolving scheduling rules for de terminis tic sequen tial CAs. Figures   4a and b presen t learning and normal opera ting modes o f the scheduler, respec tively. The experimen t is conduc ted wi thou t coevolu tion and one can see (Figure   4a) that evolving scheduling rules for de terminis tic sequen tial CAs is easy problem   for this type of a program graph.

GA discovers a rule providing an op timal scheduling wi th T = 17 a fter 20 genera tions (see, avr fin T of the best rule in Figure 4a). The average value o f initial alloca tions avr initial T of allocs genera ted randomly in each genera tion of GA oscilla tes around a value o f T_0 = 19 .2. It means that rules exposed to test problems are in the average o f the same degree o f dicul ty during the whole evolu tionary process. To see how dicul   t genera ted alloca tions are we define for each of them the average final T over all rules from a popula tion which were tested on this alloca tion. Observing the average final T of the most dicul t alloca tion avr fin T of best alloc one can see that genera ted randomly   tests become easier for rules in each genera tion. A fter genera tion 20 when the bes t rule was discovered, each s tatistical rule finds an alloca tion with final T bett er than a s tatistical ini tial alloca tion with corresponding T_0.

Figure 4b shows the normal opera ting mode o f the scheduler. In this mode each rule in the final popula tion is exposed to 1000 random ini tial alloca tions of the program graph. The figure shows   that t he near 20 the bes t rules in the sequen tial CAs find the op timal scheduling wi th T ▬ 17 in all tests.

Experimen t 2:   discovery o f rules for parallel CAs, wi thout coevolu tion

It is assumed now a parallel work o f CAs, wha t means that at a given momen t of time all cells upda te their s tates. At t his experimen t GA wi thout coevolu tion

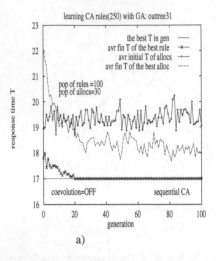
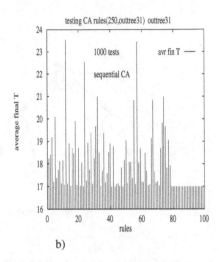

a)   b)

Fig. 4. Evolving scheduling rules for deterministic sequential CAs: learning mode of
the scheduler (a), and normal operating mode (b)

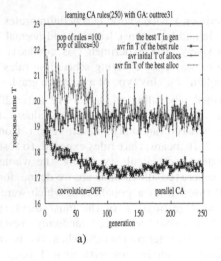
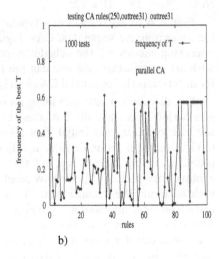

a)   b)

Fig. 5. Evolving without coevolution scheduling rules for parallel CAs: learning mode
of the scheduler (a), and normal operating mode (b)

is applied to discover rules for CAs. Figure 5 shows resul ts of a typical experimen t
with evolving scheduling rules   for parallel CAs, wi thout coevolution.

Figure 5a shows the firs t 250 genera tions of the learning mode o f the scheduler. One can see that the value o f avr fin T of the best rule   charac terizing the
best rule in each genera tion approaches to the op timal value bu t never reaches i t.

It achieves its local minimum in genera tion 125, and s tabilizes its value around 17.40 in abou t genera tion 200.

The average value avr initial T of allocs of initial alloca tions (see, Figure 5a genera ted randomly in each genera tion of GA behaves in the same manner as in the previous experimen t. However, observing the average final T of the most dicul t t est-alloca tion avr fin T of best alloc one can no tice that t he value decreases only to the genera tion 125, only as long as the bes t rule improves i ts quali ty. After this genera tion none new valuable sequence o f initial alloca tions appears in the set of tests to be exposed to the popula tion of rules. There fore, the learning process in the popula tion of rules is s topped and a be tt er rule for parallel CAs will be no t discovered. Corresponding value o f the avr initial T of allocs becomes equal to average value T_0 of initial alloca tions.

Figure 5a shows the normal opera ting mode of the scheduler. I t shows the frequency o f convergence o f CAs with a given rule to the alloca tion corresponding to the optimal value o f T = 17. One can see that t he best rules found for parallel CAs converge to the optimal T in only near 60 o f cases.

Experimen t 3:   discovery o f rules for parallel CAs, wi th coevolu tion

In this experimen t we assume that we have to do with parallel CA-based scheduler and we apply GA-based engine wi th coevolu tion to discover rules for CAs. Figure 6 shows resul ts of the experimen t.

One can see (Figure 6a) that GA with coevolu tion discovers the best rule providing convergence o f parallel CA-based scheduler   to the optimal value o f T = 17 in 35 genera tion. The dynamic o f changing the value o f avr fin T of the best rule in each genera tion is dieren t t hat one in the experimen t without coevolu tion. Also behavior o f the average value avr initial T of allocs of initial alloca tions is dieren t. One can no tice that improvemen t of avr fin T of the best rule is correla ted with changing avr initial T of allocs .

The coevolu tion mechanism which con trols changing ini tial alloca tions makes that t he average value avr initial T of allocs of initial alloca tions per forms a num-ber of hillclimbings wi th subsequen t falling down, ins tead of random oscilla tion. During hillclimbing a sequence o f initial alloca tions with increasing value o f T_0 is genera ted. These sequences make ini tial alloca tions more dicul   t and this in turn s timula tes GA to improve rules.

Figure 6b shows the normal opera ting mode of the scheduler. The figure shows that t he frequency o f convergence o f CAs with a given rule to the alloca-tion corresponding to the optimal value o f T is abou t 97. This value is much higher that in the experimen t without coevolu tion but a litt le smaller that in the sequen tial CA-based scheduler. Figure 7a shows   that t he average final T obtained wi th use of these rules is close to the optimal T. These solu tions are found in few time steps of parallel CAs (see, Figure 7b) ins tead of few hundred steps of sequen tial CAs.

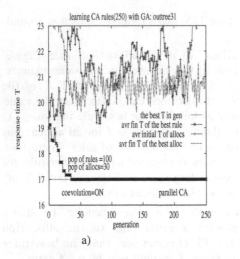
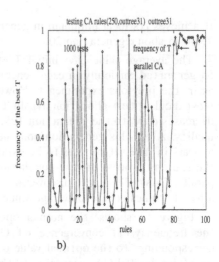

Fig. 6. Evolving with coevolution scheduling rules for parallel CA: learning mode of the scheduler (a), and normal operating mode (b)

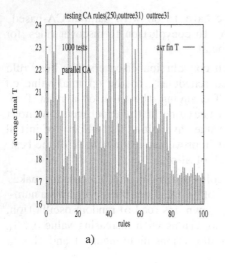
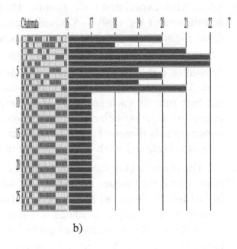

Fig. 7. Parallel CAs with coevolution: the average final   T of evolved rules in normal operating mode (a), example of time-space diagram of CAs during running (b)

5   Conclusions

We have presen ted new resul ts concerning developing CA-based algori thms o f multiprocessor scheduling. We have shown   that coevolu tionary algori thms are very promising technique s timula ting the process o f discovering eec tive rules for parallel CA-based algori thms. We compared sequen tial and parallel CA-based

scheduling algori thms and have shown advan tages of parallel approach. A number of questions in this area are s till open. One o f them is an op timal choice
of the CA s truc ture. While we have used a complex nonlinear s truc ture of CAs
based on predecessors, bro thers and successors rela tion to build a scheduler,
one of promising direc tions of research simpli fying this structure is using a linear structure based on the neighborhood o f adjacen t cells of the CA [7]. The
other impor tant question curren tly addressed is the possibili ty of wider using
and reusing discovered scheduling rules.

References

1. I. Ahmad, Y. Kwok, A parallel approach for multiprocessor scheduling, in Proceedings of the 9th Int. Parallel Processing Symposium, Santa Barbara, CA, 1995
2. R. Das, M. Mitchell, and J. P. Crutchfield, A genetic algorithm discovers particle-
   based computation in cellular automata, in Y. Davidor, H.-P. S   chwefel, R. M anner
   (eds.), Parallel Problem Solving from Nature – III   , LNCS 866, Springer, 1994
3. T. Gramb, S. Bornholdt, M. Grob, M. Mitchell, T. Pellizari,   Non - Standard Computation , Wiley-VCH, 1998
4. J. Paredis, Coevolutionary Life-Time Learning, in H. -M. Voigt, W. Ebeling, I.
   Rechenberg and H. -P. Schwefel (eds.),   Parallel Problem Solving from Nature –
   PPSN IV , LNCS 1141, Springer, 1996, pp. 72-80
5. F. Seredynski and C. Z. Janikow, Designing Cellular Automata-based Scheduling
   Algorithms, in W. Banzhaf, J. Daida, A. E. Eiben, M. H. Garzon, V. Honavar, M.
   Jakiela, R. E. Smith (eds.),   GECCO-99: Proc. of the Genetic and Evolutionary
   Comp. Conf. , 1999, Orlando, Florida, USA, Morgan Kaufmann, pp. 587-594
6. F. Seredynski, Evolving Cellular Automata-Based Algorithms for Multiprocessor
   Scheduling, in A. Z. Zomaya, F. Ercal, S. Olariu (eds.),   Solutions to Parallel and
   Distrib. Comput. Problems: Lessons from Biol. Sciences   , Wiley, 2001, pp. 179-207
7. F. Seredynski, A. Swi ecicka, Immune-like System Approach to Cellular Automata-
   based Scheduling, in   Proc. of the 4th. Int. Conf. on Parallel P   rocessing and Applied
   Math. , Naleczów, Poland, Sept. 9-12, 2001, to appear in LNCS, Springer, 2001
8. M. Tomassini, M. Sipper, M. Zolla, and M. Perrenoud, Generating high-quality
   random numbers in parallel by cellular automata,   Future Generation Computer
   Systems 16, 1999, pp. 291-305

Author Index